EIGHTH EDITION

LABORATORY MANUAL IN PHYSICAL GEOLOGY

PRODUCED UNDER THE AUSPICES OF THE
AMERICAN GEOLOGICAL INSTITUTE
http://www.agiweb.org

AND THE
NATIONAL ASSOCIATION OF GEOSCIENCE TEACHERS
http://www.nagt.org

RICHARD M. BUSCH, EDITOR
WEST CHESTER UNIVERSITY OF PENNSYLVANIA

ILLUSTRATED BY
DENNIS TASA • *TASA GRAPHIC ARTS, INC.*

PEARSON
Prentice Hall

Upper Saddle River, NJ 07458

Publisher, Geosciences and Environment: *Dan Kaveney*
Editor-in-Chief, Science: *Nicole Folchetti*
Acquisitions Editor: *Drusilla Peters*
Project Manager: *Crissy Dudonis*
Senior Managing Editor: *Kathleen Schiaparelli*
Project Manager: *Shari Toron*
Senior Operations Supervisor: *Alan Fischer*
Art Director: *Maureen Eide*
AV Project Manager: *Connie Long*
Illustrations: *Dennis Tasa*
Cover Designer: *Jonathan Boylan*
Front Cover Photo: *Canyonlands national park/Ron Niebrugge/Mountain lake*
Back Cover Photo: *Petrified wood from Arizona*

Third edition © 1993 by American Geological Institute, published by Macmillan Publishing Company.
Second edition © 1990 by Macmillan Publishing Company.
First edition © 1986 by Merrill Publishing Company.

Printed in the United States of America

10 9 8 7 6 5 4 3 2 1

ISBN-10:   0-13-600771-6
ISBN-13:   978-013-600771-5

CONTRIBUTING AUTHORS

THOMAS H. ANDERSON
University of Pittsburgh

HAROLD E. ANDREWS
Wellesley College

JAMES R. BESANCON
Wellesley College

JANE L. BOGER
SUNY–College at Geneseo

PHILLIP D. BOGER
SUNY–College at Geneseo

CLAUDE BOLZE
Tulsa Community College

JONATHAN BUSHEE
Northern Kentucky University

ROSEANN J. CARLSON
Tidewater Community College

CYNTHIA FISHER
West Chester University of
Pennsylvania

CHARLES I. FRYE
Northwest Missouri State
University

PAMELA J.W. GORE
Georgia Perimeter College

ANNE M. HALL
Emory University

EDWARD A. HAY
De Anza College

CHARLES G. HIGGINS
University of California, Davis

MICHAEL F. HOCHELLA, JR.
Virginia Polytechnic Institute and
State University

MICHAEL J. HOZIK
Richard Stockton College of
New Jersey

SHARON LASKA
Acadia University

DAVID LUMSDEN
University of Memphis

RICHARD W. MACOMBER
Long Island University, Brooklyn

GARRY D. MCKENZIE
Ohio State University

CHERUKUPALLI E. NEHRU
Brooklyn College (CUNY)

JOHN K. OSMOND
Florida State University

CHARLES G. OVIATT
Kansas State University

WILLIAM R. PARROTT, JR.
Richard Stockton College of
New Jersey

RAMAN J. SINGH
Northern Kentucky University

KENTON E. STRICKLAND
Wright State University

RICHARD N. STROM
University of South Florida,
Tampa

JAMES SWINEHART
University of Nebraska

RAYMOND W. TALKINGTON
Richard Stockton College of
New Jersey

MARGARET D. THOMPSON
Wellesley College

JAMES TITUS*
U.S. Environmental Protection
Agency

EVELYN M. VANDENDOLDER
Arizona Geological Survey

NANCY A. VAN WAGONER
Acadia University

JOHN R. WAGNER
Clemson University

DONALD W. WATSON
Slippery Rock University

JAMES R. WILSON
Weber State University

MONTE D. WILSON
Boise State University

C. GIL WISWALL
West Chester University
of Pennsylvania

*The opinions contributed by this person
do not officially represent opinions of the
U.S. Environmental Protection Agency.

CONTENTS

PREFACE

Laboratory Manual in Physical Geology is the most widely adopted, user-friendly manual available for teaching laboratories in introductory geology and geoscience. The manual has been produced under the auspices of the American Geological Institute (AGI) and the National Association of Geoscience Teachers (NAGT). It is backed up by an Internet site, GeoTools (ruler, protractor, UTM grids, sediment grain size scale, etc.), Instructor's Resource Guide, Instructor Transparency Set, and an Instructor Resource Center (IRC) on CD-ROM.

The idea for such a jointly sponsored laboratory manual was proffered by Robert W. Ridky (past president of NAGT and a member of the AGI Education Advisory Committee), who envisioned a manual made up of the "best laboratory investigations written by geology teachers." To that end, this product is the 23-year evolution of the cumulative ideas of more than 170 contributing authors, faculty peer reviewers, and students and faculty who have used past editions. Undergraduate students have field tested all parts of this eighth edition and helped make it the most student-friendly edition ever.

OUTSTANDING FEATURES

This edition contains the strengths of seven past editions published over twenty-three years and new features developed at the request of faculty and students who have used previous editions. The most outstanding features of this new edition are as follows.

16 Basic Laboratories
There are 16 laboratories on topics ranked most important by faculty peer reviewers. Each lab has 3–6 parts that can be mixed or matched at the instructor's discretion.

Consistent Focus, Pedagogy, and AGI Terminology
Each laboratory engages students in learning principles of geology and their applications to everyday life in terms of natural resources, natural hazards, and human risks. Students develop skills and infer results by analysis of maps/samples/photos, measuring, experimenting, making models, classifying, charting, graphing, and calculating. Terms are consistent with AGI's latest *Glossary of Geology*.

Materials
Laboratories are based on samples and equipment normally housed in existing geoscience teaching laboratories (page ix). No expensive items to buy. In addition, a partnership with WARD'S Natural Science, the premier provider of rock and mineral samples, has resulted in the creation of both instructional and student rock and mineral sets designed to support users of this manual. (For more information, see page xiv.)

New Geologic and Flood Hazard Maps
There are two new options for students to construct geologic cross sections from geologic maps and evaluate/modify a FEMA flood hazard map.

Introduction to Satellite Remote Sensing
There is a concise, engaging section on principles and applications of satellite remote sensing to study Earth. Students analyze and apply MODIS, Landsat, and ASTER satellite images to prospect for copper ore, evaluate volcanic activity, and predict changes in Africa's Lake Chad.

Emphasis on the Process of Geologic Inquiry
Students visualize Earth materials and processes of change using satellite imagery and infer how the geologic record is similar in some ways, yet different in others, from a book of Earth history to be "read." They explore ways that geology is a logical, testable process of scientific inquiry, "ground truthed" with data obtained by direct observation, investigation, and measurement in the field and laboratory.

Greater Visual Clarity and Appeal
The manual is more richly illustrated than any other manual on the market. Over 400 high-quality photographs, images, stereograms, maps, and charts reinforce the visual aspect of geology and enhance student learning. Many of these are revised or newly created for this edition on the basis of faculty and student feedback.

Hands-On Experimental Labs
Laboratory One engages students in geologic inquiry using satellite imagery, Earth materials, and standard laboratory equipment and techniques to measure materials, experiment with simple models, calculate numerical relationships, and evaluate how rock densities and isostasy influence global topography. Laboratory Two challenges students to explore and evaluate plate

tectonics, mantle convection, and the origin of magma using seismic tomography, lava lamps, physical and graphical models of partial melting, maps, and calculations.

More GeoTools

There are rulers, protractors, a sediment grain size scale, UTM grids, and other laboratory tools to cut from transparent sheets at the back of the manual. A new cardboard sheet of GeoTools includes a cleavage goniometer that students can use as an additional tool for mineral identification.

Emphasis on GPS and UTM

Students are introduced to these topics and their application in mapping and geology. UTM grids are provided for most scales of U.S. and Canadian maps.

Enhanced Instructor Support

Free instructor materials include a transparency set of figures in the manual plus an Instructor Resource Center (IRC) CD-ROM containing the Instructor's Resource Guide (answer key), files of all figures in the manual, PowerPoint™ presentations for each laboratory, and the Prentice Hall Geoscience Animation Library (over 100 animations illuminating the most difficult-to-visualize geological concepts and phenomena in Flash files and PowerPoint slides).

Outstanding Mineral and Rock Labs

Mineral and rock labs are better than ever with enhanced student-tested illustrations, new photographs, identification flowcharts, and a revised five-page mineral database. (Please see page xiv for more information on the rock and mineral sets available through Prentice Hall and WARD'S Natural Science.)

Internet Support at www.prenhall.com/agi

The companion web site supports all labs with additional information and links listed by laboratory topic or by state/province.

Support for Geoscience

Royalties from sales of this product support programs of the American Geological Institute and the National Association of Geoscience Teachers.

ACKNOWLEDGMENTS

We acknowledge and sincerely appreciate the assistance of many people and organizations who have helped make possible this eighth edition of *Laboratory Manual in Physical Geology.*

Revisions in this new edition are based on suggestions from faculty who used the last editon of the manual, feedback from students using the manual at West Chester University of Pennsylvania, and market research by Pearson Prentice Hall. As changes were made to the laboratories, they were field tested in Introductory Geology laboratories at West Chester University. These field tests led to final revisions that helped make the manual more practical and user friendly.

We also thank the following faculty for their independent constructive criticisms and suggestions that led to improvements in this edition of the manual:

Kurtis Burmeister, University of the Pacific
Pamela Gore, Georgia Perimeter College
Martin Helmke, West Chester University of Pennsylvania

We thank *Edwin Anderson* (Temple University), *Allen Johnson* (West Chester University of Pennsylvania), *Carrick Eggleston* (University of Wyoming), *Randall Marrett* (University of Texas at Austin), and *Paul Morin* (University of Minnesota) for the use of their personal photographs. Photographs and data related to St. Catherines Island, Georgia, were made possible by research grants to the editor from the St. Catherines Island Research Program, administered by the American Museum of Natural History and supported by the Edward J. Noble Foundation.

Maps, map data, aerial photographs, and satellite imagery have been used courtesy of the U.S. Geological Survey; Canadian Department of Energy, Mines, and Resources; Surveys and Resource Mapping Branch, Ministry of Environment, Government of British Columbia; NASA; and the U.S./Japan ASTER Science Team.

The continued success of this laboratory manual depends on criticisms, suggestions, and new contributions from persons who use it. We sincerely thank everyone who contributed to this project by voicing criticisms, suggesting changes, and conducting field tests.

Unsolicited reactions to the manual are especially welcomed as a barometer for quality control and the basis for many changes and new initiatives that keep the manual current. Please continue to submit your frank criticisms and input directly to the editor: Rich Busch, Department of Geology and Astronomy, Boucher Building, West Chester University, West Chester, PA 19383 (rbusch@wcupa.edu).

P. Patrick Leahy, *Executive Director, AGI*

Ann Benbow, *Director of Education, AGI*

Christopher M. Keane, *Director of Technology and Communications, AGI*

Richard M. Busch, *Editor*

LABORATORY EQUIPMENT

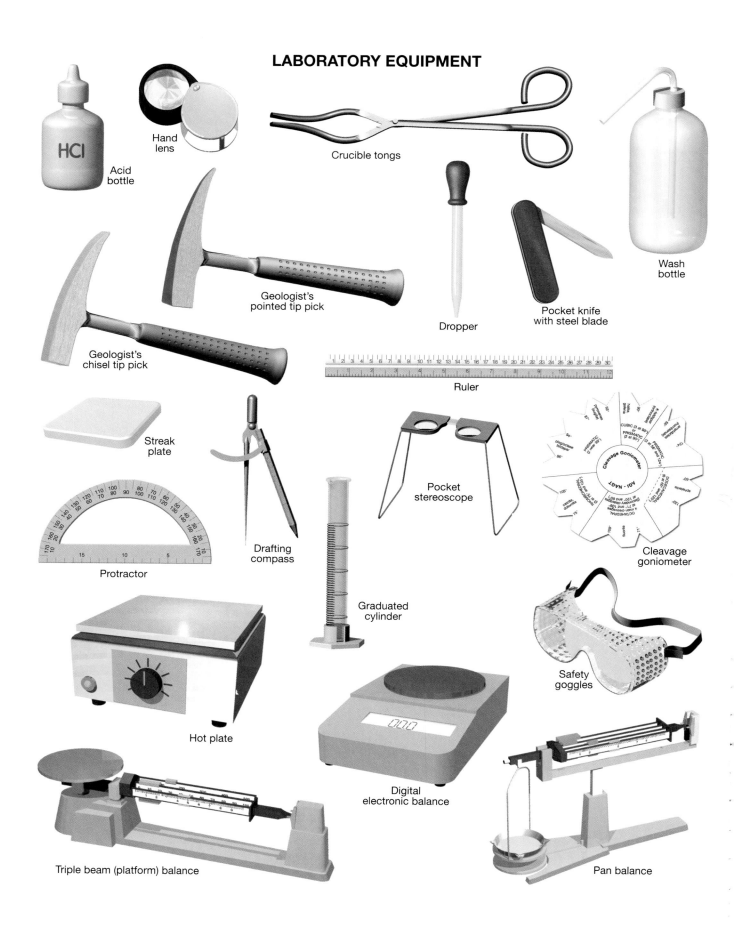

Acid bottle

Hand lens

Crucible tongs

Wash bottle

Geologist's pointed tip pick

Dropper

Pocket knife with steel blade

Geologist's chisel tip pick

Ruler

Streak plate

Drafting compass

Pocket stereoscope

Cleavage goniometer

Protractor

Graduated cylinder

Safety goggles

Hot plate

Digital electronic balance

Triple beam (platform) balance

Pan balance

LABORATORY EQUIPMENT LIST

R = Required, O = Optional

EQUIPMENT   LABORATORY NUMBERS

EQUIPMENT	1	2	3	4	5	6	7	8	9	10	11	12	13	14	15	16
Laboratory Notebook	R	R	R	R	R	R	R	R	R	R	R	R	R	R	R	R
Pencil with eraser	R	R	R	R	R	R	R	R	R	R	R	R	R	R	R	R
Calculator	R	R	R					R	R		R	R		R		R
Ruler (GeoTools Sheet 1 or 2)	R	R			R	R	R	R	R	R	R	R	R	R	R	R
Protractor (GeoTools Sheet 3)								R	R							
Colored pencils		R		R				O		R				R	R	
Scissors	R	R	R		R	R	R	R	R	R	R	R	R	R	R	
Mineral analysis tools* (steel/wire nails, glass plate, streak plate, penny, small magnet)			R	R	R	R	R									
Set of mineral samples*			R													
Set of miscellaneous rock samples*				R												
Set of igneous rock samples*					R											
Set of sedimentary rock samples*						R										
Set of metamorphic rock samples*							R									
Hand (magnifying) lens*			O	O	O	O	O									
Dropper bottle of dilute HCl*			R	R		R	R									
Small graduated cylinder (10 mL)*	R		O													
Large graduated cylinder (500 mL)*	R		O													
Basalt fragment that fits into the large graduated cylinder*	R															
Granite sample that fits into the large graduated cylinder*	R															
Small lump of modeling clay*	R															
Wood block (about 8 × 10 × 4 cm)*	R															
Gram balance*	R		O													
Small plastic bucket with water*	R																
Wash bottle with water*	O	R	O													R
Dropper with water*		R			O											
Lava lamp**		R			O											
Hot plate*		R														
Aluminum foil (roll)**		R														
Sugar cubes (2 per hot plate)*		R														
Permanent felt-tip marker*		R														
Crucible tongs*		R														
Cleavage goniometer from GeoTools Sheet 1			R													
Visual estimation of percent chart from GeoTools Sheet 1 or 2		R			R											
Sediment grain size scale from GeoTools Sheet 1 or 2						R										
UTM grids (GeoTools Sheets 2–4)										R						
Topographic quadrangle map*										O						
Geologic map*											O					
Pocket stereoscope										R		O		O	R	O
Cardboard models 1–6 cut from the back of this manual											R					
String (about 30 cm long)												R				R
Drafting compass																R
2 small plastic cups with dry sand*																R
Several coins per cup of sand																R

*Per group. **Per class.

MEASUREMENT UNITS

People in different parts of the world have historically used different systems of measurement. For example, people in the United States have historically used the English system of measurement based on units such as inches, feet, miles, acres, pounds, gallons, and degrees Fahrenheit. However, for more than a century, most other nations of the world have used the metric system of measurement. In 1975 the U.S. Congress recognized that global communication, science, technology, and commerce were aided by use of a common system of measurement, and they made the metric system the official measurement system of the United States. This conversion is not yet complete, so most Americans currently use both English and metric systems of measurement.

The International System (SI)

The International System of Units (SI) is a modern version of the metric system adopted by most nations of the world, including the United States. Each kind of metric unit can be divided or multiplied by 10 and its powers to form the smaller or larger units of the metric system. Therefore, the metric system is also known as a "base-10" or "decimal" system. The International System of Units (SI) is the official system of symbols, numbers, base-10 numerals, powers of 10, and prefixes in the modern metric system.

SYMBOL	NUMBER	NUMERAL	POWER OF 10	PREFIX
T	one trillion	1,000,000,000,000	10^{12}	tera-
G	one billion	1,000,000,000	10^9	giga-
M	one million	1,000,000	10^6	mega-
k	one thousand	1,000	10^3	kilo-
h	one hundred	100	10^2	hecto-
da	ten	10	10^1	deka-
	one	1	10^0	
d	one-tenth	0.1	10^{-1}	deci-
c	one-hundredth	0.01	10^{-2}	centi-
m	one-thousandth	0.001	10^{-3}	milli-
μ	one-millionth	0.000,001	10^{-6}	micro-
n	one-billionth	0.000,000,001	10^{-9}	nano-
p	one-trillionth	0.000,000,000,001	10^{-12}	pico-

Examples

1 meter (1 m) = 0.001 kilometers (0.001 km), 10 decimeters (10 dm), 100 centimeters (100 cm), or 1000 millimeters (1000 mm)
1 kilometer (1 km) = 1000 meters (1000 m)
1 micrometer (1 μm) = 0.000,001 meter (.000001 m) or 0.001 millimeters (0.001 mm)
1 kilogram (kg) = 1000 grams (1000 g)
1 gram (1 g) = 0.001 kilograms (0.001 kg)
1 metric ton (1 t) = 1000 kilograms (1000 kg)
1 liter (1 L) = 1000 milliliters (1000 mL)
1 milliliter (1 mL) = 0.001 liter (0.001 L)

Abbreviations for Measures of Time

A number of abbreviations are used in the geological literature to refer to time. Some of these abbreviations combine SI symbols with "yr" (*years*). Some abbreviations combine SI symbols with "a," for annum (*years before the present*).

yr (or y) = year
kyr = kiloyear—thousand years
Myr (or m.y.) = megayear—million years
Gyr (or Byr or b.y.) = gigayear—billion years
ka = kiloannum—thousand years before present or thousand years ago
Ma = megannum—million years before present or million years ago
Ga = gigannum—billion years before present or billion years ago

MATHEMATICAL CONVERSIONS

To convert:	To:	Multiply by:	
kilometers (km)	meters (m)	1000 m/km	**LENGTHS AND DISTANCES**
	centimeters (cm)	100000 cm/km	
	miles (mi)	0.6214 mi/km	
	feet (ft)	3280.83 ft/km	
meters (m)	centimeters (cm)	100 cm/m	
	millimeters (mm)	1000 mm/m	
	feet (ft)	3.2808 ft/m	
	yards (yd)	1.0936 yd/m	
	inches (in.)	39.37 in./m	
	kilometers (km)	0.001 km/m	
	miles (mi)	0.0006214 mi/m	
centimeters (cm)	meters (m)	0.01 m/cm	
	millimeters (mm)	10 mm/cm	
	feet (ft)	0.0328 ft/cm	
	inches (in.)	0.3937 in./cm	
	micrometers (μm)*	10000 μm/cm	
millimeters (mm)	meters (m)	0.001 m/mm	
	centimeters (cm)	0.1 cm/mm	
	inches (in.)	0.03937 in./mm	
	micrometers (μm)*	1000 μm/mm	
	nanometers (nm)	1000000 nm/mm	
micrometers (μm)*	millimeters (mm)	0.001 mm/μm	
nanometers (nm)	millimeters (mm)	0.000001 mm/nm	
miles (mi)	kilometers (km)	1.609 km/mi	
	feet (ft)	5280 ft/mi	
	meters (m)	1609.34 m/mi	
feet (ft)	centimeters (cm)	30.48 cm/ft	
	meters (m)	0.3048 m/ft	
	inches (in.)	12 in./ft	
	miles (mi)	0.000189 mi/ft	
inches (in.)	centimeters (cm)	2.54 cm/in.	
	millimeters (mm)	25.4 mm/in.	
	micrometers (μm)*	25,400 μm/in.	
square miles (mi^2)	acres (a)	640 acres/mi^2	**AREAS**
	square km (km^2)	2.589988 km^2/mi^2	
square km (km^2)	square miles (mi^2)	0.3861 mi^2/km^2	
acres	square miles (mi^2)	0.001563 mi^2/acre	
	square km (km^2)	0.00405 km^2/acre	
gallons (gal)	liters (L)	3.78 L/gal	**VOLUMES**
fluid ounces (oz)	milliliters (mL)	30 mL/fluid oz	
milliliters (mL)	liters (L)	0.001 L/mL	
	cubic centimeters (cm^3)	1.000 cm^3/mL	
liters (L)	milliliters (mL)	1000 mL/L	
	cubic centimeters (cm^3)	1000 cm^3/mL	
	gallons (gal)	0.2646 gal/L	
	quarts (qt)	1.0582 qt/L	
	pints (pt)	2.1164 pt/L	
grams (g)	kilograms (kg)	0.001 kg/g	**WEIGHTS AND MASSES**
	pounds avdp. (lb)	0.002205 lb/g	
pounds avdp. (lb)	kilograms (kg)	0.4536 kg/lb	
kilograms (kg)	pounds avdp. (lb)	2.2046 lb/kg	

To convert from degrees Fahrenheit (°F) to degrees Celsius (°C), subtract 32 degrees and then divide by 1.8
To convert from degrees Celsius (°C) to degrees Fahrenheit (°F), multiply by 1.8 and then add 32 degrees.

*Formerly called microns

National Association of Geoscience Teachers

Journal of Geoscience Education

Supporting this generation of educators and the next

For subscription & membership information:
www.nagt.org

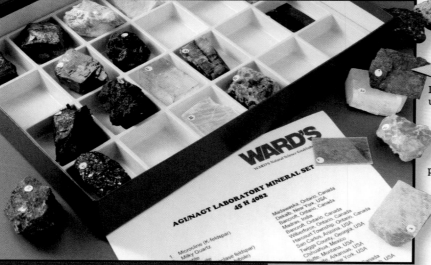

LABORATORY ONE

Observing and Measuring Earth Materials and Processes

•CONTRIBUTING AUTHORS•

Cynthia Fisher • *West Chester University of Pennsylvania*

C. Gil Wiswall • *West Chester University of Pennsylvania*

OBJECTIVES

A. Know what the "geologic record" is and how it is similar to, yet different from, a book.

B. Understand basic principles and tools of satellite remote sensing used by geoscientists and apply these principles and tools with your skills of observation to identify Earth materials, observe and describe processes of change, and make predictions.

C. Know that geology is based on a logical, testable process of science that is "ground truthed" with data obtained by direct observation, investigation, and measurement in the field (out of doors, in natural context) and in the laboratory.

D. Measure or calculate length, area, volume, mass, and density of Earth materials using basic scientific equipment and techniques.

E. Develop and test physical and quantitative models of isostasy based on floating wood blocks and icebergs. Then apply your quantitative model and your measurements of basalt and granite density to calculate the isostasy of average blocks of oceanic and continental crust.

F. Analyze Earth's global topography in relation to your work and a hypsographic curve, and infer how Earth's global topography may be related to isostasy.

MATERIALS

Part 1A: Pencil, eraser, laboratory notebook.
Parts 1B–1D: Pencil, eraser, laboratory notebook, metric ruler, small (10 mL) graduated cylinder, large (500 mL) graduated cylinder, pieces of basalt and granite that will fit into the large graduated cylinder, small lump of modeling clay (marble size), water, wood block (about 8 cm × 10 cm × 4 cm), small bucket to float wood block, gram balance, wash bottle or dropper (optional), and calculator (optional).

INTRODUCTION

Exploring and interacting with the natural world leads one to fundamental questions about the natural world and how it operates. **Science** is a way of answering these questions by *gathering evidence* (from careful observations), *running tests* to verify or falsify tentative ideas, *engaging in discourse* (discussing, organizing, and evaluating the evidence and test results with other scientists), and *making inferences* (conclusions based on reasonable interpretation of the evidence and test results). **Geology** is the branch of science that deals with Earth: its rocky body, 4.55 billion-year-long history, and environmental changes that affect humans. Its name comes from two Greek words, *geo* = Earth and *logos* = discourse. So geologists are also Earth scientists or geoscientists.

Geologists use their senses and tools (microscopes, rock hammers, rulers, etc.) to make direct observations of Earth materials and processes of change.

They also use computer-based technologies to make exact, automated, and even remote observations. These collective observations serve as a growing body of *data* (information, evidence) that enables geologists to characterize and classify Earth materials, identify relationships of cause (process) and effect (product), design models (physical, conceptual, mathematical, graphical, or artistic representations of something to test or demonstrate how it works), and publish inferences about Earth and how it operates. Geologists also apply their information and inferences to locate and manage resources, identify hazards, predict change, and help communities plan for the future.

PART 1A (PRELAB.): OBSERVING EARTH MATERIALS AND PROCESSES OF CHANGE THROUGH TIME

As you complete exercises in this laboratory manual, think of yourself as a geologist. Conduct tests and make careful observations. Record your observations carefully so you have a body of information (data, evidence) to justify your ideas (hypotheses, inferences). The quality of your ideas depends on your logic (method of thinking) and the information that you use to justify them. Your ideas may change as you make new observations, locate new information, or apply a different method of thinking. Your instructor will not accept simple yes or no answers to questions. S/he will expect your answers to be complete inferences justified with information and an explanation of your logic. Show your work whenever you use mathematics to solve a problem so your method of thinking is obvious.

When making observations, you should observe and record **qualitative information,** by *describing* how materials look, feel, smell, sound, taste, or behave. You should also collect and record **quantitative information,** by *measuring* materials, energy levels, and processes of change in time and space. This will require you to understand and use some scales and tools of measurement that professional geologists use in their work. You will also be expected to infer and quantify relationships by comparing one set of measurements to another.

Scales of Earth Observation

The most widely known geologic feature in the United States is undoubtedly the Grand Canyon (Figure 1.1). This canyon cuts a mile deep, through millions of rock layers that are like pages of an immense stone book of geologic history called the **geologic record.** These

FIGURE 1.1 Photograph of a portion of the Grand Canyon, Arizona. Rocks exposed at the base of the canyon are more than a billion years old, yet some layers of sand along the Colorado River that runs through the canyon may have formed just seconds ago. (Photo by Allen Johnson)

stone pages vary in thickness from millimeters to meters. Each page has distinguishing features—some as tiny as microscopic fossils or grains of sand and some as large as fossil trees, dinosaur skeletons, or ancient stream channels. Each successive stone page, from the bottom (oldest page) to the top (youngest page) of the canyon, is but one of millions of recorded events and times in Earth's long geologic history.

Geologists study all of Earth's materials, from the spatial scale of atoms (atomic scale) to the scale of our entire planet (global scale). At each spatial scale of observation, they identify materials and characterize relationships. Each scale is also related to the others. You should familiarize yourself with these **spatial scales of observation** as they are summarized in Figure 1.2 and in the summary table of quantitative units of measurement, symbols, abbreviations, and conversions on pages x and xi at the front of this manual.

Geologists also think about **temporal scales of observation.** As geologic detectives, they analyze the stone pages of the geologic record for evidence of events and relationships. As geologic historians, they group the events and relationships into paragraphs, chapters, sections, and parts of geologic history that occurred over epochs, periods, eras, and eons of time. The index to this book of geologic history is called the **geologic time scale** (Figure 1.3). Notice that the geologic time scale is a chart showing named intervals of the geologic record (rock units), the sequence in which they formed (oldest at the bottom), and their ages in millions of years. The eonothems, erathems, systems, and series of rock are the physical record of what happened during eons, eras, periods, and epochs of time. The intervals have been named and dated on the basis of more than a century of cooperative work among scientists of different nations, races, religions, genders, classes, and ethnic groups from throughout the world. What all of these scientists have had in common is the ability to do science and an intense desire to decipher Earth's long and complex history based on evidence contained in the stone pages that record geologic history.

SPATIAL SCALES OF OBSERVATION USED BY GEOLOGISTS

Scale of observation		Used to study things like...	Measured in...
Global	Macroscopic: visible with the naked eye	Entire planet and its interactive "spheres"	Thousands of kilometers (km) or miles (mi)
Regional		Portions of oceans, continents, countries, provinces, states, islands	Kilometers (km), miles (mi)
Local (outcrop or field site)		Specific locations that can be "pin-pointed" on a map	Meters (m), feet (ft)
Hand sample (field/lab. sample)		Sample of a mineral, a rock, air, water, or an organism that can be held in your hand	Centimeters (cm), millimeters (mm), inches (in.) 0 1 cm 0 10 mm
Microscopic		Features of a hand sample that can only be seen with a hand lens (magnifier) or microscope	Fractions of millimeters (mm), micrometers (m)
Atomic (or molecular)		Arrangements of the atoms or molecules in a substance	Nanometers (nm), angstroms (Å)

FIGURE 1.2 Spatial scales of observation used by geologists in their work.

Processes and Cycles of Change

Earth is characterized by energy flow and processes of change at every spatial and temporal scale of observation. Earth's surface is energized by geothermal energy (from inside the planet) and solar energy (from outside the planet). The energy flows from *sources* to *sinks* (materials that store or convert energy) and drives processes of change like the examples in

THE GEOLOGIC TIME SCALE

A chart showing the sequence, names, and ages of Earth's rock layers (oldest at the bottom)

Eon of time Eonothem of rock	Era of time Erathem of rock	Period of time System of rock**		Epoch of time Series of rock	Millions of years ago (Ma)	Some notable fossils in named rock layers
Phanerozoic	Cenozoic: (new life) Age of Mammals 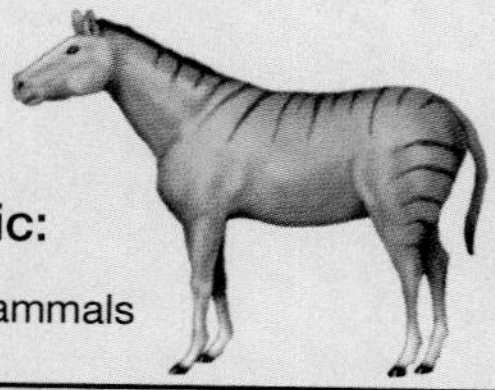	Q	Neogene (N)	Recent	.01	Human fossils in addition to abundant fossils of mammals
				Pleistocene	1.8	
				Pliocene	2.6	
		Tertiary		Miocene	5.3	
					23	
			Paleogene (P_G)	Oligocene	34	Abundant fossils of mammals
				Eocene	56	
				Paleocene		
					65	
	Mesozoic: (middle life) Age of Reptiles 	Cretaceous (K)				Last dinosaur fossils: including *Tyrannosaurus rex*
					145	
		Jurassic (J)				First bird fossil: *Archaeopteryx*
					200	
		Triassic (Ŧ)				First dinosaur, mammal, turtle, and crocodile fossils
					251	
	Paleozoic: (old life) Age of Trilobites 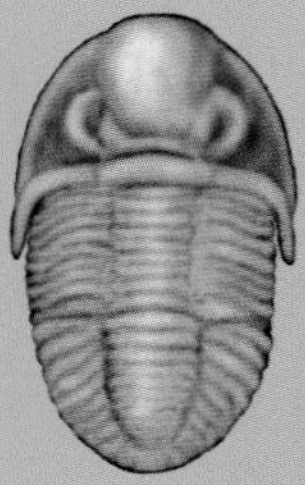	Permian (P)				Last (youngest) trilobite fossils
					299	
		Carboniferous (C)*	Pennsylvanian (ℙP)			First reptile fossils
					318	
			Mississippian (M)			First fossil conifer trees
					359	
		Devonian (D)				First amphibian, insect, tree, and shark fossils
					416	
		Silurian (S)				First true land plant fossils
					444	
		Ordovician (O)				First fossils of coral and fish
					488	
		Cambrian (Ꞓ)				First trilobite fossils First abundant visible fossils
					542	
Proterozoic	Precambrian: An informal name for all of this time and rock.				2500	Oldest fossils: mostly microscopic life, visible fossils rare
Archean						
		Acasta Gneiss, northwestern Canada			4030	
Hadean						
	Oldest meteorites	Zircon mineral crystals in the Jack Hills Metaconglomerate, Western Australia			4400	
					4550	

*European name
**Symbols in parentheses are abbreviations commonly used to designate the age of rock units on geologic maps.
Q = Quaternary

FIGURE 1.3 The geologic time scale. Absolute ages in millions of years ago (Ma) were provided courtesy of the International Commission on Stratigraphy, 2004. See their web site for more detailed versions and recent updates of the geological time scale (**http://www.stratigraphy.org**).

Figure 1.4. Most of these processes involve organic (biological; parts of living or once living organisms) and inorganic (non-biological) materials in solid, liquid, and gaseous states, or *phases* (Figure 1.5). Note that many of the processes have opposites depending on the flow of energy to or from a material: melting and freezing, evaporation and condensation, sublimation and deposition, dissolution and chemical precipitation, photosynthesis (food energy storage) and respiration (food energy release or "burning" without flames). And while some chemical reactions are irreversible, most are reversible (as in the process of dissociation). Thus opposing processes of change cause chemical materials to be endlessly cycled and recycled between two or more phases. One of these cycles is the *hydrologic cycle*, or "water cycle" (Figure 1.6).

The hydrologic cycle involves several processes and changes in relation to all three phases of water and all of Earth's spheres (global subsystems). It is one of the most important cycles that geologists routinely consider in their work. The hydrologic cycle is generally thought to operate like this: water (hydrosphere) evaporating from Earth's surface produces water vapor (atmospheric gas). The water vapor eventually condenses in the atmosphere to form aerosol water droplets (clouds). The droplets combine to form raindrops or snowflakes (atmospheric precipitation). Snowflakes can accumulate to form ice (cryosphere) that sublimates back into the atmosphere or melts back into water. Both rainwater and meltwater soak into the ground (to form groundwater), evaporate back into the atmosphere, drain back into the ocean, or are consumed by plants and animals (which release the water back to the atmosphere via the process of transpiration).

In addition to water that is moving about the Earth system, there is also water that is stored and not circulating at any given time. For example, a very small portion of Earth's water (about 2% of the water volume in oceans) is currently stored in snow and glacial ice at the poles and on high mountaintops. Additional water (perhaps as much as 80% of the water now in oceans) is also stored in *"hydrous"* (water-bearing) minerals inside Earth. When glaciers melt, or rocks melt, the water can return to active circulation.

The endless exchange of energy and recycling of water undoubtedly has occurred since the first water bodies formed on Earth billions of years ago. Your next drink may include water molecules that once were part of a hydrous (water-bearing) mineral inside Earth or that once were consumed by a thirsty dinosaur!

Relating Scales of Understanding

The hydrologic cycle is a reminder that each thing on Earth is somehow related to everything else in space, time, or process. Geologists seek to understand these complex relationships relative to human lifetimes and the geologic time scale. For you to think like a geologist, you must consider many materials and processes over a broad range of temporal and spatial scales of observation. One way to simplify this thinking process is to develop hierarchical levels (domains) of understanding, from the most general to the most specific. The most generalized way of thinking about Earth is to think of it as a single planet (global scale). A more specific and complex way to think about Earth is to think about individual locations, samples, or even atoms. Of course, materials and processes observed at one scale might not be visible at another scale. Therefore, a central tenet of geological analysis and understanding is that geologic materials and processes should be addressed first and foremost at their characteristic scales (i.e., their obvious spatial, temporal, or spacio-temporal domains).

At the global scale of observation, geologists conceptualize Earth as a dynamic planetary system comprised of interacting *spheres* (subsystems). They are lithosphere (rock), hydrosphere (water), atmosphere (air), cryosphere (ice), and biosphere (life).

The rocky body of Earth consists of a core, mantle, and crust (Figure 1.7A). The outer edge of this rocky body is the rigid *lithosphere,* which consists of Earth's crust and lithospheric mantle (i.e., the rigid outermost part of the mantle). The lithosphere (Figure 1.7B) is about 100 km (60 mi) thick on average and rests on the *asthenosphere,* a relatively weak, soft zone of the upper mantle where some of the rocks flow very slowly due to the intense heat and pressure. The lithosphere is not a single eggshell-like covering. It is a mosaic of about a dozen large, rigid "plates" that are moved by the slow plastic flow of the asthenosphere. Zones between the plates, called *plate boundaries,* are regions where common earthquakes and volcanic activity occur. This is because the plates are sliding past each other (to form faults), colliding with each other (to form mountains), or separating apart from each other (to form valleys and oceans).

The most rigid part of the lithosphere is Earth's crust. Geologists have not directly observed the base of the crust (called the Mohorovicic discontinuity or "Moho", Figure 1.7B) or the subjacent asthenosphere. Even the deepest exploration wells have only been drilled to a depth of about 12 km (7.5 mi) (Figure 1.7C).

COMMON PROCESSES OF CHANGE

Process	Kind of Change	Example
Melting	Solid phase changes to liquid phase.	Water ice turns to water.
Freezing	Liquid phase changes to solid phase.	Water turns to water ice.
Evaporation	Liquid phase changes to gas (vapor) phase.	Water turns to water vapor or steam (hot water vapor).
Condensation	Gas (vapor) phase changes to liquid phase.	Water vapor turns to water droplets.
Sublimation	Solid phase changes directly to a gas (vapor) phase.	Dry ice (carbon-dioxide ice) turns to carbon dioxide gas.
Deposition	The laying down of solid material as when a gas phase changes into a solid phase or solid particles settle out of a fluid.	Frost is the deposition of ice (solid phase) from water vapor (gas). There is deposition of sand and gravel on beaches.
Dissolution	A substance becomes evenly dipersed into a liquid (or gas). The dispersed substance is called a solute, and the liquid (or gas) that causes the dissolution is called a solvent.	Table salt (solute) dissolves in water (solvent).
Vaporization	Solid or liquid changes into a gas (vapor), due to evaporation or sublimation.	Water turns to water vapor or water ice turns directly to water vapor.
Reaction	Any change that results in formation of a new chemical substance (by combining two or more different substances).	Sulfur dioxide (gas) combines with water vapor in the atmosphere to form sulfuric acid, one of the acids in rain.
Decomposition	An irreversible reaction. The different elements in a chemical compound are irreversibly split apart from one another to form new compounds.	Feldspar mineral crystals decompose to clay minerals and metal oxides (rust).
Dissociation	A reversible reaction in which some of the elements in a chemical compound are temporarily split up. They can combine again under the right conditions to form back into the starting compound.	The mineral gypsum dissociates into water and calcium sulfate, which can recombine to form gypsum again.
Chemical precipitation	A solid that forms when a liquid solution evaporates or reacts with another substance.	Salt forms as ocean water evaporates. Table salt forms when hydrochloric acid and sodium hydroxide solutions are mixed.
Photosynthesis	Sugar (glucose) and oxygen are produced from the reaction of carbon dioxide and water in the presence of sunlight (solar energy).	Plants produce glucose sugar and oxygen.
Respiration	Sugar (glucose) and oxygen undergo combustion (burning) without flames and change to carbon dioxide, water, and heat energy.	Plants and animals obtain their energy from respiration.
Transpiration	Water vapor is produced by the biological processes of animals and plants (respiration, photosynthesis).	Plants release water vapor to the atmosphere through their pores.
Evolution	Change in a specific direction (gradually or in stages).	Biological evolution, change in the shape of Earth's landforms through time.
Crystallization	Atoms, ions, or molecules arrange themselves into a regular repeating 3-dimensional pattern. The formation of a crystal.	Water vapor freezes into snowflakes. Liquid magma cools into a solid mass of crystals.
Weathering	Materials are fragmented, worn, or chemically decomposed.	Rocks break apart, get worn into pebbles or sand, dissolve, rust, or decompose to mud.
Transportation	Materials are pushed, bounced, or carried by water, wind, ice, or organisms.	Sand and soil are blown away. Streams push, bounce, and carry materials downstream.
Convection	Current motion (and heat transfer) within a body of material (gas, liquid, or soft solid). As part of the material is heated and rises, a cooler part of the material descends to replace it and form a cycle of convection (convection cell).	Warm air in the atmosphere rises and cooler air descends to replaces it; water boiling in a pot.

FIGURE 1.4   Some common processes of change on Earth.

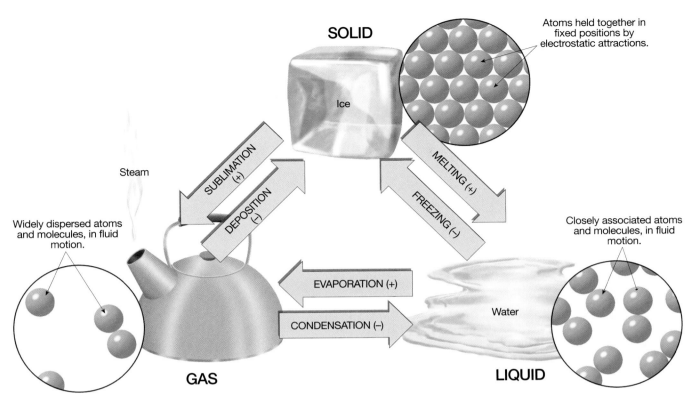

FIGURE 1.5 Ternary diagram showing the three states (*phases*) of water, plus six common processes that change states of matter by heating (+) and cooling (−). Note the distribution and packing of atoms and molecules in fluid (liquid and gas) versus solid states.

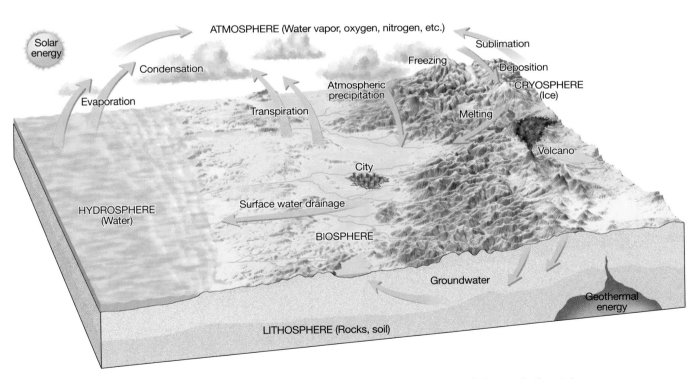

FIGURE 1.6 The hydrologic cycle (water cycle). Note the relationship of processes of change in the states of water (evaporation, condensation, etc.) to Earth's spheres (lithosphere, cryosphere, hydrosphere, atmosphere, biosphere). Also note that the hydrologic cycle is driven (forced to operate) by energy from the Sun (solar energy), energy from Earth's interior (volcanoes and geothermal energy), and gravity.

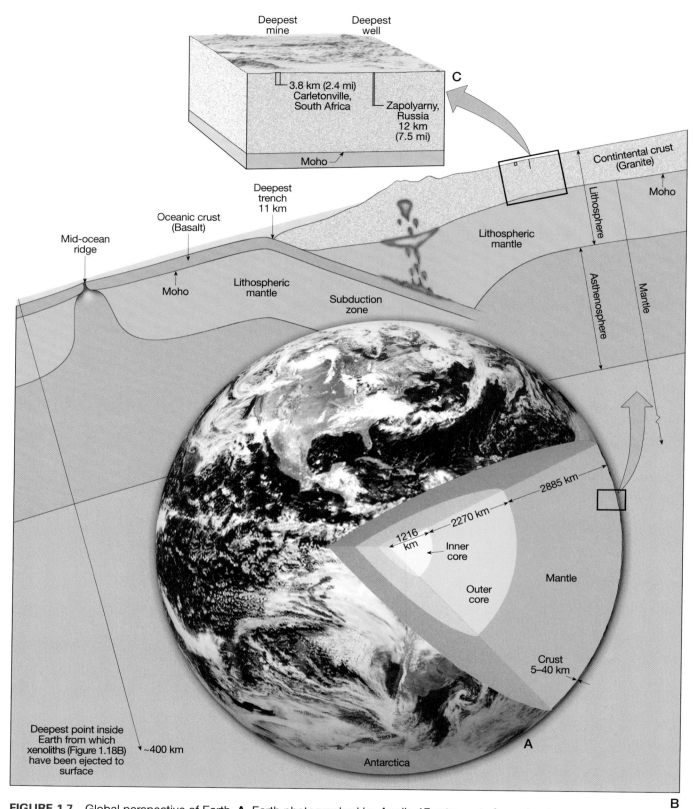

Deepest
mine

Deepest
well

C

3.8 km (2.4 mi)
Carletonville,
South Africa

Zapolyarny, Russia
12 km
(7.5 mi)

Moho

Contintental crust
(Granite)

Moho

Lithosphere

Lithospheric
mantle

Deepest
trench
11 km

Oceanic crust
(Basalt)

Mid-ocean
ridge

Moho

Lithospheric
mantle

Subduction
zone

Asthenosphere

Mantle

2885 km

2270 km

1216
km

Inner
core

Outer
core

Mantle

Crust
5–40 km

Deepest point inside
Earth from which
xenoliths (Figure 1.18B)
have been ejected to
surface

~400 km

Antarctica

A

B

FIGURE 1.7 Global perspective of Earth. **A.** Earth photographed by *Apollo 17* astronauts from about 37,000 km (23,000 mi) away in 1972 (Courtesy of NASA). Note compositional divisions of Earth's rocky body in cut-away view (inner core, outer core, mantle, crust). **B.** Hypothetical cross section of the edge of Earth's rocky body. Note locations of continental crust, oceanic crust, Moho (base of the crust), lithosphere, and asthenosphere. **C.** Depths of the deepest mine and well ever drilled into Earth.

All of Earth's other visible spheres rest on its rocky body. The *hydrosphere* is all of the liquid water on Earth's surface and in the ground (groundwater). Most of the hydrosphere is salt water in the oceans. The *atmosphere* is the gaseous envelope that surrounds the Earth. It consists of about 75% nitrogen, 20% oxygen, and small amounts of other gases like argon, carbon dioxide, water vapor, and methane. The *cryosphere* is the snow and ice that forms from freezing parts of the hydrosphere or atmosphere. (Since ice is a mineral, the cryosphere can also be thought of as part of the lithosphere.) Most of it exists in the polar ice sheets (continental glaciers), permafrost (permanently frozen moisture in the ground), and sea ice (ice on the oceans). The *biosphere* is the living part of Earth, the part that is organic and self-replicating. It is comprised of all of the plants and animals, so you are a member of the biosphere.

Observation of materials and processes at more specific scales of observation provides more specific data and levels of understanding. Any comprehensive study of geology involves data collected from several spatio-temporal domains, each one linked to the others at some level of significance.

Ground Truth: Field Geology and Lab Work

The most reliable information about Earth is obtained by direct observation, investigation, and measurement in the field (out of doors, in natural context) and laboratory. Most geologists study *outcrops*—field sites where rocks *crop out* (stick out of the ground). The outcrops are comprised of rocks, and rocks are comprised of minerals.

Samples obtained "in the field" (from outcrops at field sites) are often removed to the laboratory for further analysis using basic science. Careful observation (use of your senses, tactile abilities, and tools to gather information) and critical thought lead to questions and hypotheses (tentative ideas to test). Investigations are then designed and carried out to test the hypotheses and gather data (information, evidence). Results of the investigations are analyzed to answer questions and justify logical conclusions.

Refer to the example of field and laboratory analysis in Figure 1.8. Observation 1 (in the field) reveals that Earth's rocky lithosphere crops out at the surface of the land. Observation 2 reveals that outcrops are comprised of rocks. Observation 3 reveals that rocks are comprised of mineral crystals such as the mineral *chalcopyrite*. This line of reasoning leads to the next **logical question:** *What is chalcopyrite comprised of?* Let us consider the two most logical possibilities, or **working hypotheses** (tentative ideas to investigate, test). It is always best to have more than one working hypothesis.

1. Chalcopyrite may be a pure substance, or chemical *element*, so chalcopyrite is made of chalcopyrite. What investigating and gathering of evidence could we do to reasonably determine if this is true or false?

2. Chalcopyrite may be a *compound* comprised of two or more elements. What investigating and gathering of evidence could we do to reasonably determine if this is true or false? If true, then how could we find out what elements make up chalcopyrite?

Let us conduct two **investigations** (activities planned and conducted to test hypotheses, gather and record data, make measurements, or control and explore variables). In Investigation 1, the chalcopyrite is ground to a powder and heated. This investigation reveals the presence of sulfur and at least one other substance. The remaining substance is attracted to a magnet, so it may be iron, or a compound containing iron. When the powder is leached (dissolved in acidic water) and subjected to electrolysis (Investigation 2), copper separates from the powder. The remaining powder is attracted to a magnet, indicating the presence of iron. **Analysis of the results** of these two investigations leads us to the **logical conclusion** that Hypothesis 2 was correct (i.e., chalcopyrite is a compound). The results are also evidence that chalcopyrite is comprised of three different elements: sulfur (S), iron (Fe), and copper (Cu). Chemists call chalcopyrite, copper-iron sulfide ($CuFeS_2$). Since chalcopyrite contains a significant proportion of copper, it is also a *copper ore* (natural material from which copper can be extracted at a reasonable profit).

This same laboratory procedure is applied on a massive scale at copper mines. Because most copper-bearing rock contains only a few percent of chalcopyrite or another copper-bearing mineral, the rock is mined, crushed, and powdered. It is then mixed with water, detergents, and air bubbles that float the chalcopyrite grains to the surface of the water. When these grains are removed, they are smelted (roasted) to separate the copper from the other parts of the chalcopyrite and melted rock (that cool to form *slag*). The remaining copper powder is then leached in sulfuric acid and subjected to electrolysis, whereupon the copper is deposited as a mass of pure copper on the positive electrode (cathode).

Example of Geologic Field and Laboratory Investigation

OBSERVATION 1:
Earth's lithosphere crops out in surface exposures called *outcrops*.

OBSERVATION 2:
Outcrops are comprised of *rocks*.

OBSERVATION 3:
Rocks are comprised of *mineral crystals*. Chalcopyrite is a kind of mineral crystal found in some rocks.

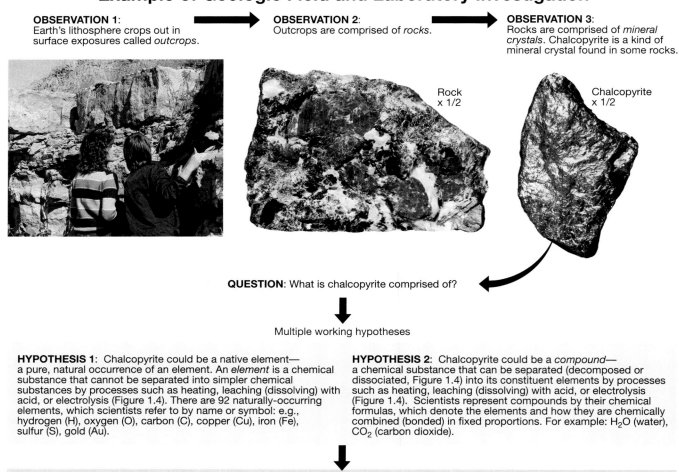

Rock
x 1/2

Chalcopyrite
x 1/2

QUESTION: What is chalcopyrite comprised of?

Multiple working hypotheses

HYPOTHESIS 1: Chalcopyrite could be a native element— a pure, natural occurrence of an element. An *element* is a chemical substance that cannot be separated into simpler chemical substances by processes such as heating, leaching (dissolving) with acid, or electrolysis (Figure 1.4). There are 92 naturally-occurring elements, which scientists refer to by name or symbol: e.g., hydrogen (H), oxygen (O), carbon (C), copper (Cu), iron (Fe), sulfur (S), gold (Au).

HYPOTHESIS 2: Chalcopyrite could be a *compound*— a chemical substance that can be separated (decomposed or dissociated, Figure 1.4) into its constituent elements by processes such as heating, leaching (dissolving) with acid, or electrolysis (Figure 1.4). Scientists represent compounds by their chemical formulas, which denote the elements and how they are chemically combined (bonded) in fixed proportions. For example: H_2O (water), CO_2 (carbon dioxide).

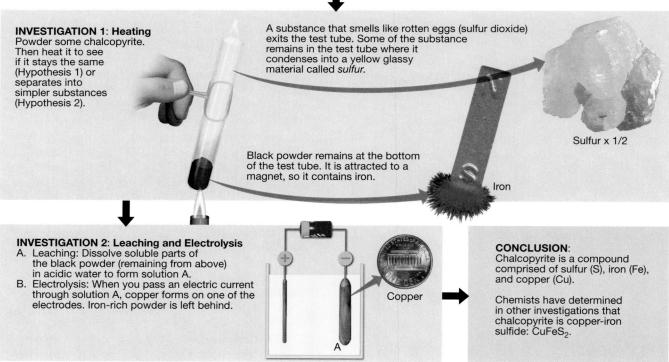

INVESTIGATION 1: Heating
Powder some chalcopyrite. Then heat it to see if it stays the same (Hypothesis 1) or separates into simpler substances (Hypothesis 2).

A substance that smells like rotten eggs (sulfur dioxide) exits the test tube. Some of the substance remains in the test tube where it condenses into a yellow glassy material called *sulfur*.

Sulfur x 1/2

Black powder remains at the bottom of the test tube. It is attracted to a magnet, so it contains iron.

Iron

INVESTIGATION 2: Leaching and Electrolysis
A. Leaching: Dissolve soluble parts of the black powder (remaining from above) in acidic water to form solution A.
B. Electrolysis: When you pass an electric current through solution A, copper forms on one of the electrodes. Iron-rich powder is left behind.

Copper

A

CONCLUSION:
Chalcopyrite is a compound comprised of sulfur (S), iron (Fe), and copper (Cu).

Chemists have determined in other investigations that chalcopyrite is copper-iron sulfide: $CuFeS_2$.

FIGURE 1.8 Example of geologic field and laboratory investigation.

Satellite Remote Sensing of Geology

There are times when geologists cannot make direct observations of Earth and must rely on a technology to acquire and record information remotely (from a distance, without direct contact). This is called *remote sensing*. One of the most common kinds of remote sensing used by geologists is satellite remote sensing. Some satellites, such as the space shuttle and International Space Station, are manned by astronauts who take photographs of Earth using cameras. However, most satellite remote sensing is done by scanners mounted on unmanned environmental satellites. These satellites scan Earth to obtain a digital data set, and the data are sent to ground stations that convert it into visual satellite images of Earth. Instruments aboard the satellites scan information from not only the visible part of the electromagnetic spectrum, but also parts of the spectrum that are not visible to humans (e.g., infrared).

The electromagnetic (EM) spectrum of radiation (Figure 1.9) is a spectrum of electric and magnetic waves that travel at the speed of light (300,000,000 meters/second, or 3×10^8 m/s). The spectrum is subdivided into **bands**—parts of the EM spectrum that are defined and named according to their wavelength (distance between two adjacent wave crests or troughs). Waves in the gamma ray and X-ray bands have the shortest wavelengths, which are measured in billionths of a meter (called nanometers, nm). Waves in the ultraviolet, visible, and infrared bands have wavelengths measured in millionths of a meter (called micrometers, μm). The long waves of the microwave and broadcast radio bands are measured in meters (m) and kilometers (km).

Earth materials may *emit* (send out) EM waves in the thermal (heat) infrared band (3–15 μm). Earth materials also *reflect* EM waves (like a mirror), *absorb* EM waves and convert them to another wavelength or form of energy (like sunlight converted to heat), or *transmit* EM waves (allow them to pass through, like a glass window). Many Earth materials can be identified from space by their characteristic wavelengths and intensities of emitted and reflected EM radiation. However, Earth's atmosphere is very effective at absorbing (not transmitting) EM radiation in the gamma ray, X-ray, most of the ultraviolet, and some parts of the infrared wavelengths. These absorbed bands cannot be used in satellite remote sensing. (Note how this is shown graphically at the top of Figure 1.9, which shows the percentage of EM radiation transmitted through Earth's atmosphere.) Therefore, satellite remote sensing is based on the use of EM radiation that is transmitted through Earth's atmosphere and provides the most useful information to identify and characterize Earth materials.

The most intense EM radiation that reaches Earth's surface is short wavelengths of EM radiation from our Sun. This includes some ultraviolet bands (0.3−0.4 μm), all of the visible bands (0.4−0.7 μm), and the reflected infrared bands [both NIR—near infrared (0.7−0.9 μm) and SWIR—short wave infrared (0.9−3.0 μm)]. The most intense solar radiation is the green visible band. Some solar EM radiation is absorbed by Earth materials or transmitted through them. What is not absorbed or transmitted is reflected back into space, where it is detected by satellite instruments that are remotely sensing the reflected bands of EM radiation. This is done during the day, when the reflections occur.

Some solar radiation is absorbed by Earth materials, converted to heat, and then emitted from those materials as long wavelengths of invisible, thermal (heat) infrared radiation (TIR radiation, 3−30 μm). Earth materials warmed in other ways (e.g., fire, body heat produced by respiration, volcanic heat from Earth's interior) also emit thermal (heat) infrared radiation (TIR). It is normally detected and measured by satellite instruments operating at night, when there is no sunlight to heat up surfaces. This allows geologists to see objects that emit heat on their own, even when they are not absorbing solar radiation and converting it to heat.

Microwaves (0.1–30 cm) are also invisible to humans. However, they are used in imaging radar instruments that measure the distance between a satellite (or aircraft) and points on Earth. Imaging radar is used to make images with a three-dimensional perspective.

Data from environmental satellite instruments must be rendered into an image that humans can see, either by giving objects in the image their true color or a false color. **True color** photographs and satellite images show objects in the colors that they would appear to be if viewed by the human eye. They are comprised of the colors we detect within the visible spectrum of red, orange, yellow, green, blue, indigo, and violet colors. Note examples (Figure 1.10) of a true color photograph obtained by astronauts on the International Space Station and a true-color satellite image produced with data from the MODIS instrument (currently mounted on unmanned Aqua and Terra satellites). Instruments aboard environmental satellites also detect and measure EM radiation in the bands that humans cannot normally see. Because they are invisible, they have no true color. These normally invisible bands must be given a **false color** in satellite images.

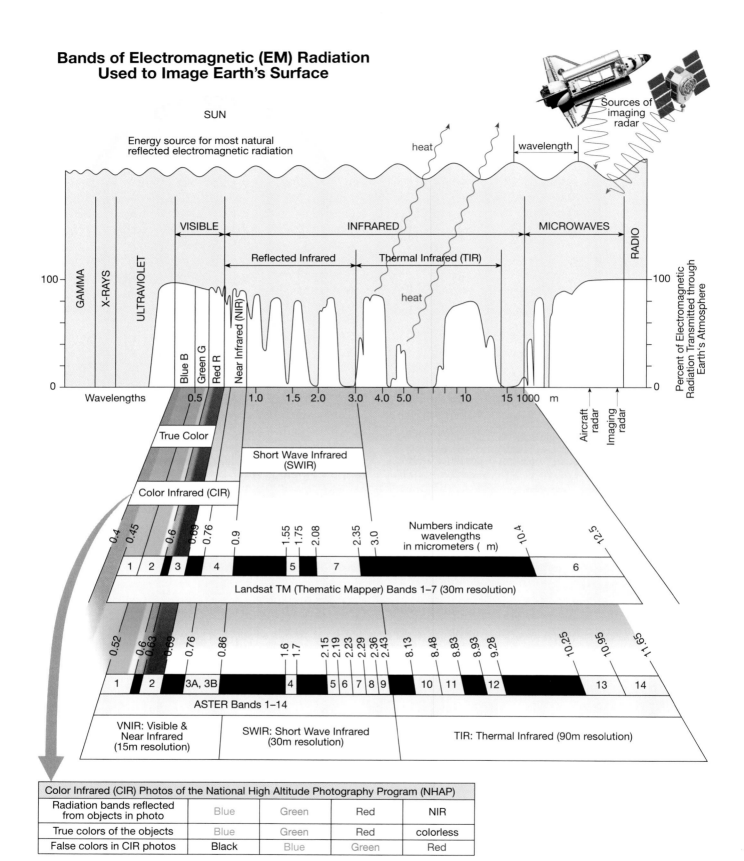

Bands of Electromagnetic (EM) Radiation Used to Image Earth's Surface

SUN

Energy source for most natural reflected electromagnetic radiation

heat

wavelength

Sources of imaging radar

VISIBLE   INFRARED   MICROWAVES

Reflected Infrared   Thermal Infrared (TIR)

heat

GAMMA   X-RAYS   ULTRAVIOLET   Blue B   Green G   Red R   Near Infrared (NIR)   RADIO

Percent of Electromagnetic Radiation Transmitted through Earth's Atmosphere

100   100

0   0

Wavelengths   0.5   1.0   1.5   2.0   3.0   4.0   5.0   10   15   1000   µm

Aircraft radar   Imaging radar

True Color

Short Wave Infrared (SWIR)

Color Infrared (CIR)

Numbers indicate wavelengths in micrometers (µm)

0.4   0.45   0.6   0.69   0.76   0.9   1.55   1.75   2.08   2.35   3.0   10.4   12.5

| 1 | 2 | 3 | 4 | | 5 | 7 | | | 6 |

Landsat TM (Thematic Mapper) Bands 1–7 (30m resolution)

0.52   0.6   0.63   0.69   0.76   0.86   1.6   1.7   2.15   2.19   2.23   2.29   2.36   2.43   8.13   8.48   8.83   8.93   9.28   10.25   10.95   11.65

| 1 | 2 | 3A, 3B | 4 | 5 | 6 | 7 | 8 | 9 | 10 | 11 | 12 | 13 | 14 |

ASTER Bands 1–14

VNIR: Visible & Near Infrared (15m resolution)   SWIR: Short Wave Infrared (30m resolution)   TIR: Thermal Infrared (90m resolution)

Color Infrared (CIR) Photos of the National High Altitude Photography Program (NHAP)				
Radiation bands reflected from objects in photo	Blue	Green	Red	NIR
True colors of the objects	Blue	Green	Red	colorless
False colors in CIR photos	Black	Blue	Green	Red

FIGURE 1.9 Bands of electromagnetic (EM) radiation detected with the human eye (true color), NHAP color-infrared film, and the Landsat TM and ASTER instruments on environmental satellites. Refer to text for discussion (page 11).

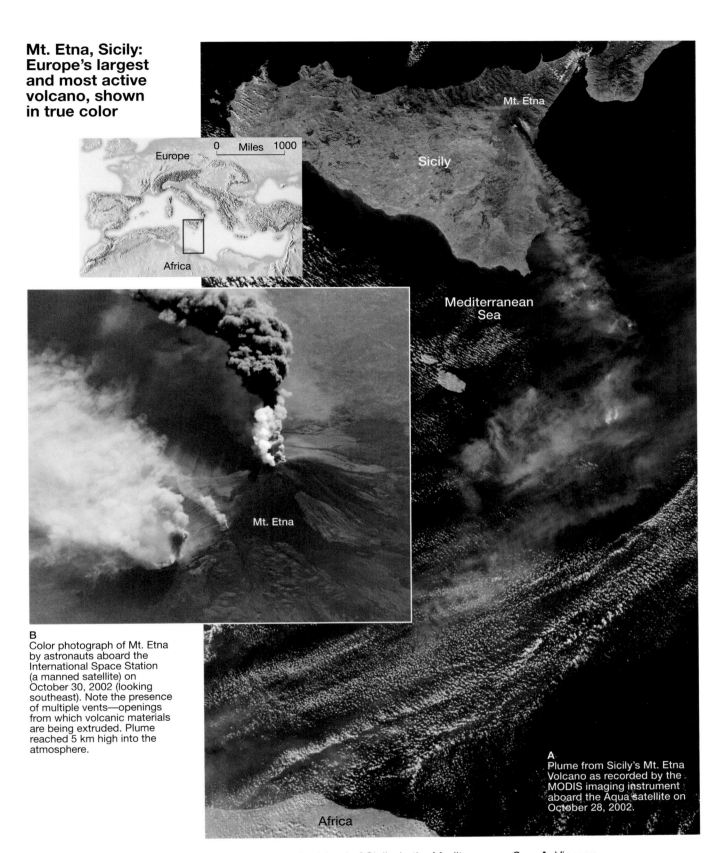

Mt. Etna, Sicily: Europe's largest and most active volcano, shown in true color

Europe

0 Miles 1000

Africa

Mt. Etna

Sicily

Mediterranean Sea

Mt. Etna

B
Color photograph of Mt. Etna by astronauts aboard the International Space Station (a manned satellite) on October 30, 2002 (looking southeast). Note the presence of multiple vents—openings from which volcanic materials are being extruded. Plume reached 5 km high into the atmosphere.

A
Plume from Sicily's Mt. Etna Volcano as recorded by the MODIS imaging instrument aboard the Aqua satellite on October 28, 2002.

Africa

FIGURE 1.10   True color pictures of Mt. Etna, on the island of Sicily, in the Mediterranean Sea. **A.** View on October 28, 2002, by the MODIS instrument on the Aqua satellite (Courtesy of Jacques Descloitres, MODIS Rapid Response Team, NASA GSFC). **B.** View on October 30, 2002, photographed by astronauts aboard the International Space Station.   (Image ISS05-E-19024: Courtesy of Earth Science & Image Analysis Laboratory, NASA Johnson Space Center)

Some of the false color images in this manual are *color infrared* (CIR, Figure 1.9) photographs taken from airplanes by the National High Altitude Photography Program (NHAP, Figure 1.9). They are taken with cameras that record information from the blue, green, red, and near infrared bands. Because the near infrared band is normally invisible to humans, these NHAP–CIR photographs must be false colored. Blue/gray objects are colored black, green objects are colored blue, red objects are colored green, and the near infrared objects (objects reflecting near infrared wavelengths) are colored red. Healthy vegetation generally reflects near infrared radiation and is false colored red in the NHAP–CIR images. Where vegetation is very sparse and the true color of the land is brown (green + red), the NHAP–CIR landscape appears as a false blue-green color. This is because the visible green is false colored blue and the visible red is false colored green.

Because Earth materials can be identified from space by the characteristic EM wavelengths that they emit and reflect, instruments on environmental satellites scan both visible and invisible bands of EM radiation. For example, the MODIS instrument on Terra and Aqua satellites scans 36 different bands of data! At the same time, however, the resolution of MODIS is not as good as that of some other instruments like the Landsat TM (Thematic Mapper) and ASTER (Advanced Spaceborne Thermal Emission and Reflection Radiometer) instruments.

All instruments that measure spatial or temporal information have a **resolution**—ability to resolve and measure detail. The smaller the increments of measurement or imaging, the greater the resolution. Therefore, one way to tell the resolution of a photograph or image is to determine the smallest object that it reveals. Image resolution also depends on the resolution of the digital scanning instrument that collected the data for the image.

Scanners on environmental satellites are digital scanners that subdivide the field of view into rows of square picture elements, or *pixels*. The pixels are assembled row-by-row via computer to produce an image. All of the light (EM radiation) entering the scanner for the area represented by a single pixel is averaged into a single reading for that pixel. Therefore, satellite images are not exact copies of what was in the satellite's field of view. They are a grid of information averaged into pixels at some resolution. The easiest way to express the resolution of a satellite image is to determine the surface area that one of its pixels represents. If each pixel in an image represents a 100 m by 100 m square of land, then the satellite image is said to have a resolution of 100 m. At this resolution, each pixel is bigger than a house, so homes would not be visible in the image.

MODIS satellite images have a resolution of 250 m (bands 1–2, Figure 1.10A), 500 m (bands 3–7), and 1000 m (bands 8–36). This makes them useful for viewing large features of Earth at a regional scale of observation. MODIS instruments scan the entire Earth every 1–2 days.

Landsat TM instruments collect digital data at a resolution of 30 m from seven different bands (Figure 1.9). Bands 1, 2, and 3 are blue, green, and red bands useful for true color images. Band 4 is near infrared (NIR), useful for detecting vegetation that reflects NIR. Bands 5, 6, and 7 are thermal infrared (TIR) bands that can be used to measure and produce false color images of the temperature of water, clouds, volcanoes, or other objects of the landscape. The Landsat TM images in Figure 1.11 are comprised of bands 2 (green), 3 (red), and 4 (NIR), shown in false color as blue, green, and red.

The ASTER instrument scans 14 bands of electromagnetic radiation at three different resolutions (Figure 1.9). Bands 1, 2, and 3 are visible (green, red) and NIR bands, collectively called VNIR, obtained at a resolution of 15 m. Bands 4–9 are short wave infrared (SWIR) obtained at a resolution of 30 m. Bands 10–14 are TIR obtained at a resolution of 90 m. The examples of ASTER satellite images in Figures 1.12 and 1.13 show how they can be used to study volcanoes and prospect for copper ore.

Satellite remote sensing is a tool used by geologists to help them view regions that are not accessible to them, connect sample and outcrop data to regional and global perspectives, or survey regions to devise a strategy for field work. Inferences made from satellite images must be ground truthed as much as possible with field geology and laboratory work. The more they are ground truthed, the more reliable they are as tools for understanding Earth.

Questions

1. Recall that the "geologic record" is the millions of layers of rock that record events and times in geologic history (e.g., Figure 1.1). Also recall that the geologic time scale (Figure 1.4) is a chart showing named intervals of the geologic record (rock units), the sequence in which they formed (oldest at the bottom), and their ages in millions of years.

   a. How is the geologic record like a book?

   b. How is reading the geologic record different from reading a book?

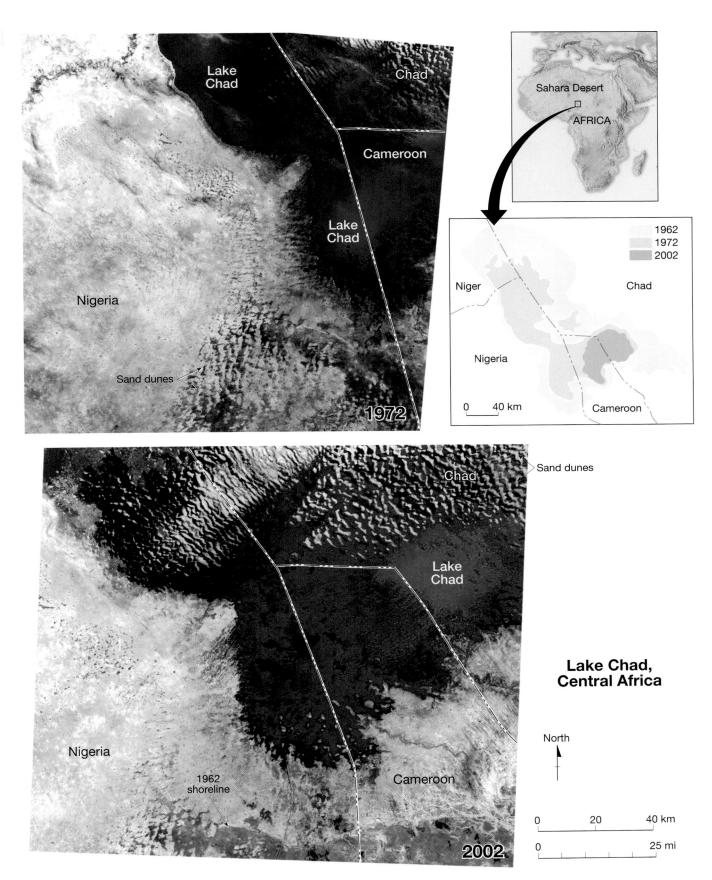

FIGURE 1.11 Lake Chad region, west central Africa, observed by the Landsat 1 satellite in 1972 and the Landsat 7 satellite in 2002. Both images are equivalent to Landsat TM bands 2, 3, and 4 (Figure 1.9), which are colored blue, green, and red in these images. (Images courtesy of U.S.G.S., Landsat Project)

15

Chiliques Volcano, Chile: Dormant or active?

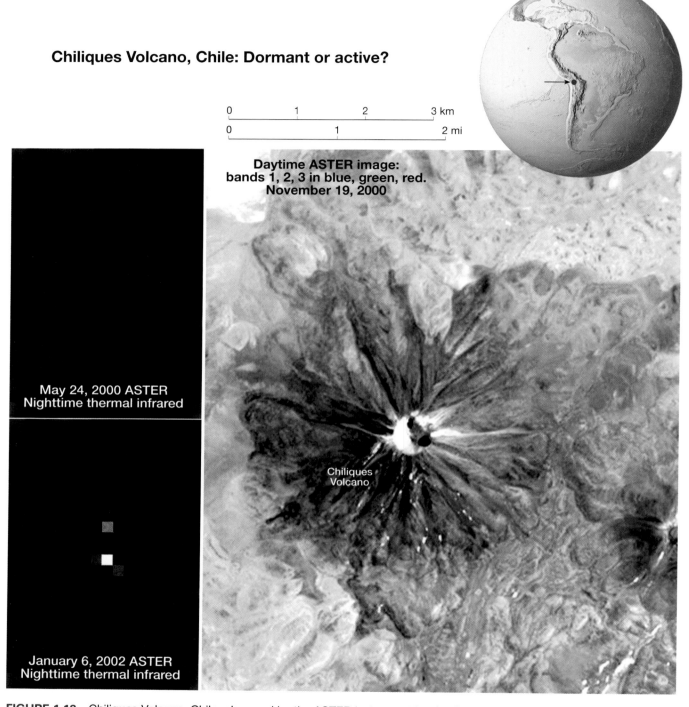

0   1   2   3 km

0   1   2 mi

**Daytime ASTER image:
bands 1, 2, 3 in blue, green, red.
November 19, 2000**

Chiliques
Volcano

May 24, 2000 ASTER
Nighttime thermal infrared

January 6, 2002 ASTER
Nighttime thermal infrared

FIGURE 1.12 Chiliques Volcano, Chile, observed by the ASTER instrument by day (bands 1, 2, 3—colored blue, green, and red) and night (a thermal infrared band). In the nighttime images, the lighter red pixels are hotter than dark red pixels, and white pixels are hotter than red pixels. Refer to Figure 1.9 to review what kind of EM radiation is detected by each of the ASTER bands.   (Images courtesy of NASA/GSFC/METI/ERSDAC/JAROS, and U.S./Japan ASTER Science Team: **asterweb.jpl.nasa.gov**)

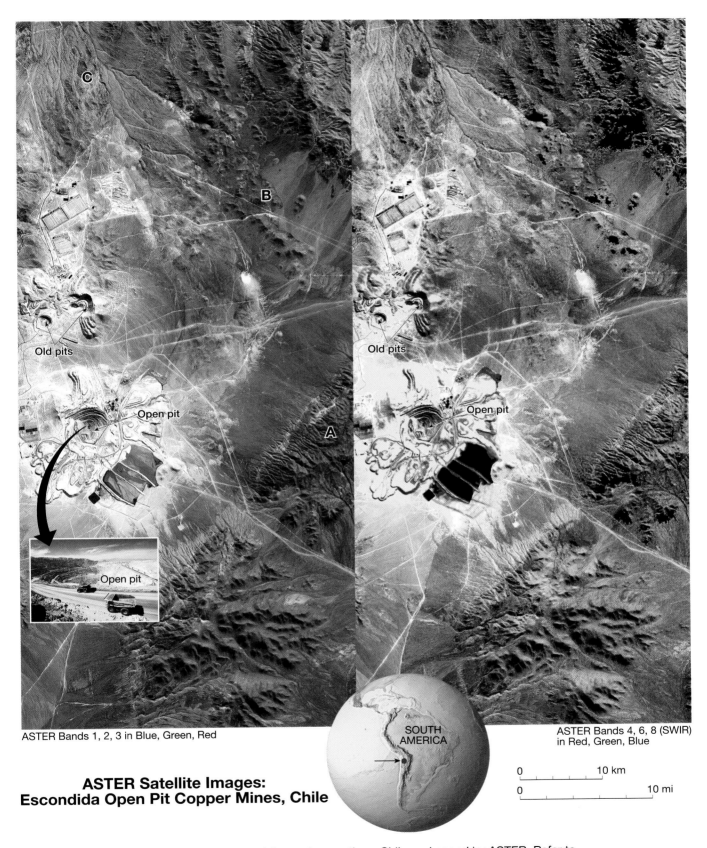

ASTER Bands 1, 2, 3 in Blue, Green, Red

ASTER Bands 4, 6, 8 (SWIR) in Red, Green, Blue

0   10 km

0   10 mi

SOUTH AMERICA

**ASTER Satellite Images:
Escondida Open Pit Copper Mines, Chile**

FIGURE 1.13 Escondida open-pit copper mining region, northern Chile, as imaged by ASTER. Refer to Figure 1.9 to review what kind of EM radiation is detected by each of the ASTER bands. (Images courtesy of NASA/GSFC/METI/ERSDAC/JAROS, and U.S./Japan ASTER Science Team: asterweb.jpl.nasa.gov; photograph courtesy of Rio Tinto: **www.riotinto.com**)

2. Analyze Figure 1.10, an astronaut's photograph and MODIS satellite image of the eruption of Sicily's Mt. Etna in the Mediterranean Sea in 2002.

   a. Volcanic "vents" are openings from which volcanic materials are extruded. Analyze Figure 1.10B (October 30, 2002 eruption of Mt. Etna). How many vents are extruding volcanic materials, and what two kinds of materials do you think are being extruded from them?

   b. What distance has the brown extruded material traveled so far, and where do you predict it will land?

   c. How did this eruption affect the atmosphere and hydrosphere?

3. Analyze Figure 1.11, two Landsat satellite images of the Lake Chad region of Africa. Lake Chad lies on the southern edge of the Sahara Desert and was once the world's sixth largest lake. In 1962, the lake covered about 16,000 square kilometers. But the lake has decreased in size since 1962 due to climate change and human diversion of river water that used to enter the lake. These 1972 and 2002 images show the lake in Landsat TM bands 2, 3, and 4 colored blue, green, and red. Refer to Figure 1.9 to determine what EM radiation is detected by these Landsat TM bands.

   a. Are these true color or false color images? Explain your answer.

   b. What material is probably represented in these images by each of these colors?
   - black/blue:
   - bright red:
   - white/blue-gray:

   c. Between 1972 and 2002, did the land near the label "Nigeria" get wetter or drier? How can you tell?

   d. Notice the white/blue-gray features in the northern part of the 2002 image. What do you think these features might be, and how could you test your hypothesis?

   e. Predict what a Landsat TM satellite image of this area will look like (common colors, features) if one is obtained in 2012 using the same bands as the images in Figure 1.11.

4. Analyze Figure 1.12, ASTER satellite images of Chile's Chiliques Volcano. Refer to Figure 1.9 to review what kind of EM radiation is detected by each of the ASTER bands.

   a. The daytime image is ASTER bands 1, 2, and 3 shown in blue, green, and red. The volcano is blue-green in this image. What is the true color of the volcano?

   b. Notice the thermal infrared images of this volcano. The lighter the color, the hotter the material. This volcano has never erupted in historic time and was classified as dormant (showing no signs of erupting). Would you still classify this volcano as dormant? Explain.

5. Analyze Figure 1.13, false colored, ASTER satellite images of Chile's Escondida Mine and vicinity. This is primarily a copper mine, but it also produces some silver and gold. The copper ore is mined from large open pits. Notice how these pits appear in the images.

   a. Imagine that you have been hired by Escondida Mine to find the best location for a new pit. Which location, A, B, or C, is probably the best site for a new pit? How can you tell?

   b. What plan of scientific investigation would you carry out to see if the location you chose above is actually a good source of more copper ore?

PART 1B: MEASURING EARTH MATERIALS AND RELATIONSHIPS

Every material has a *mass* that can be weighed and a *volume* of space that it occupies. An object's mass can be measured by determining its weight under the pull of Earth's gravity (using a balance). An object's volume can be calculated by determining the multiple of its linear dimensions (measured using a ruler) or directly measured by determining the volume of water that it displaces (using a graduated cylinder). In this laboratory, you will use metric balances, rulers, and graduated cylinders to analyze and evaluate the dimensions and density of Earth materials. Refer to page viii at the front of this manual for illustrations of this basic laboratory equipment.

Metric System of Measurement

People in different parts of the world have historically used different systems of measurement. For example, people in the United States have historically used the English system of measurement based on units such as inches, feet, miles, pounds, gallons, and degrees Fahrenheit. However, for more than a century, most nations of the world have used the metric system of measurement based on units such as meters, liters, and degrees Celsius. In 1975, the U.S. Congress recognized the value of a global system of measurement and adopted the metric system as the official measurement system of the United States. This conversion is not yet complete, so Americans currently use both English and metric systems of measurement. In this laboratory we will only use the metric system.

Each kind of metric unit can be divided or multiplied by 10 and its powers to form the smaller or larger units of the metric system. Therefore, the metric system is also known as a base-10 or decimal system. The International System of Units (SI) is the modern version of metric system symbols, numbers, base-10 numerals, powers of ten, and prefixes (see page x).

Linear Measurements and Conversions

You must be able to use a metric ruler to make exact measurements of **length** (how long something is). This is called *linear measurement*. Most rulers in the United States are graduated in English units of length (inches) on one side and metric units of length (centimeters) on the other. For example, notice that one side of the ruler in Figure 1.14A is graduated in numbered inches, and each inch is subdivided into

eighths and sixteenths. The other side of the ruler is graduated in numbered centimeters (hundredths of a meter), and each centimeter is subdivided into ten millimeters. The ruler provided for you in GeoTools Sheets 1 and 2 at the back of this manual are graduated in exactly the same way.

Review the examples of linear metric measurement in Figure 1.14A to be sure that you understand how to make *exact* metric measurements. Note that the length of an object may not coincide with a specific centimeter or millimeter mark on the ruler, so you may have to estimate the fraction of a unit as exactly as you can. The length of the red rectangle in Figure 1.14A is between graduation marks for 106 and 107 millimeters (mm), so the most exact measurement of this length is 106.5 mm. Also be sure that you measure lengths starting from the zero point on the ruler and *not from the end of the ruler*.

There will be times when you will need to convert a measurement from one unit of measure to another. This can be done with the aid of the mathematical conversions chart on page xi at the front of the manual. For example, you convert millimeters (mm) to meters (m) by multiplying the measurement in mm by 0.001 m/mm. Thus,

$$\frac{106.5 \text{ mm}}{1} \times \frac{0.001 \text{ m}}{1 \text{ mm}} = 0.1065 \text{ m}$$

so 106.5 millimeters is the same as 0.1065 meters.

Area and Volume

An **area** is a two-dimensional space, such as the surface of a table. The long dimension is the *length*, and the short dimension is the *width*. If the area is square or rectangular, then the size of the area is the product of its length multiplied times its width. For example, the blue rectangular area in Figure 1.14A is 7.3 cm long and 3.8 cm wide. So the size of the area is 7.3 cm × 3.8 cm, which equals 27.7 cm^2. This is called 27.7 square centimeters. Using this same method, the yellow front of the box in Figure 1.14B has an area of 9.0 cm × 4.0 cm, which equals 36.0 cm^2. The green side of this same box has an area of 4.0 cm × 4.0 cm, which equals 16.0 cm^2.

Three-dimensional objects are said to occupy a **volume** of space. Box shaped objects have *linear volume* because they take up three linear dimensions of space: their length (longest dimension), width (or depth), and height (or thickness). So the volume of a box shaped object is the product of its length, width, and height. For example, the box in Figure 1.14B has a length of 9.0 cm, a width of 4.0 cm, and a height of

A. LINEAR MEASUREMENT

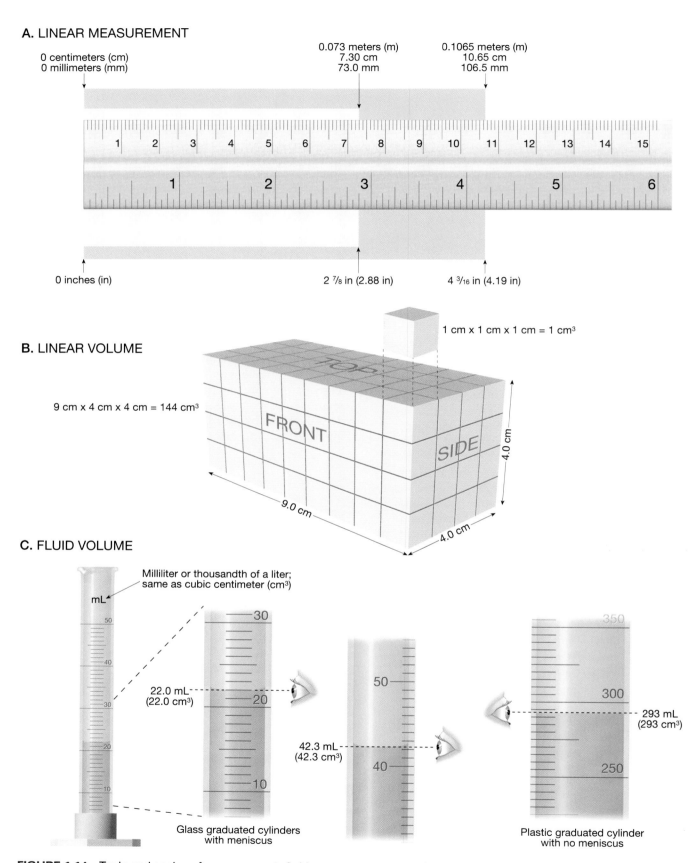

FIGURE 1.14 Tools and scales of measurement. **A.** Linear measurement using a ruler. **B.** Linear volume measured in cubic centimeters. **C.** Fluid volume measured with graduated cylinder (at base of meniscus). A milliliter (mL) is the same as a cubic centimeter (cm^3).

4.0 cm. Its volume is 9.0 cm × 4.0 cm × 4.0 cm, which equals 144 cm³. This is called 144 cubic centimeters.

Most natural materials such as rocks do not have linear dimensions, so their volumes cannot be calculated from linear measurements. However, the volumes of these odd-shaped materials can be determined by measuring the volume of water they displace. This is often done in the laboratory with a *graduated cylinder* (Figure 1.14C), an instrument used to measure volumes of fluid (fluid volume). Most graduated cylinders are graduated in metric units called milliliters (mL), which are thousandths of a liter. *You should also note that 1 mL of fluid volume is exactly the same as 1 cm³ of linear volume.*

When you pour water into a graduated cylinder, the surface of the liquid is usually a curved *meniscus*, and the volume is read at the bottom of the curve (Figure 1.14C: middle and left-hand examples). In some plastic graduated cylinders, however, there is no meniscus. The water level is flat (Figure 1.14C: right-hand example).

If you drop a rock into a graduated cylinder full of water, then it takes up space previously occupied by water at the bottom of the graduated cylinder. This displaced water has nowhere to go except higher into the graduated cylinder. Therefore, the volume of an object such as a rock is exactly the same as the volume of fluid (water) that it displaces.

The water displacement procedure for determining the volume of an object is illustrated in Figure 1.15. First place water in the bottom of a graduated cylinder. Choose a graduated cylinder into which the rock will fit easily, and add enough water to be able to totally immerse the rock. It is also helpful to use a dropper or wash bottle and bring the volume of water (before adding the rock) up to an exact graduation mark (5.0 mL mark in Figure 1.15A). Record this starting volume of water. Then carefully slide the rock sample down into the same graduated cylinder and record this ending level of the water (7.8 mL mark in Figure 1.15B). Subtract the starting volume of water from the ending volume of water, to obtain the displaced volume of water (2.8 mL, which is the same as 2.8 cm³). This volume of displaced water is also the volume of the rock sample.

Mass

Earth materials do not just take up space (volume). They also have a mass of atoms that can be weighed. You will use a gram balance to measure the **mass** of materials (by determining their weight under the pull of Earth's gravity). The gram (g) is the basic unit of mass in the metric system, but instruments used to

VOLUME DETERMINATION BY WATER DISPLACEMENT

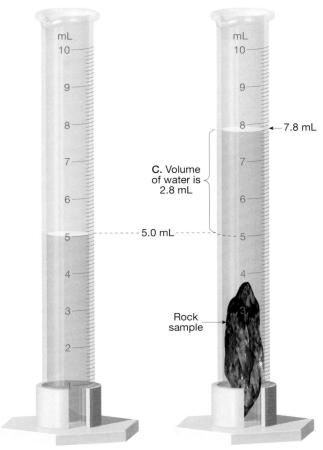

A. Starting volume of water   B. Ending volume of water

FIGURE 1.15 Procedure for determining volume of a rock sample by water displacement. **A.** Place water in the bottom of a graduated cylinder. Choose a graduated cylinder into which the rock will fit easily and add enough water to be able to totally immerse the rock. It is also helpful to use a dropper or wash bottle and bring the volume of water (before adding the rock) up to an exact graduation mark like the 5.0 mL mark. Record this starting volume of water. **B.** Carefully slide the rock sample down into the same graduated cylinder and record this ending level of the water (7.8 mL mark). Subtract the starting volume of water from the ending volume of water to obtain the displaced volume of water (2.8 mL, which is the same as 2.8 cm³). This volume of displaced water is also the volume of the rock sample.

measure grams vary from triple-beam balances to spring scales to digital balances (page viii). Consult with your laboratory instructor or other students to be sure that you understand how to read the gram balance provided in your laboratory.

Density

Every material has a *mass* that can be weighed and a *volume* of space that it occupies. However, the relationship between a material's mass and volume tends to vary from one kind of material to another. For example, a bucket of rocks has much greater mass than an equal-sized bucket of air. Therefore a useful way to describe an object is to determine its mass per unit of volume, called **density**. *Per* refers to division, as in miles *per* hour (distance divided by time). So density is the measure of an object's mass divided by its volume (density = mass ÷ volume). Scientists and mathematicians use the Greek character rho (ρ) to represent density. Also, the gram (g) is the basic metric unit of mass, and the cubic centimeter is the basic unit of metric volume (cm^3), so density (ρ) is usually expressed in grams per cubic centimeter (g/cm^3).

Questions

6. Make the following unit conversions using the Mathematical Conversions chart on page xi.

   a. 10.0 miles = _____ kilometers.

   b. 1.0 foot = _____ meters.

   c. 16 kilometers = _____ meters.

   d. 25 meters = _____ centimeters.

   e. 25.4 mL = _____ cm^3.

   f. 1.3 liters = _____ cm^3.

7. Use a ruler to help you draw a line segment that has a length of exactly 1 cm (1 centimeter). A line occupies only one dimension of space, so a line that is 1 cm long is 1 cm^1.

8. Use a ruler to help you draw a square area that has a length of exactly 1 cm and a width of exactly 1 cm. An area occupies two dimensions of space, so a square that is 1 cm long and 1 cm wide is 1 cm^2 of area (1 cm × 1 cm = 1 cm^2).

9. Use a ruler to help you draw a cube that has a length of 1 cm, width of 1 cm, and height of 1 cm. This cube made of centimeters occupies three dimensions of space, so it is 1 cm^3 (1 cubic centimeter) of volume.

10. Explain how you could use a small graduated cylinder and a gram balance to determine the density of water (ρ_{water}) in g/cm^3. Then use your procedures to calculate the density of water as exactly as you can. Show your data and calculations.

11. Obtain a small lump of clay (grease-based modeling clay) and determine its density (ρ_{clay}) in g/cm^3. There is more than one way to do this, so develop and apply a procedure that makes the most sense to you. Explain the procedure that you use, show your data, and show your calculations.

12. Reconsider your answers to Questions 10 and 11 and the fact that modeling clay sinks in water.

    a. Why does modeling clay sink in water?

    b. What could you do to a lump of modeling clay to get it to float in water? Try your hypothesis and experiment until you get the clay to float.

    c. When you got the clay to float, why did it float?

13. Compared to your answer to Question 10, what is the density of Earth's atmosphere ($\rho_{atmosphere}$), cryosphere ($\rho_{cryosphere}$), hydrosphere ($\rho_{hydrosphere}$), and lithosphere ($\rho_{lithosphere}$)?

    a. ($\rho_{atmosphere}$) = _____ g/cm^3

    b. ($\rho_{cryosphere}$) = _____ g/cm^3

    c. ($\rho_{hydrosphere}$) = _____ g/cm^3

    d. ($\rho_{lithosphere}$) = _____ g/cm^3

14. How is the distribution of Earth's spheres related to their relative densities?

PART 1C: DENSITY, GRAVITY, AND ISOSTASY

Scientists have wondered for centuries about how the distribution of Earth materials is related to their density and gravity. For example, Greek scientist and mathematician, Archimedes, experimented with floating objects around 225 B.C. When he placed a block of wood in a bucket of water, he noticed that the block floated and the water level rose (Figure 1.16A). When he pushed down on the wood block, the water level rose even more. And when he removed his fingers from the wood block, the water pushed it back up to its original level of floating. Archimedes eventually realized that every floating object is pulled down (toward Earth's center) by gravity, so the object displaces fluid and causes the fluid level to rise. However, Archimedes also realized that every floating object is also pushed upward by a buoyant force that is equal to the weight of the displaced fluid. This is now called Archimedes' Principle.

Buoyant force (buoyancy) is caused as gravity pulls on the mass of a fluid, causing it to exert a *fluid pressure* on submerged objects that increases steadily with increasing depth in the fluid. The deeper (greater amount of) a fluid, the more it weighs, so deep water exerts greater fluid pressure than shallow fluid. Therefore, the lowest surfaces of a submerged object are squeezed more (by the fluid pressure) than the upper surfaces. This creates the wedge of buoyant force that pushes the object upward and opposes the downward pull of gravity (white arrows in Figure 1.16B). An object will sink if it is heavier than the fluid it displaces (is denser than the fluid it displaces). An object will rise if it is lighter than the fluid it displaces (is less dense than the fluid it displaces). But a floating object is balanced between sinking and rising. The object sinks until it displaces a volume of fluid that has the same mass as the entire floating object. When the object achieves a motionless floating condition, it is balanced between the downward pull of gravity and the upward push of the buoyant force.

Isostasy

In the 1880s, geologists began to realize the abundant evidence that levels of shoreline along lakes and oceans had changed often throughout geologic time in all parts of the world. Geologists like Edward Suess hypothesized that changes in sea level can occur if *the volume of ocean water changes* in response to climate. Global atmospheric warming leads to sea level rise caused by melting of glaciers (cryosphere), and global atmospheric cooling leads to a drop in sea level as more of Earth's hydrosphere gets stored in thicker glaciers. However, an American geologist named Clarence Dutton suggested that shorelines can also change *if the level of the land changes* (and the volume of water remains the same).

Dutton reasoned that if blocks of Earth's crust are supported by fluid materials beneath them, then they must float according to Archimedes' Principle (like wood blocks, icebergs, and boats floating in water). Therefore, he proposed that Earth's crust consists of buoyant blocks of rock that float in gravitational balance on top of the mantle. He called this floating condition **isostasy** (Greek for "equal standing"). Loading a crustal block (by adding lava flows, sediments, glaciers, water, etc.) will decrease its buoyancy, and the block will sink (like pushing down on a floating wood block). Unloading materials from a crustal block will increase its buoyancy, and the block will rise. Therefore, you can also think of isostasy as the equilibrium (balancing) condition between any floating object (such as the iceberg in Figure 1.16) and the more dense fluid in which it is floating (such as the water in Figure 1.16). Gravity pulls the iceberg down toward Earth's center (this is called *gravitational force*), so the submerged root of the iceberg displaces water. At the same time, gravity also tries to pull the displaced water back into its original place (now occupied by the iceberg's root). This creates fluid pressure that increases with depth along the iceberg's root, so the iceberg is squeezed and wedged (pushed) upward. This squeezing and upward-pushing force is called *buoyant force*. **Isostatic equilibrium** (balanced floating) occurs when the buoyant force equals (is in equilibrium with) the gravitational force that opposes it. An **equilibrium line** (like the waterline on a boat) separates the iceberg's submerged root from its exposed top.

Questions

15. Obtain one of the wood blocks provided at your table. Determine the density of the wood block (ρ_{wood}) in g/cm^3. Show your calculations.

16. Float the same wood block in a bowl of water (like Figure 1.16A) and mark the equilibrium line (waterline).

    a. Draw an exact sketch of the side view of the wood block and show the exact position of the waterline (equilibrium line). Label the total height of the wood block (H_{block}), the height of the wood block that is submerged below the waterline (H_{below}), and the height of the wood block that is exposed above the waterline (H_{above}).

A. FLOATING WOOD BLOCK

B. ICEBERG

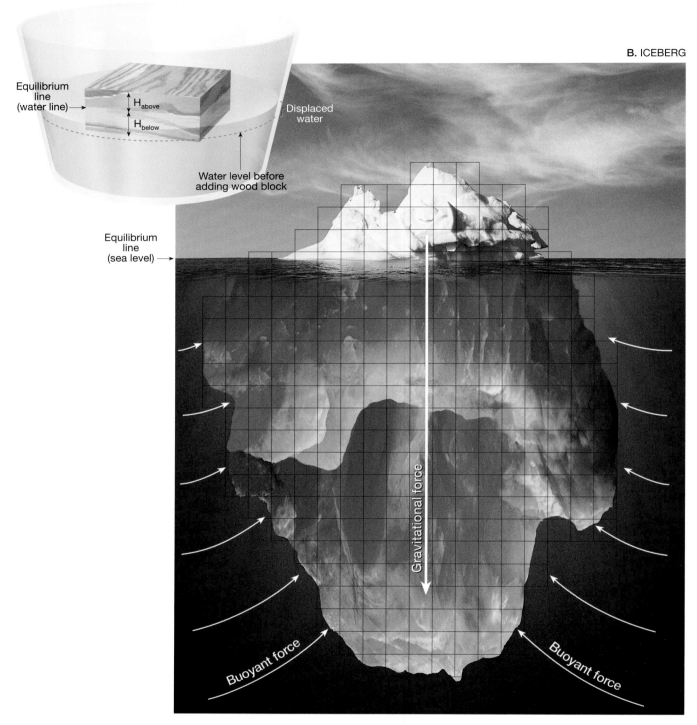

Equilibrium line (water line)

H_{above}

H_{below}

Displaced water

Water level before adding wood block

Equilibrium line (sea level)

Gravitational force

Buoyant force

Buoyant force

FIGURE 1.16 Isostasy relationships of a floating wood block (**A**) and iceberg (**B**). Refer to text for discussion. (Iceberg image © Ralph A. Clavenger/CORBIS. All rights reserved.)

b. Measure and record H_{block}: _____ cm

c. Measure and record H_{below}: _____ cm

d. Measure and record H_{above}: _____ cm

17. Write an isostasy equation (mathematical model) that expresses how the density of the wood block (ρ_{wood}) compared to the density of the water (ρ_{water}) is related to the height of the wood block that floats *below* the equilibrium line (H_{below}).

    [*Hint:* Recall that the wood block achieves isostatic equilibrium (motionless balanced floating) when it displaces a volume of water that has the same mass as the entire wood block. For example, if the wood block is 80% as dense as the water, then only 80% of the wood block will be below the equilibrium line (water line). Therefore, the portion of the wood block's height that is below the equilibrium line (H_{below}) is equal to the total height of the wood block (H_{block}) times the ratio of the density of the wood block (ρ_{wood}) to the density of water (ρ_{water}).]

18. Change your answer in Question 17 to an equation (mathematical model) that expresses how the density of the wood block (ρ_{wood}) compared to the density of the water (ρ_{water}) is related to the height of the wood block that floats *above* the equilibrium line (H_{above}).

19. The density of water ice (in icebergs) is $0.917 \, g/cm^3$. The average density of (salty) ocean water is $1.025 \, g/cm^3$.

    a. Use your isostasy equation for H_{below} (Question 17) to calculate how much of an iceberg is exposed below sea level. Show your work.

    b. Use your isostasy equation for H_{above} (Question 18) to calculate how much of an iceberg is exposed above sea level. Show your work.

    c. Notice the graph paper grid overlay on the picture of an iceberg in Figure 1.16B. Use this grid to determine and record the cross-sectional area of this iceberg that is below sea level and the cross-sectional area that is above sea level (by adding together all of the whole boxes and fractions of boxes that overlay the root of the iceberg or the exposed top of the iceberg). Use this data to calculate what proportion of the iceberg is below sea level (the equilibrium line) and what proportion is above sea level. How do your results compare to your calculations in Questions 19a and 19b?

    d. What will happen as the top of the iceberg melts?

    e. Clarence Dutton proposed his isostasy hypothesis to explain how some ancient shorelines have been elevated to where they now occur on the slopes of adjacent mountains. Use *your* understanding of isostasy and icebergs to explain how this may happen.

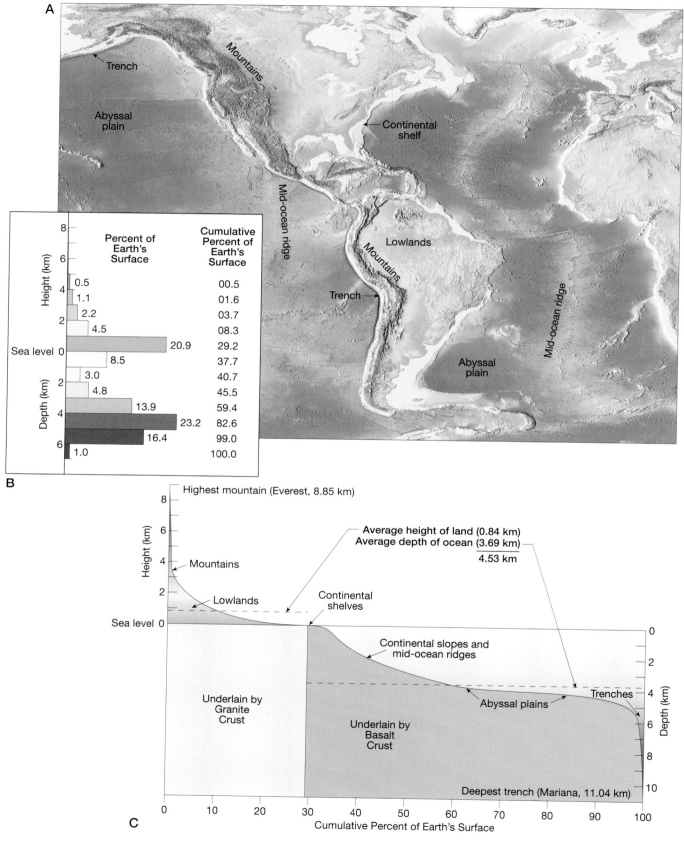

FIGURE 1.17 Global topography of Earth. **A.** Portion of Earth with ocean removed, based on satellite-based radar and laser technologies. **B.** Histogram of global topography. **C.** Hypsographic curve of Earth's global topography. (Refer to text for discussion.)

PART 1D: ISOSTASY AND EARTH'S GLOBAL TOPOGRAPHY

Clarence Dutton applied his isostasy hypothesis in 1889 to explain how the shorelines of lakes or oceans could be elevated by vertical motions of Earth's crust. At that time, little was known about Earth's mantle or topography of the seafloor. Modern data show that Dutton's isostasy hypothesis has broader application for understanding global topography.

Global Topography

Radar and laser imaging technologies carried aboard satellites now measure Earth's topography very exactly, and the data can be used to form very precise relief images of the height of landforms and depths of ocean basins. For example, satellite data was used to construct the image in Figure 1.17A of Earth with ocean water removed. The seafloor is shaded blue and includes features such as shallow continental shelves, submarine mountains (mid-ocean ridges), deep abyssal plains, and even deeper trenches. Land areas (continents) are shaded green (lowlands) and brown (mountains).

The histogram (bar diagram) of Earth's topography in Figure 1.17B shows the percentage of Earth's surface for each depth or height class (bar) in kilometers. Notice that the histogram is bimodal (shows two levels of elevation that are most common on Earth). One of the elevation modes occurs above sea level and corresponds to the continents. The other elevation mode occurs below sea level and corresponds to the ocean floor.

Figure 1.17C is called a *hypsographic curve* and shows the cumulative percentage of Earth's spherical surface that occurs at specific elevations or depths in relation to sea level. This curve is not the profile of a continent, because it represents Earth's entire spherical surface. Notice that the cumulative percentage of land is only 29.2% of Earth's surface, and most of the land is lowlands. The remaining 70.8 cumulative percent of Earth's surface is covered by ocean, and most of the seafloor is more than 3 km deep.

Global Isostasy

The average elevation of the continents is about 0.84 km above sea level (+0.84 km), but the average elevation of the ocean basins is 3.87 km below sea level (−3.87 km). Therefore the difference between the average continental and ocean basin elevations is 4.71 km! If the continents did not sit so much higher than the floor of the ocean basins, then Earth would have no dry land and there would be no humans. What could account for this elevation difference?

One clue may be the difference between crustal granite and basalt in relation to mantle peridotite.

Granite (light-colored, coarse-grained igneous rock) and basalt (dark-colored, fine-grained igneous rock) make up nearly all of Earth's crust (Figure 1.18). *Basaltic rocks* form the crust of the oceans, beneath a

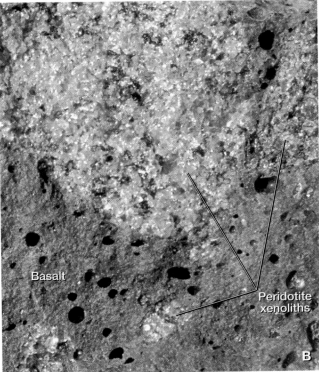

FIGURE 1.18 The most abundant rocks of Earth's crust and mantle. **A. Granite** is a light-colored igneous rock that forms the crust of continents, beneath layers of sedimentary and metamorphic rocks like those shown in Figure 1.1. **B. Basalt** is an igneous rock that forms the crust of all oceans, beneath layers of sand and mud. The upper mantle consists of **peridotite** rock like these *xenoliths*—pieces of rock carried to Earth's surface by magma in a volcanic eruption. The magma cooled to form a body of basalt with the xenoliths (and gas bubbles) trapped inside.

thin veneer of sediment. *Granitic rocks* form the crust of the continents, usually beneath a thin veneer of sediment and other rock types. Therefore, you can think of the continents (green and brown) in Figure 1.17A as granitic islands surrounded by a low sea of basaltic ocean crust (blue). All of these rocky bodies rest on mantle rock called *peridotite* (Figure 1.18B). Could differences among the three rock types making up Earth's outer edge explain Earth's bimodal global topography? Let's investigate. You will need a 500 mL or 1000 mL graduated cylinder, small samples (about 30–50 g) of basalt and granite that fit into the graduated cylinder, a gram balance, and water.

Questions

20. *As exactly as you can*, weigh (grams) and determine the volume (by water displacement) of a sample of basalt. Add your data to the basalt density chart in Figure 1.19A. Calculate the density of your sample

A. BASALT DENSITY CHART

Basalt Sample Number	Sample Weight (g)	Sample Volume (cm³)	Sample Density (g/cm³)
1	40.5	13	3.1
2	29.5	10	3.0
3	46.6	15	3.0
4	31.5	10	3.2
5	37.6	12	3.1
6	34.3	11	3.1
7	78.3	25	3.1
8	28.2	9	3.1
9	55.6	18	3.1
10			

Average density of basalt = _____

B. GRANITE DENSITY CHART

Granite Sample Number	Sample Weight (g)	Sample Volume (cm³)	Sample Density (g/cm³)
1	32.1	12	2.7
2	27.8	10	2.8
3	27.6	10	2.8
4	31.1	11	2.8
5	58.6	20	2.9
6	62.1	22	2.8
7	28.8	10	2.9
8	82.8	30	2.8
9	52.2	20	2.6
10			

Average density of granite = _____

FIGURE 1.19 Charts for recording data (Questions 20, 21) and calculating the average density of basalt (**A**) and granite (**B**).

of basalt to tenths of a g/cm^3. Then determine the average density of basalt using all ten lines of sample data in the basalt density chart.

21. *As exactly as you can*, weigh (grams) and determine the volume (by water displacement) of a sample of granite. Add your data to the granite density chart in Figure 1.19B. Calculate the density of your sample of granite to tenths of a g/cm^3. Then determine the average density of granite using all ten lines of sample data in the granite density chart.

22. Seismology (the study of Earth's structure and composition using earthquake waves), mantle xenoliths (Figure 1.18B), and laboratory experiments indicate that the upper mantle is peridotite rock. The peridotite has an average density of about $3.3\,g/cm^3$ and is capable of slow flow. Seismology also reveals the thicknesses of crust and mantle layers.

    a. Seismology indicates that the average thickness of basaltic ocean crust is about 5.0 km. Use the average density of basalt (Figure 1.18A) and your isostasy equation (Question 18) to calculate how high (in kilometers) basalt floats in the mantle. Show your work.

    b. Seismology indicates that the average thickness of granitic continental crust is about 30.0 kilometers. Use the average density of granite (Figure 1.19B) and your isostasy equation (Question 18) to calculate how high (in kilometers) granite floats in the mantle. Show your work.

    c. What is the difference (in km) between your answers in a and b?

    d. How does this difference between a and b compare to the actual difference between the average height of continents and average depth of oceans on the hypsographic curve (Figure 1.17C)?

23. Reflect on all of your work in this laboratory. Explain why Earth has a bimodal global topography.

24. How is a mountain like the iceberg in Figure 1.16B?

25. Clarence Dutton was not the first person to develop the concept of a floating crust in equilibrium balance with the mantle, which he called *isostasy* in 1889. Two other people proposed floating crust (isostasy) hypotheses in 1855 (Figure 1.20). John Pratt (a British physicist and Archdeacon of Calcutta) studied the Himalaya Mountains and hypothesized that floating blocks of Earth's crust have different densities, but they all sink to the same *compensation level* within the mantle. The continental blocks are higher because they are less dense. George Airy (a British astronomer and mathematician) hypothesized that floating blocks of Earth's crust have the same density but different thicknesses. The continental blocks are higher because they are thicker. Do you think that one of these two hypotheses (Pratt vs. Airy) is correct, or would you propose a compromise between them? Explain.

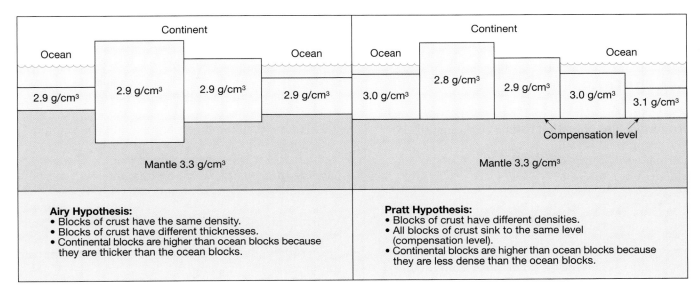

FIGURE 1.20 Two hypotheses proposed to explain the isostasy of Earth's crust. Refer to text and Question 25 for discussion.

Plate Tectonics and the Origin of Magma

•CONTRIBUTING AUTHORS•

Edward A. Hay • *De Anza College*

Cherukupalli E. Nehru • *Brooklyn College (CUNY)*

C. Gil Wiswall • *West Chester University of Pennsylvania*

OBJECTIVES

A. Analyze Earth forces, faults, and plate boundaries to determine if Earth's volume (size) is decreasing, increasing, or staying about the same.

B. Use seismic tomography to evaluate a lava lamp model of Earth's mantle.

C. Use physical and graphical models of rock melting to infer how magma forms in relation to pressure, temperature, water, and plate tectonics.

D. Measure and calculate rates of plate tectonic processes.

MATERIALS

Pencil, eraser, ruler, calculator, hot plate, aluminum foil or two aluminum foil baking cups, sugar cubes (2), dropper with water, red and blue colored pencils, permanent felt-tipped marker.

INTRODUCTION

Ever since the first reasonably accurate world maps were constructed in the 1600s, people have proposed models to explain the origin of Earth's mountain belts, continents, ocean basins, rifts, and trenches. For example, some people proposed that surficial processes, such as catastrophic global floods, had carved our ocean basins and deposited mountains of gravel. Others proposed that global relief was the result of what is now called **tectonism:** large-scale movements and deformation of Earth's crust. What kinds of tectonic movements occur on Earth, and what process(es) could cause them?

German scientist, Alfred Wegener, noticed that the shapes of the continents matched up like pieces of a global jig-saw puzzle. In 1915, he hypothesized that all continents were once part of a single supercontinent, *Pangea,* parts of which drifted apart to form the smaller modern continents. However, most scientists were immediately skeptical of Wegener's **Continental Drift Hypothesis**, because he could not think of a natural process that could force the continents to drift apart. These "anti-drift" scientists viewed continents as stationary landforms that could rise and fall but not drift sideways.

The anti-drift scientists argued that it was impossible for continents to drift or plow through solid oceanic rocks. They also reasoned that Earth was cooling from an older semi-molten state, so it must be shrinking. Their **Shrinking Earth Hypothesis** suggested that the continents were moving together, rather than drifting apart. As Earth's crust shrank into less space, flat rock layers in ocean basins would have been squeezed and folded between the continents (as observed in the Alps).

Two other German scientists, Bernard Lindemann (in 1927) and Otto Hilgenberg (in 1933), independently evaluated the Continental Drift and Shrinking Earth

Hypotheses. Both men agreed with Wegener's notion that the continents seemed to fit together like a jig-saw puzzle, but they also felt that the ocean basins were best explained by a new **Expanding Earth Hypothesis** (that they developed and published separately). According to this hypothesis, Earth was once much smaller (about 60% of its modern size) and covered entirely by granitic crust. As Earth expanded, the granitic crust split apart into the shapes of the modern continents and basaltic ocean crust was exposed between them (and covered by ocean).

During the 1960s more data emerged in favor of the Continental Drift Hypothesis. For example, geologists found that it was not only the shapes (outlines) of the continents that matched up like pieces of a Pangea jig-saw puzzle. Similar bodies of rock and the patterns they make at Earth's surface also matched up like a picture on the puzzle pieces. Abundant studies also revealed that ocean basins were generally younger than the continents. An American Geologist, Harry Hess, even developed a *Seafloor-Spreading Hypothesis* to explain this. According to Hess' hypothesis, seafloor crust is created along mid-ocean ridges above regions of upwelling magma from Earth's mantle. As new magma rises, it forces the old seafloor crust to spread apart on both sides of the ridge and cools to form new crust. The seafloor spreads apart in this way until it encounters a trench, whereupon it descends back into the mantle. Hess' hypothesis was supported by studies showing that Earth has a thin, rigid *lithosphere* underlain by a plastic *asthenosphere* (see Figure 1.7). Earthquakes occur below Earth's surface wherever ocean crust is created and wherever it descends back into the mantle. Zones of abundant earthquake and volcanic activity are also concentrated along cracks in the lithosphere that are boundaries (plate boundaries) between rigid stable sheets of lithosphere, called *lithospheric plates*. Thus, by the end of the 1960s a new hypothesis of global tectonics had emerged called the **Plate Tectonics Hypothesis.** It is now the prevailing model of Earth's global tectonism.

According to the developing Plate Tectonics Model, the continents are parts of rigid lithospheric plates that move about relative to one another. Plates form and spread apart along **divergent boundaries** such as mid-ocean ridges (Figure 2.1), where magma rises up between two plates, forces them to spread apart, and cools to form new rock on the edges of both plates. Plates are destroyed along **convergent boundaries**, where the edge of one plate may *subduct* (descend beneath the edge of another plate) back into the mantle (Figure 2.1) or both plates may crumple and merge to form a mountain belt. Plates slide past one another along **transform fault boundaries**, where plates are neither formed nor destroyed (Figure 2.1). The Plate Tectonics Model does not require the Earth to shrink or expand in size. Earth's size can remain constant, because there are processes that simultaneously form and destroy crust (lithospheric plates).

Most evidence for plate tectonics has come from the detailed observations, maps, and measurements made by field geologists studying Earth's surface directly. However, some of the best modern evidence of lithospheric plate motions is now obtained remotely with satellites orbiting thousands of kilometers above Earth's surface. Several different kinds of satellite technologies and measurement techniques are used, but the most common is the Global Positioning System (GPS).

The Global Positioning System (GPS) is a constellation of 24 satellites in orbit above Earth. These satellites transmit their own radio signals that can be detected by a fixed or hand-held GPS receiver. A receiver can simultaneously detect radio transmissions from four or more GPS satellites, then triangulate its exact position and elevation on Earth's surface. Because GPS receivers can be purchased for as little as $100, they are widely used by airplane navigators, automated vehicle navigation systems, ship captains, and hikers to map locations on Earth (see Laboratory 9). More expensive GPS receivers are used to measure plate motions within a few millimeters of accuracy.

PART 2A: IS EARTH'S SIZE INCREASING, DECREASING, OR STAYING ABOUT THE SAME?

The possibility that Earth's size might be changing brings up three fundamental questions about Earth's global rocky sphere. Is Earth shrinking (decreasing) in size? Is Earth expanding (increasing) in size? Or is Earth staying about the same size? These questions can be evaluated by studying Earth's natural forces and faults in relation to those that you predict on the basis of models of a shrinking, expanding, or constant Earth volume.

Earth Forces and the Faults They Produce

Three kinds of directed force (stress) can be applied to a solid mass of rock (Figure 2.2). **Compression** compacts a block of rock and squeezes it into less space. This can cause *reverse faulting*, in which the hanging wall block is forced up the footwall block in opposition to the pull of gravity (Figure 2.3). **Tension** (also

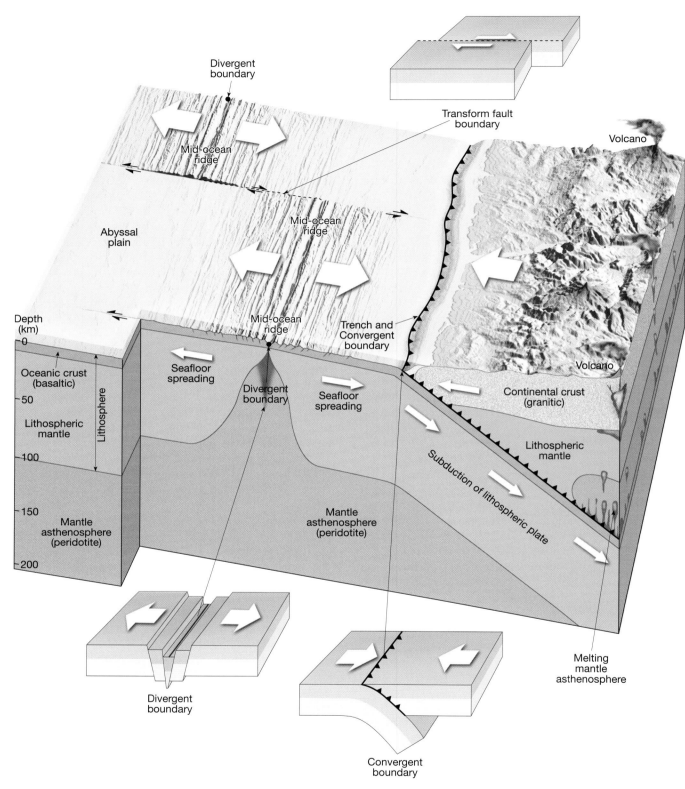

Divergent
boundary

Transform fault
boundary

Volcano

Mid-ocean
ridge

Abyssal
plain

Mid-ocean
ridge

Trench and
Convergent
boundary

Depth
(km)

Mid-ocean
ridge

Seafloor
spreading

Divergent
boundary

Seafloor
spreading

Subduction of lithospheric plate

Volcano

0

Oceanic crust
(basaltic)

Lithosphere

Continental crust
(granitic)

50

Lithospheric
mantle

Lithospheric
mantle

100

150

Mantle
asthenosphere
(peridotite)

Mantle
asthenosphere
(peridotite)

200

Melting
mantle
asthenosphere

Divergent
boundary

Convergent
boundary

FIGURE 2.1   Three kinds of plate boundaries: divergent, convergent, and transform fault boundaries.
White arrows indicate motions of lithospheric plates.

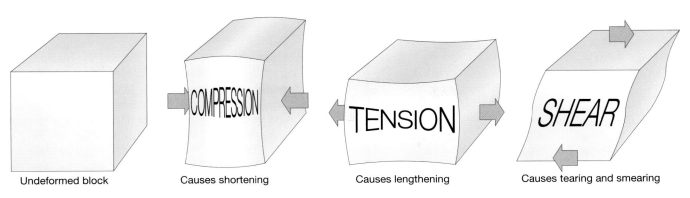

FIGURE 2.2 Three kinds of stress (applied force, as indicated by arrows) and the strain (deformation) that they cause in an undeformed block of rock.

A. Block diagram	B. Fault type	C. Has the crust: • Shortened? • Lengthened? • Neither?	D. Was the stress: • Shear? • Compression? • Tension?	E. Is the plate boundary type: • Transform? • Divergent? • Convergent? See Figure 2.1
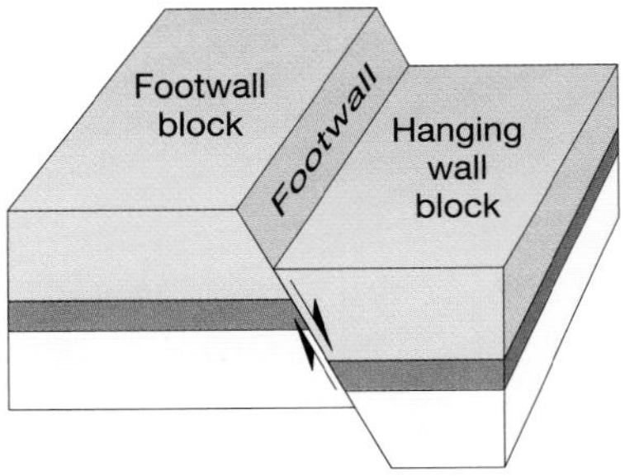 Footwall block / Footwall / Hanging wall block	Normal fault			
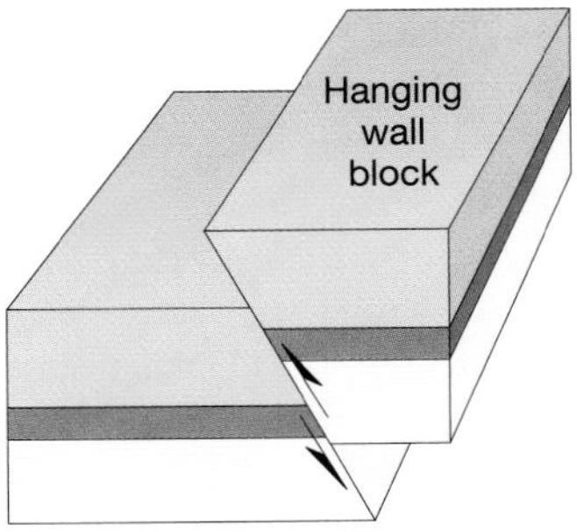 Hanging wall block	Reverse fault			
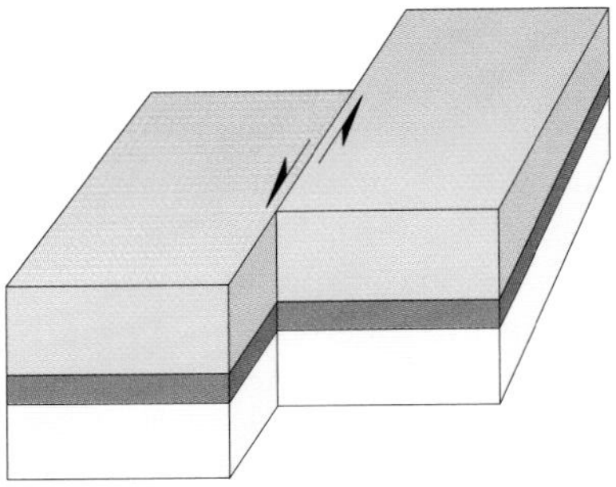	Strike-slip fault (lateral fault)			

FIGURE 2.3 Chart for comparing fault types (columns A and B) to stress (column D), strain (column C), and plate boundary types (column E).

called *dilation*) pulls a block of rock apart and increases its length. This can cause *normal faulting*, in which gravity pulls the hanging wall block down and forces it to slide down off of the footwall block (see Figure 2.3). **Shear** smears a block of rock from side to side and may eventually tear it apart into two blocks of rock that slide past each other along a lateral or *strike-slip fault* (Figure 2.3).

Questions

1. Analyze the three block diagrams in Figure 2.2, the three kinds of faults shown in Figure 2.3, and the three kinds of plate boundaries shown in Figure 2.1. Complete columns C, D, and E in Figure 2.3 to infer what kind of stress and strain is associated with each kind of faulting and plate boundary.

2. Refer to your completed chart in Figure 2.3 and the models of Earth volume in Figure 2.4. Think of Earth's lithosphere as the shell of an egg. What main kind of faulting and plate boundary would you expect to occur in Earth's thin, eggshell-like lithosphere if:

   a. Earth is expanding in size?

   b. Earth is shrinking in size?

   c. Earth's outer shell fractures but there is no expansion or shrinking of its size?

3. Refer to Figure 2.5, a map showing the distribution of Earth's lithospheric plates and their rates of motion in centimeters per year. Study this figure to determine the location, distribution, and length of Earth's three kinds of plate boundaries: divergent (red), convergent (hachured), and transform (dashed). Then visually estimate the following data.

   a. What percentage (based on length) of Earth's plate boundaries are transform fault boundaries?

   b. What percentage (based on length) of Earth's plate boundaries are divergent boundaries?

   c. What percentage (based on length) of Earth's plate boundaries are convergent boundaries?

4. Do you think Earth's size is increasing (expanding), decreasing (shrinking), or staying about the same? Justify your answer by citing evidence from your answers to Questions 1–3.

PART 2B: WHAT DRIVES PLATE TECTONICS?

Modern data and scientific logic suggest that Earth's volume has changed very little in the past 4 billion years. Plate tectonics has gained widespread acceptance in the scientific community and is now a widely used model or theory.

According to the Plate Tectonics Model, Earth's lithosphere is broken into at least a dozen major plates that move about on the weak asthenosphere and are separated by linear plate boundaries. Zones of crustal deformation, earthquakes, and volcanoes develop along the plate boundaries, because they are contacts between the mobile lithospheric plates. At each plate boundary, lithospheric plates are either forming and spreading apart (diverging), sliding past one another (shearing), or converging.

While much is known about plate tectonics, there has been uncertainty about how mantle rocks beneath the asthenosphere may influence this process. In the 1930s, an English geologist named Arthur Holmes speculated that the mantle may experience circular (convection cell) flow like a boiling pot of soup. He proposed that such flow could carry continents about the Earth like a giant conveyor belt. This idea was also adapted in the 1960s by Harry Hess, who suggested that mantle flow is the driving mechanism of plate tectonics. Yet other geologists have proposed that magma erupting along plate boundaries pushes plates along (a *slab push* hypothesis) while gravity pulls dense edges of plates down into the mantle (a *slab pull* hypothesis).

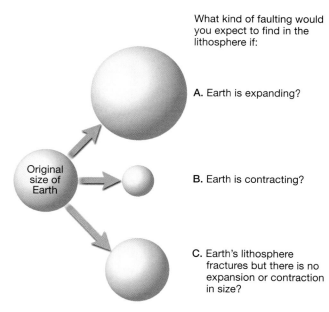

FIGURE 2.4 Diagram for predicting the effects of changing Earth's size (Question 2).

Now new technologies provide data directly related to these ideas. For example, seismic tomography now provides sound evidence that processes at least 660 kilometers deep inside the mantle may have dramatic effects on plate tectonics at the surface.

Seismic Tomography

Earth's mantle is nearly 3000 km thick and occurs between the crust and the molten outer core. Although mantle rocks behave like a brittle solid on short timescales associated with earthquakes, they seem to flow like a very thick (viscous) fluid on longer timescales of days to years. Geologists use a technique called *seismic tomography* to detect this mantle flow.

The word *tomography* (Greek: *tomos* = slice, *graphe* = drawing) refers to the process of making drawings of slices through an object or person. Geologists use seismic tomography to view slices of Earth's interior similar to the way that medical technologists view slices of the human body. The human body slices are known as CAT (computer axial tomography) scans and are constructed using X rays to penetrate and image the human body. The tomography scans of Earth's interior are constructed using seismic waves to penetrate and image the body of Earth.

In seismic tomography, geologists collect data on the velocity (rate and direction) of many thousands of seismic waves as they pass through Earth. The waves travel fastest through rocks that are the most dense and presumed to be coolest. The waves travel slower through rocks that are less dense and presumed to be warmer. When a computer is used to analyze all of the data, from all directions, it is possible to generate seismic tomography images of Earth. These images can be viewed individually or combined to form three-dimensional perspectives. The computer can also assist in false-coloring seismic tomography images to show bodies of mantle rock that are significantly warmer (red) and cooler (blue) than the rest of the mantle (Figure 2.6).

Questions

5. A "lava lamp" is inactive when the light is off, but a lighted lava lamp is dynamic and ever-changing. Observe the rising and sinking motion of the lava-like wax in a lighted lava lamp.

    a. Describe (or sketch and label) the motions of the lava-like wax that occur over one full minute of time.

    b. What causes the "lava" to move from the base of the lamp to the top of the lamp? (Be as specific and complete as you can.)

    c. What causes the "lava" to move from the top of the lamp to the base of the lamp? (Be as specific and complete as you can.)

    d. What is the name applied to this kind of cycle of change? (*Hint:* Refer to Figure 1.4)

6. Observe the seismic tomography image in Figure 2.6: a slice through Earth's mantle at a depth of 350 kilometers. Unlike the lava lamp that you viewed in a vertical profile from the side of the lamp, this image is a horizontal slice of Earth's mantle viewed from above. This image is also false-colored to show where rocks are significantly warmer and less dense (colored red) versus cooler and more dense (colored blue).

    a. How is Earth's mantle like a lava lamp?

    b. How is Earth's mantle different from a lava lamp?

7. Compare the tectonic plates and plate boundaries in Figure 2.5 to the red and blue regions of the seismic tomography image in Figure 2.6.

    a. Under what kind of plate tectonic feature do the warm, less dense rocks (red) occur most often?

    b. Under what kind of plate tectonic feature do the cool, more dense rocks (blue) occur most often?

8. Based on your work in Questions 6 and 7, draw a vertical cross section (vertical slice) of Earth that shows how mantle convection may be related to plate tectonics. Include and label the following features in your drawing: lithospheric plates, a mid-ocean ridge (divergent plate boundary), a subduction zone (convergent plate boundary), continental crust, ocean crust, and locations of slab pull and slab push. Use colored pencils to show where the mantle rocks in your vertical cross section would be red and blue like the false-colored mantle rocks in Figure 2.6.

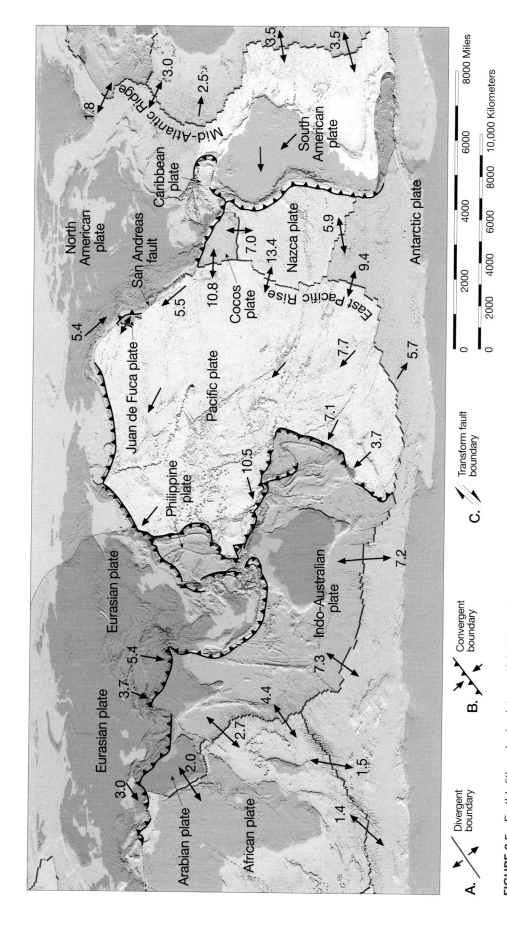

FIGURE 2.5 Earth's lithospheric plates and their boundaries. Numerals indicate rates of plate motion in centimeters per year (cm/yr) based on satellite measurements (Courtesy of NASA). Divergent boundaries (red) occur where two adjacent plates form and move apart (diverge) from each other. Convergent boundaries (hachured) occur where two adjacent plates move together (converge). Transform fault boundaries (dashed) occur along faults where two adjacent plates slide past each other. Refer back to Figure 2.1 for another perspective of the three kinds of plate boundaries.

A. Divergent boundary

B. Convergent boundary

C. Transform fault boundary

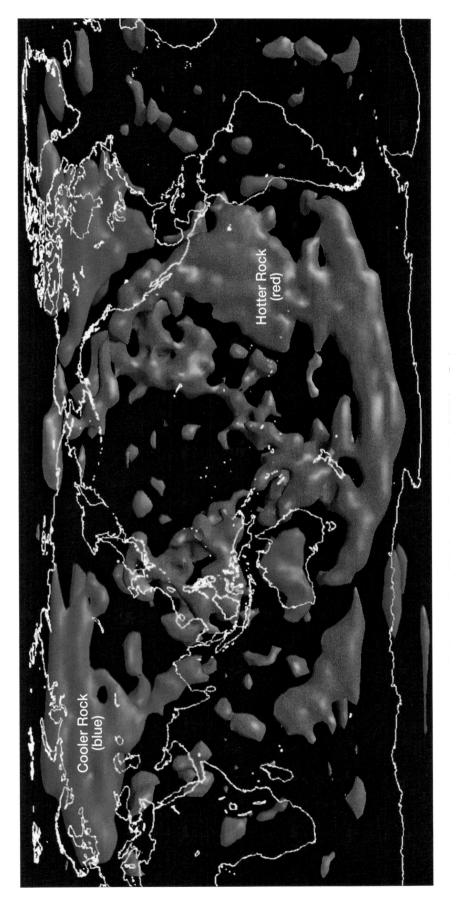

FIGURE 2.6 Seismic tomography image (horizontal slice) of Earth's mantle at a depth of 350 km. Red false-coloring indicates hot rock that is less dense and ascending in comparison to the blue-colored cooler rock that is static or descending. See text for discussion. (Courtesy of Paul J. Morin, University of Minnesota)

PART 2C: THE ORIGIN OF MAGMA

Seismic studies indicate that Earth's outer core contains a substantial proportion of melted rock called *magma*. However, this magma is so deep inside Earth that it cannot erupt to the surface. Seismic studies also indicate that nearly all of Earth's mantle and crust are solid rock—not magma. Therefore, except for some specific locations where active volcanoes occur, there is no reservoir or layer of magma beneath Earth's surface just waiting to erupt. On a global scale, the volume of magma that feeds active volcanoes is actually very small. What, then, are the special conditions that cause these rare bodies of upper mantle and lower crust magma to form?

Magma generally forms in three plate tectonic settings (divergent plate boundaries, convergent plate boundaries, and hot spots). Its origin (rock melting) is also influenced by underground temperature, underground pressure (lithostatic pressure), and the kind of minerals that comprise underground rocks.

Temperature (T)

Rocks are mostly masses of solid mineral crystals. Therefore, some or all of the mineral crystals must melt to form magma. According to the Kinetic Theory, a solid mineral crystal will melt if its kinetic energy (motion of its atoms and molecules) exceeds the attractive forces that hold together its orderly crystalline structure. Heating a crystal is the most obvious way to melt it. If enough heat energy is applied to the crystal, then its kinetic energy level may rise enough to cause melting. The specific temperature at which crystals of a given mineral begin to melt is the mineral's **melting point.**

All minerals have different melting points. So when heating a rock comprised of several different kinds of mineral crystals, one part of the rock (one kind of mineral crystal) will melt before another part (another kind of mineral crystal). Geologists call this **partial melting** of rock. But where would the heat come from to begin melting rocks below the ground?

Unless you live near a volcano or hot spring, you probably are not aware of Earth's body heat. But South African gold miners know all about it. The deeper they mine, the hotter it gets. In the deepest mine (see Figure 1.7), 3.8 kilometers below ground, temperatures reach 60 °C (140 °F) and the mines must be air conditioned. This gradient of increasing temperature with depth is called the **geothermal gradient.** This gradient also varies between ocean crust and continental crust, but the global average for all of Earth's crust is about 25 °C (77 °F) per kilometer. In other words, rocks located 1 kilometer below your house are about 25 °C warmer than the foundation of your house. If the geothermal gradient continued at this rate through the mantle, then the mantle would eventually melt at depths of 100–150 kilometers. Seismology shows that this does not occur, so temperature is not the only factor that determines whether a rock melts or remains solid. Pressure is also a factor.

Pressure (P)

When you press your hand against something like a bookshelf, you can apply all of your body *weight* against the surface *area* under your hand. Therefore, **pressure** is expressed as amount of weight applied per unit of area. For example, imagine that you weigh 100 pounds and that your hand is 5 inches long and 4 inches wide. If you exert all of your weight against a wall by leaning against the wall with one hand, then you are exerting 100 pounds of weight over an area of 20 square inches (5 inches × 4 inches = 20 square inches). This means that you are exerting 5 pounds of pressure per square inch of your hand.

Atoms and molecules of air (atmosphere) are masses of matter that are pulled by gravity toward the center of Earth. But they cannot reach Earth's center because water, rocks, and your body are in their path. As a result, the weight of the air presses against surfaces of water, rocks, and your body. If you stand at sea level, then your body is confined by 14.7 pounds of weight pressing on every square inch of your body (14 lbs./inch^2). This is called atmospheric **confining pressure.** Scientists also refer to this as one *atmosphere* (1 atm) of pressure.

You do not normally feel one atmosphere of confining pressure, because your body exerts the same pressure to keep you in equilibrium (balance) with your surroundings. But if you ever dove into the deep end of a swimming pool, then you experienced the confining pressure exerted by the water plus the confining pressure of the atmosphere. The deeper you dove, the more pressure you felt. It takes 10 m (33.9 ft) of water to exert another 1 atm of confining pressure on your body.

Rocks are about three times denser than water, so it takes only about 3.3 m of rock to exert a force equal to that of 10 m of water or the entire thickness of the atmosphere! 100 m of rock exert a confining pressure of about 30 atm, and 1 km (1000 m) of rock exerts a confining pressure of about 300 atm. At 300 atm/km, a rock buried 5 km underground is confined by 1500 atm of pressure!

The confining pressure under kilometers of rock is so great that a mineral crystal cannot melt at its "normal" melting point observed on Earth's surface. The pressure confines the atoms and molecules and prevents them from flowing apart. More heat is required to raise the kinetic energy level of atoms and molecules in the crystal enough to melt the crystal. Consequently, an increase in confining pressure causes an increase in the melting point of a mineral. Reducing confining pressure lowers the melting point of a mineral. This means that if a mineral is already near its melting point, and its confining pressure decreases enough, then it will melt. This is called **decompression melting.**

Pressure-Temperature (P-T) Diagrams

Geologists understand that rock melting (the origin of magma) is related to both temperature and pressure. Therefore they heat and pressurize rock samples under controlled conditions in geochemical laboratories to determine how rock melting is influenced by specific combinations of both pressure and temperature. Samples are pressurized and heated to specific P-T points to determine if they remain solid, undergo partial melting, or melt completely. The data are then plotted as specific points on a **pressure-temperature (P-T) diagram** such as the one in Figure 2.7 for mantle peridotite. Mantle peridotite is made of olivine,

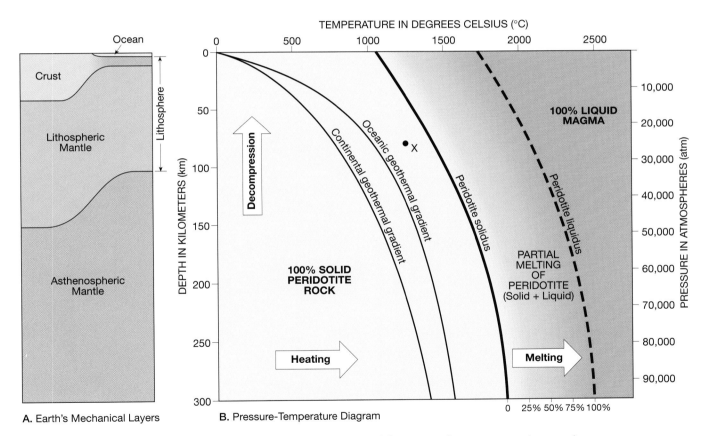

FIGURE 2.7 Mechanical layers at the edge of Earth's rocky body **(A)** compared to a pressure-temperature (P-T) diagram of environmental conditions that exist there **(B)**. The pressure-temperature diagram shows how P-T conditions affect peridotite rock (made of olivine, pyroxene, amphibole, and garnet mineral crystals). At P-T points below (to the left of) the *peridotite solidus,* all mineral crystals in the rock remain solid. At P-T points above (to the right of) the *peridotite liquidus,* all mineral crystals in the rock melt to liquid. At P-T points between the solidus and liquidus, the rock undergoes partial melting—one kind of mineral at a time, so solid and liquid are present. Continental and oceanic geothermal gradients are curves showing how temperature normally varies according to depth below the continents and ocean basins. Temperatures along both of these geothermal gradients are too cool to begin partial melting of peridotite. Both gradients occur below (to the left of) the peridotite solidus. (1 atm = about 1 bar)

pyroxene, amphibole, and garnet mineral crystals. Therefore, this diagram also shows the combined effects of pressure and temperature on a rock made of several different minerals. At P-T points below (to the left of) the *solidus,* all mineral crystals in the rock remain solid. At P-T points above (to the right of) the *liquidus,* all mineral crystals in the rock melt to liquid. At P-T points between the solidus and liquidus, the rock undergoes partial melting—one kind of mineral at a time. Therefore, a P-T diagram also reveals stability fields for states (phases) of matter. In this case (see Figure 2.7) there are stability fields for solid, solid + liquid, and liquid.

Notice that lines for the continental and oceanic geothermal gradients are also plotted on Figure 2.7. They show how temperature normally varies according to depth below the continents and ocean basins. Temperatures along both of these geothermal gradients are not great enough to begin melting peridotite. Both gradients occur along temperatures below (to the left of) the peridotite solidus.

Questions

9. Examine the pressure-temperature (P-T) diagram for mantle peridotite in Figure 2.7, and locate point **X**. This point represents a mass of peridotite buried 80 km underground.

   a. According to the continental geothermal gradient, rocks buried 80 km beneath a continent would normally be heated to what temperature?

   b. According to the oceanic geothermal gradient, rocks buried 80 km beneath an ocean basin would normally be heated to what temperature?

   c. Is the peridotite at point **X** a mass of solid, a mixture of solid and liquid, or a mass of liquid? How do you know?

   d. What would happen to the mass of peridotite at point **X** if it were heated to 1750 °C? How do you know?

   e. What would happen to the mass of peridotite at point **X** if it were heated to 2250 °C? How do you know?

10. At its current depth, the peridotite at point **X** in Figure 2.7 is under about 25,000 atm of pressure.

    a. At what depth and pressure will this peridotite begin to melt if it is uplifted closer to Earth's surface and its temperature remains the same?

    b. What is the name applied to this kind of melting?

    c. Name and describe a process that could uplift mantle peridotite to start it melting in this way,

and name a specific plate tectonic setting where this may be happening now. (*Hint:* Study Figures 2.1, 2.5, and 2.6.)

11. Based on your answers in Questions 9 and 10, what are two environmental changes that can cause the peridotite at point **X** (see Figure 2.7) to begin partial melting?

12. Obtain the materials shown in Figure 2.8. Turn the hot plate on a low setting (about 3 on most commercial hot plates) and allow it to heat up in a safe location. Next place two sugar cubes on a flat piece of aluminum foil. Label (on the foil) one sugar cube "dry." Moisten the second sugar cube with a few drops of water, and label it "wet." Carefully place the aluminum foil (with the sugar cubes) onto the hot plate and observe what happens. (*Note:* Turn off the hot plate when one cube begins to melt.)

    a. Which sugar cube melted first?

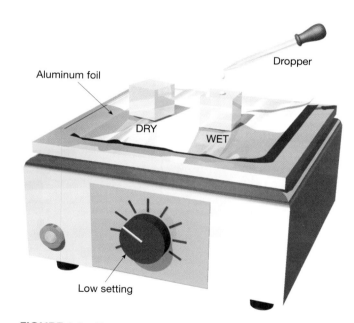

FIGURE 2.8 Procedures for melting experiment in Question 12. Turn the hot plate on a low setting (about 2 or 3 on most commercial hot plates) and allow it to heat up in a safe location (be careful not to touch hot surfaces directly). Next, place two sugar cubes on a flat piece of aluminum foil or in aluminum foil baking cups. Label (on the foil) one sugar cube "dry." Moisten the second sugar cube with about 4 or 5 drops of water and label it "wet." Carefully place the aluminum foil with the labeled sugar cubes onto the hot plate and observe what happens. When one of the sugar cubes begins to melt, use crucible tongs and/or hot pads to remove the foil and sugar cubes from the hot plate and avoid burning the sugar. Turn off the hot plate!

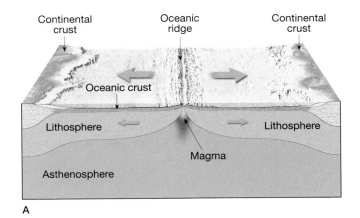

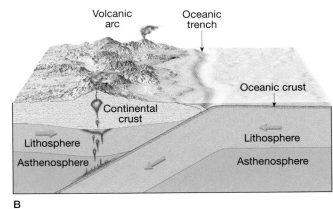

FIGURE 2.9 Plate boundary block diagrams to evaluate in Questions 13 and 14.

b. The rapid melting that you observed in 12a is called "**flux melting**," because flux is something that speeds up a process. What was the flux?

c. How would the P-T diagram in Figure 2.7 change if all of the peridotite in the diagram was "wet" peridotite?

d. In what specific kind of plate tectonic setting could water enter Earth's mantle and cause flux melting of mantle peridotite?
(*Hint:* Figure 2.1)

13. Examine the cross section of a plate boundary in Figure 2.9A.

a. What kind of plate boundary is this?

b. Name the specific process that led to formation of magma in this cross section.

c. Describe the sequence of plate tectonic and magma generating processes that led to formation of the volcanoes (oceanic ridge) in this cross section.

14. Examine the cross section of a plate boundary in Figure 2.9B.

a. What kind of plate boundary is this?

b. Name the specific process that led to formation of magma in this cross section.

c. Describe the plate tectonic and magma generating processes that led to formation of the volcanoes in this cross section.

PART 2D: MEASURING AND EVALUATING PLATE TECTONICS

Hot spots are centers of volcanic activity that persist in a stationary location for tens-of-millions of years. Geologists think they are either: a) the result of long-lived narrow *plumes* of hot rock rising rapidly from Earth's mantle and forming magma by decompression melting (like a stream of heated lava rising in a lava lamp), or b) the slow decompression melting of a large mass of hot mantle rock in the upper mantle that persists for a long interval of geologic time.

As a lithospheric plate migrates across a stationary hot spot, a volcano develops directly above the hot spot. When the plate slides on, the volcano that was over the hot spot becomes dormant, and over time it migrates many kilometers from the hot spot. Meanwhile, a new volcano arises as new lithosphere passes over the hot spot. The result is a string of volcanoes, with one end of the line located over the hot spot and quite active, and the other end distant and inactive. In between is a succession of volcanoes that are progressively older with distance from the hot spot. The Hawaiian Islands and Emperor Seamount chain are thought to represent such a line of volcanoes that formed over the Hawaiian hot spot (Figure 2.10).

Question

15. Figure 2.10 shows the distribution of the Hawaiian Island chain and Emperor Seamount chain. The numbers indicate the age of each island in millions of years (m.y.), obtained from the basaltic igneous rock of which each island is composed.

   a. What was the rate in centimeters per year (cm/yr) and direction of plate motion in the Hawaiian region from 4.7 to 1.6 million years (m.y.) ago?

   b. What was the rate in centimeters per year (cm/yr) and direction of plate motion from 1.6 million years ago to the present time?

   c. How does the rate and direction of Pacific plate movement during the past 1.6 million years differ from the older rate and direction (4.7–1.6 m.y.) of plate motion?

   d. Locate the Hawaiian Island chain and the Emperor Seamount chain (submerged volcanic islands) in the top part of Figure 2.10. How are the two island chains related?

   e. Based on the distribution of the Hawaiian Islands and Emperor Seamount chains, suggest how the direction of Pacific plate movement has generally changed over the past 60 million years.

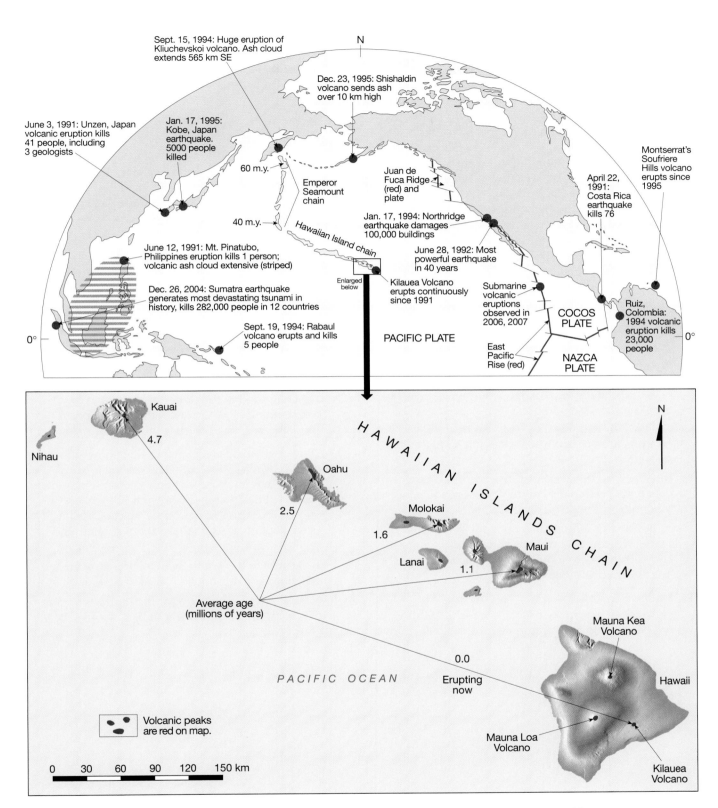

FIGURE 2.10 Map of the northern Pacific Ocean (top) and adjacent land masses showing some notable geologic hazards (natural disasters) from the 1990s, the Hawaiian Islands Chain, and the Emperor Seamount Chain. Lower map shows details of the Hawaiian Islands Chain, including locations of volcanic peaks.

Another hot spot is beneath Yellowstone National Park in the United States (Figure 2.11). There are no erupting volcanoes on the Yellowstone hot spot today, but there are hot springs and geysers. The high heat flow also causes buckling and faulting of Earth's crust.

Geologist Mark Anders has observed that as the North American Plate slides over this hot spot, it causes development of a U-shaped set of faults (with the closed end of the U pointing northeast). Also, as layers of volcanic ash and lava flows accumulate on deforming crust, they tilt. By dating the layers of tilted rocks, and mapping the U-shaped fault systems that moved beneath them, Anders has been able to compose a map of circular regions that were once centered over the hot spot at specific times. This line of deformation circles and their ages are shown in Figure 2.11.

Questions

16. Examine the part of Figure 2.11 showing the distribution of circular areas that were centers of crustal faulting and buckling when they were located over the Yellowstone hot spot. The numbers indicate the ages of deformation, as determined by Mark Anders.

    a. What direction is the North American Plate moving, according to Anders' data? Explain your reasoning.

    b. What was the average rate in centimeters per year (cm/yr) that the North American Plate has moved over the past 11 million years?

    c. Beside the Yellowstone hot spot in Figure 2.11, place an arrow and rate of motion (from items 16a and 16b above) to indicate the velocity of the North American Plate.

17. Notice the ages of seafloor volcanic rocks in Figure 2.11. The modern seafloor rocks of this region (located at 0 million years old) are forming along a divergent plate boundary called the Juan de Fuca Ridge. The farther one moves away from the plate boundary (B in Figure 2.11), the older are the seafloor rocks.

    a. What has been the average rate of seafloor spreading in centimeters per year (cm/yr) east of the Juan de Fuca Ridge (along line B–D) over the past 8 million years?

    b. What has been the average rate of seafloor spreading in centimeters per year (cm/yr) west of the Juan de Fuca Ridge (along line A–B) over the past 8 million years?

    c. Notice that seafloor rocks older than 8 million years are present west of the Juan de Fuca Ridge but not east of the ridge. What could be happening to the seafloor rocks along line C–D that would explain their absence from the map?

    d. Based on your reasoning in item 17c, what kind of plate boundary is represented by the red line running through location C on Figure 2.11?

    e. Notice the line of volcanoes in the Cascade Range located at the center of the map. How could magma form beneath these volcanoes? (Be as specific as you can.)

18. Draw a geologic cross section that shows all of the plate tectonic features developed along line A–D. Be sure that your sketch shows ocean lithosphere (Juan de Fuca Plate), continental lithosphere (North American Plate), and a volcano of the Cascade Range. Add labels, arrows, rates, and brief descriptions for all.

    a. Be sure to draw and label all of the processes described in Questions 17c and 17e.

    b. Add arrows and rates for the motions of the plates (from Questions 16a, 16b, 17a, and 17b) to your geologic cross section.

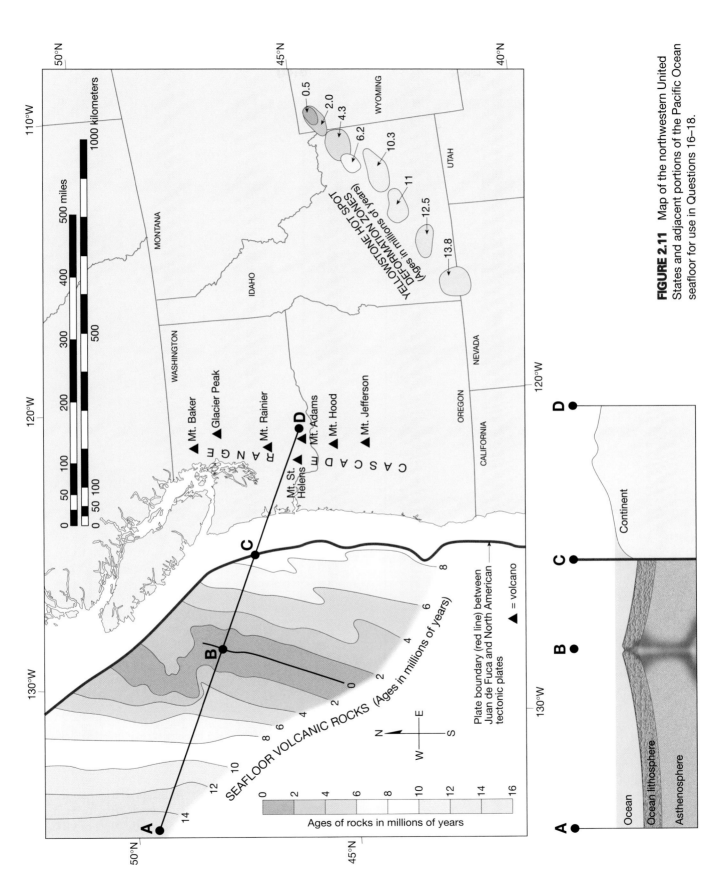

FIGURE 2.11 Map of the northwestern United States and adjacent portions of the Pacific Ocean seafloor for use in Questions 16–18.

45

FIGURE 2.12 Generalized geologic map of southern California. Half-arrows indicate relative motions along the San Andreas Fault. The fault is also a boundary between two of Earth's lithospheric plates. The relative motion of the Pacific Plate (under the Pacific Ocean) is northwest. The North American Plate is located east of the fault and is moving relatively southeast.

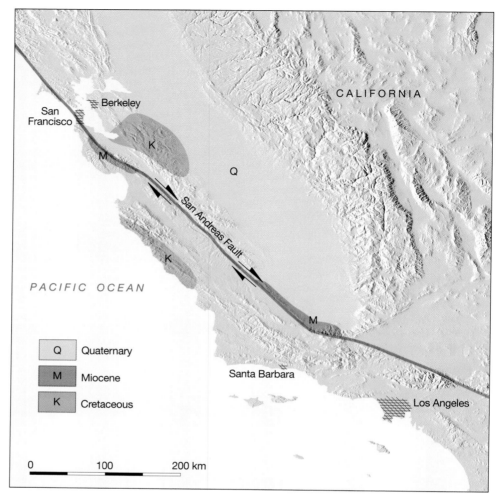

San Andreas Fault Hazards

Study the geologic map of southern California in Figure 2.12, showing the position of the famous San Andreas Fault. The east side of this fault is rocks of the North American lithospheric plate. The west side of the fault is the Pacific lithospheric plate, which is moving northwest. It is well known to all who live in southern California that plate motions along the fault cause frequent earthquakes that place at risk humans and their properties.

Question

19. The two bodies of Late Miocene rocks (about 25 million years old) located along either side of the San Andreas Fault (Figure 2.12) were one body of rock that has been separated by motions along the fault. Note that arrows have been placed along the sides of the fault to show the relative sense of movement.

    a. The San Andreas Fault is what kind of plate boundary?

    b. You can estimate the average annual rate of movement along the San Andreas Fault by measuring how much the Late Miocene rocks have been offset by the fault and by assuming that these rocks began separating soon after they formed. What is the average annual rate of fault movement in centimeters per year (cm/yr)?

    c. The average yearly rate of movement on the San Andreas Fault is very small. Does this mean that the residents of southern California have nothing to worry about from this fault? Explain.

    d. An average movement of about 5 m (16 ft) along the San Andreas Fault was associated with the devastating 1906 San Francisco earthquake that killed people and destroyed properties. Assuming that all displacement along the fault was produced by Earth motions of this magnitude, how often must such earthquakes have occurred in order to account for the total displacement?

Mineral Properties, Uses, and Identification

•CONTRIBUTING AUTHORS•

Jane L. Boger • *SUNY, College at Geneseo*

Philip D. Boger • *SUNY, College at Geneseo*

Roseann J. Carlson • *Tidewater Community College*

Charles I. Frye • *Northwest Missouri State University*

Michael F. Hochella, Jr. • *Virginia Polytechnic Institute*

OBJECTIVES

A. Analyze minerals for seven common properties (color, crystal form, luster, streak, hardness, cleavage, fracture) and six other properties (tenacity, reaction with acid, magnetism, striations, exsolution lamellae, specific gravity).

B. Be able to identify common minerals (in hand samples) on the basis of their properties and then infer how they are or could be used by people.

C. Use Internet resources to learn more about minerals upon which you depend and where those minerals are mined.

MATERIALS

Pencil, eraser, laboratory notebook; mineral analysis tools: pocket knife or steel masonry nail, wire nail, glass plate, streak plate, copper penny, small magnet, dilute (1–3%) hydrochloric acid (HCl) in a dropper bottle; cleavage goniometer (cut from GeoTools Sheet 1 at the back of the manual), mineral and rock samples (purchased or provided by your instructor and marked with an identifying number or letter); mineral data charts (Figure 3.22).

INTRODUCTION

You may know minerals as the beautiful gemstones in jewelry or masses of natural crystals displayed in museums. However, most **rocks** are aggregates of one or more kinds of minerals. If you look closely at rocks, you will see the minerals that comprise them (Figure 3.1). Minerals are also the natural materials from which every inorganic item in our industrialized society has been manufactured. Therefore, minerals are the physical foundation of both our planet and our human societies.

All **minerals** are inorganic, naturally occurring substances that have a characteristic chemical composition, distinctive physical properties, and crystalline structure. *Crystalline structure* is an orderly three-dimensional arrangement of atoms or molecules (Figure 3.2), and materials with crystalline structure form *crystals* (Figures 3.1A, 3.2A, 3.3). A few "minerals," such as limonite (rust) and opal (Figure 3.4), do not have crystalline structure and never form crystals, so they are not true minerals. They are sometimes called *mineraloids*.

More than 2000 different kinds of minerals have been identified and named on the basis of their characteristic chemical composition and physical properties. Some are best known as *rock-forming minerals,*

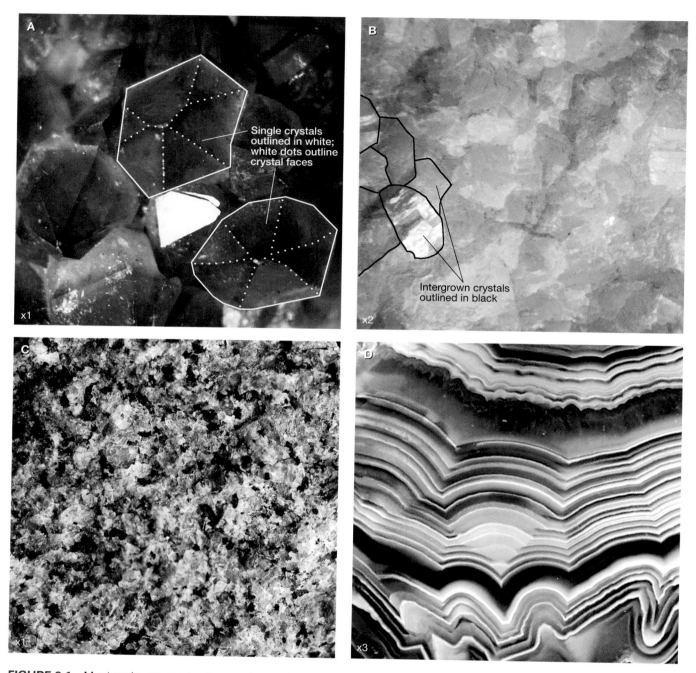

FIGURE 3.1 Most rocks are made of mineral crystals, as in these four examples. **A.** Rock comprised of numerous mineral crystals of amethyst, a purple variety of quartz. Two crystals have been outlined to emphasize their crystal form, and dots show the boundaries between their crystal faces (the flat surfaces on the outside of a crystal). **B.** Rock (called marble) comprised of intergrown calcite mineral crystals. The calcite crystals grew together, so their crystal form is not visible. **C.** Rock comprised of red-brown potassium-feldspar crystals, white plagioclase feldspar crystals, gray quartz crystals, and black biotite crystals. Like rock B, the crystals grew together so closely that their crystal form is obscured. **D.** A variety of quartz called agate. It is comprised of variously colored quartz crystals that are *cryptocrystalline* (so tiny that they are not visible in hand sample) in most of the bands of the agate.

FIGURE 3.2 Galena is lead sulfide—PbS. This mineral forms cubic (cube-shaped) crystals (**A**) that are the color of dull tarnished lead metal. Galena crystals are brittle. When struck with a hammer, they shatter into silvery gray fragments with cubic shapes (**B**). The orderly arrangement of lead and sulfur atoms in galena can be seen in an atomic-resolution image of galena (**C**) taken with a scanning tunneling microscope (STM) by C. M. Eggleston, University of Wyoming. Each *sulfur* atom is bonded to four lead atoms in the image, plus another lead atom beneath it. Similarly, each *lead* atom is bonded to four sulfur atoms in the image, plus a sulfur atom beneath it. Galena is an ore of lead.

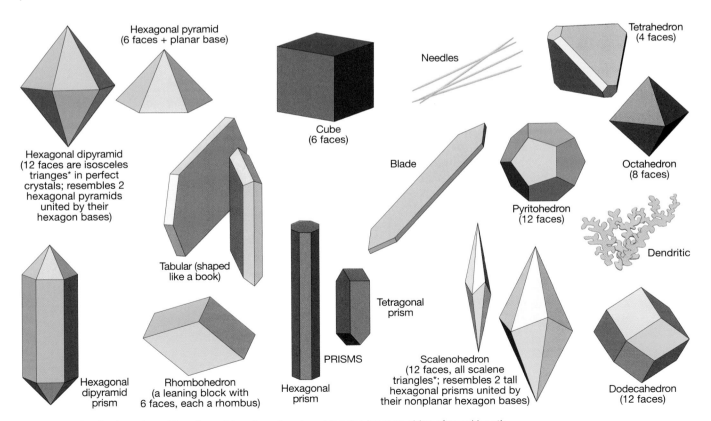

*Isosceles triangles have two sides of equal length and scalene triangles have no sides of equal length.

FIGURE 3.3 Some *crystal forms* (geometric shapes) or habits. The flat outer surfaces of these forms are called *crystal faces*. Crystal form is an external feature of mineral crystals. *Massive* form refers to cases where mineral crystals are so tightly intergrown that no distinguishing crystal form is visible.

FIGURE 3.4 Opal is hydrated silicon dioxide—$SiO_2 \cdot nH_2O$. Opal is not a true mineral because it is an *amorphous* (without crystalline structure) solid. It is sometimes referred to as a mineraloid (mineral-like) material. **A.** This sample of Australian opal formed on brown rock and ranges from colorless to translucent white and blue. **B.** This *precious opal* from Mexico has fire (internal flashes of color) and has been artificially shaped into a cabochon (oval shape with a rounded top and flat back) to be set in jewelry.

meaning that they are the main minerals observed in rocks. Others are best known as *industrial minerals,* meaning that they are the main minerals used to manufacture physical materials of industrialized societies. Although all minerals are inorganic, *biochemical minerals* can be manufactured by organisms. A good example is the mineral aragonite, which clams and other molluscs use to construct their shells.

PART 3A: MINERAL PROPERTIES AND USES

Seven properties are commonly used to identify minerals and determine how they may be used. You should be able to analyze minerals for all of these common properties (color, crystal form, luster, streak, hardness, cleavage, fracture). It will also be useful for you to be familiar with ways to analyze minerals for six other properties that may be used to distinguish specific minerals or specific groups of minerals (tenacity, reaction with acid, magnetism, striations, exsolution lamellae, specific gravity).

Color and Clarity

Color of a mineral is usually its most noticeable property and may be a clue to its identity. A rock made up of one color of mineral crystals is usually made up of one kind of mineral (Figure 3.1A, 3.1B), and a rock made of more than one color of mineral crystals is usually made up of more than one kind of mineral (Figure 3.1C). However, there are exceptions. The rock in Figure 3.1D has many colors, but they are simply *varieties* (var.)—different forms or colors—of the mineral quartz. This means that a mineral cannot be identified solely on the basis of its color. The mineral's other properties must also be observed, recorded, and used collectively to identify it. Most minerals also tend to exhibit one color on freshly broken surfaces and a different color on tarnished or weathered surfaces. Be sure to note this difference, if present, to aid your identification.

Mineral crystals may vary in the **clarity** of their color (or lack of color). They may be *transparent* (clear and see-through, like window glass), *translucent* (foggy, like looking through a steamed-up shower door), or *opaque* (impervious to light, like concrete and metals). It is good practice to record not only a mineral's color, but also its clarity. For example, the crystals in Figure 3.1A are purple in color and have transparent to translucent clarity. Galena mineral crystals (Figure 3.2) are opaque.

Crystal Form (Habit)

When you think of mineral crystals, you probably imagine museum displays of crystals having perfect **crystal forms** (geometric shapes) like cubes, pyramids, or prisms (see Figure 3.3). The characteristic crystal form (or combination of forms) of a mineral is that mineral's *habit*.

Note that crystal form is an external feature of mineral crystals. Perfect crystal forms can only develop if a mineral crystal is unrestricted as it grows. This is rare. It is more common for mineral crystals to crowd together as they grow, resulting in a network of intergrown crystals that do not exhibit their crystal form. For example, calcite normally forms 12-sided scalenohedrons (see Figure 3.3). But the calcite crystals in Figure 3.1B crowded together so much as they grew that none of them exhibits its crystal form. Most crystalline rocks form in this way. Even if crystal forms develop, the crystals may be *cryptocrystalline*—too small to see with the naked eye (Figure 3.1D). Most of the laboratory samples of minerals that you will analyze do not exhibit their crystal forms either, because they are small broken pieces of larger crystals. However, if

FIGURE 3.5 Native copper—Cu (naturally-occurring pure copper) has a metallic luster until it tarnishes. **A.** Copper crystals tarnished brown and nonmetallic. **B.** Tarnished copper crystals with a thin coating of green malachite (copper carbonate) also have a nonmetallic luster. **C.** A freshly minted copper coin has a bright metallic luster (until it tarnishes).

the form of a crystal is visible, then it should be noted and used as evidence for mineral identification.

Luster

Luster is a description of how light reflects from the surface of an object, such as a mineral. Luster is of two main types—metallic and nonmetallic—that vary in intensity from bright (very reflective, shiny, polished)

to dull (not very reflective, not very shiny, not polished). For example, if you make a list of objects in your home that are made of metal (e.g., coins, knives, keys, jewelry, door hinges, aluminum foil), then you are already familiar with metallic luster. Yet the metallic objects can vary from bright (very reflective—like polished jewelry, the polished side of aluminum foil, or new coins) to dull (non-reflective—like unpolished jewelry or the unpolished side of aluminum foil).

Minerals with a **metallic luster (M)** reflect light just like the metal objects in your home—they have opaque, reflective surfaces with a silvery, gold, brassy, or coppery sheen (Figures 3.2B, 3.5C, 3.6). All other minerals have a **nonmetallic luster (NM)**—a luster unlike that of the metal objects in your home (Figures 3.1, 3.2A, 3.4). The luster of nonmetallic minerals can also be described with the more specific terms below:

- Vitreous—resembling the luster of freshly broken glass or a glossy photograph

- Waxy—resembling the luster of a candle

- Pearly—resembling the luster of a pearl

- Satiny—resembling the luster of satin or silk cloth

- Earthy—lacking reflection, completely dull, like dry soil

- Greasy—resembling the luster of grease, oily

- Porcelaneous—resembling the luster of porcelain (translucent white ceramic ware)

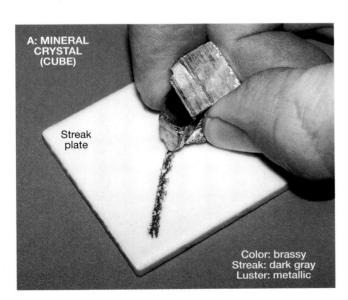

FIGURE 3.6 Streak tests. **A.** This mineral has a brassy color and metallic *luster* (surface reflection), but its *streak* (color in powdered form) is dark gray. **B.** This mineral has a reddish silver color and silvery or steel-gray metallic luster, but its streak is red-brown. If you do not have a streak plate, then determine the streak color by crushing or scratching part of the sample to see the color of its powdered form.

Exposed surfaces of most minerals, especially metallic minerals, will normally tarnish or weather to a more dull or earthy nonmetallic luster. Notice how the exposed metallic copper crystals in Figure 3.5A, 3.5B, and the galena crystals in Figure 3.2A, have tarnished to a nonmetallic luster. Always observe freshly broken surfaces of a mineral to determine whether it has a metallic or nonmetallic luster. It is also useful to note a mineral's luster on fresh versus tarnished surfaces when possible.

Streak

Streak is the color of a substance after it has been ground to a fine powder (so fine that you cannot see the grains of powder). The easiest way to do this is simply by scratching the mineral back and forth across a hard surface such as concrete, or a square of unglazed porcelain (called a *streak plate*). The color of the mineral's fine powder is its streak. Note that the brassy mineral in Figure 3.6 has a dark gray streak, but the reddish silver mineral has a red-brown streak. A mineral's streak is usually similar even among all of that mineral's varieties.

If you encounter a mineral that is harder than the streak plate it will scratch the streak plate and make a white streak of powder from the streak plate. The streak of such hard minerals can be determined by crushing a tiny piece of them with a hammer (if available). Otherwise, record the streak as unknown.

Hardness (H)

Hardness is a measure of resistance to scratching. A harder substance will scratch a softer one (Figure 3.7). German mineralogist Friedrich Mohs (1773–1839) developed a quantitative scale of relative mineral hardness on which the softest mineral (talc) has a hardness of 1 and the hardest mineral (diamond) has a hardness of 10. Higher-numbered minerals will scratch lower-numbered minerals (e.g., diamond will scratch talc, but talc cannot scratch diamond).

Mohs Scale of Hardness (Figure 3.8) is now widely used by geologists and engineers. When identifying a mineral, you should mainly be able to distinguish minerals that are relatively hard (6.0 or higher on Mohs Scale) from minerals that are relatively soft (less than or equal to 5.5 on Mohs Scale). You can use common objects such as a glass plate (Figure 3.7), pocket knife (steel), or masonry (steel) nail to make this distinction as follows.

- **Hard minerals:** Will scratch glass. Cannot be scratched with a knife blade or masonry nail (or glass).

- **Soft minerals:** Will not scratch glass. Can be scratched with a knife blade or masonry nail (or glass).

You can determine a mineral's hardness number on Mohs Scale by comparing the mineral to common objects shown in Figure 3.8 or pieces of the minerals in Mohs Scale. Commercial *hardness kits* contain a set of all of the minerals in Figure 3.8 or a set of metal scribes of known hardnesses. When using such kits to make hardness comparisons, remember that the harder mineral/object is the one that scratches, and the softer mineral/object is the one that is scratched.

Cleavage and Fractures

Cleavage is the tendency of some minerals to break (*cleave*) along flat, parallel surfaces (**cleavage planes**) like the flat surfaces on broken pieces of galena (Figure 3.2B). The cleavage can be described as excellent, good, or poor (Figures 3.9, 3.10). An *excellent cleavage* direction reflects light in one direction from a set of obvious, large, flat, parallel surfaces. A *good cleavage* direction reflects light in one direction from a set of many small, obvious, flat, parallel surfaces. A *poor cleavage* direction reflects light from a set of small, flat, parallel surfaces that are difficult to detect. Some of the light is reflected in one direction from the small cleavage surfaces, but most of the light is scattered randomly by fracture surfaces separating the cleavage surfaces. **Fracture** refers to any break in a mineral that does not occur along a cleavage plane. Therefore, fracture surfaces are normally not flat and they never occur in parallel sets. Fracture can be described as *uneven* (rough, like the milky quartz in Figure 3.9A), *splintery* (like splintered wood), or *hackly* (having jagged edges). Opal (Figure 3.4A) and pure quartz (Figure 3.10B) both tend to fracture like glass—along ribbed, smoothly curved surfaces called *conchoidal fractures*.

The cleavage planes are parallel surfaces of weak chemical bonding (attraction) between repeating, parallel layers of atoms in a crystal, and more than one set of cleavage planes can be present in a crystal. Each different set has an orientation relative to the crystalline structure and is referred to as a **cleavage direction** (Figure 3.11). For example, muscovite (Figure 3.12) has one excellent cleavage direction and splits apart like pages of a book (book cleavage). Halite (Figure 3.13) and galena (Figure 3.2B) both have three cleavage directions developed at right angles to one another, so they have cubic cleavage (break either into cubes or fragments that have sides at right angles to one another).

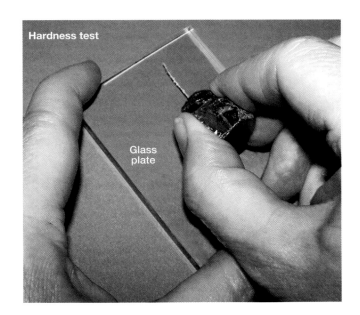

FIGURE 3.7 *Hardness* (resistance to scratching) test using a glass plate, which has a hardness of 5.5 on Mohs Scale of Hardness (Figure 3.8). Be sure the edges of the glass have been dulled. If not, then wrap the edges in masking tape or duct tape. Hold the glass plate firmly against a flat table top, then try to scratch the glass with the mineral sample. A mineral that scratches the glass (this image) is a *hard* mineral (i.e., harder than 5.5 on Mohs Scale of Hardness). A mineral that does not scratch the glass is a *soft* mineral (i.e., less than or equal to 5.5 on Mohs Scale of Hardness).

Mohs Scale of Hardness*		Hardness of Some Common Objects (Harder objects scratch softer objects)
HARD	10 Diamond	
	9 Corundum	
	8 Topaz	
	7 Quartz	6.5 Streak plate
	6 Orthoclase Feldspar	
SOFT	5 Apatite	5.5 Glass, Masonry nail, Knife blade
	4 Fluorite	4.5 Wire (iron) nail
	3 Calcite	3.5 Brass (wood screw, washer)
		3.0 Copper coin (penny)
	2 Gypsum	2.5 Fingernail
	1 Talc	

* A scale for measuring relative mineral hardness (resistance to scratching).

FIGURE 3.8 Mohs Scale of Hardness (resistance to scratching) and the hardness of some common objects. *Hard minerals* have a Mohs hardness number greater than 5.5, so they scratch glass and cannot be scratched with a knife blade or masonry (steel) nail. *Soft minerals* have a Mohs hardness number of 5.5 or less, so they do not scratch glass and are easily scratched by a knife blade or masonry (steel) nail. Mohs hardness numbers can be determined for hard or soft minerals by comparing them to the hardness of other common objects or minerals of Mohs Scale of Hardness.

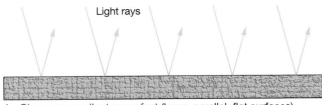

A. Cleavage excellent or perfect (large, parallel, flat surfaces)

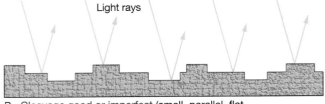

B. Cleavage good or imperfect (small, parallel, flat, stair-like surfaces)

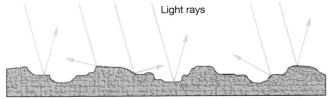

C. Cleavage poor (a few small, flat surfaces difficult to detect)

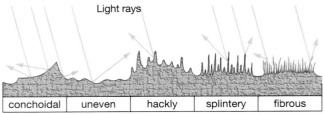

| conchoidal | uneven | hackly | splintery | fibrous |

D. Fractures (broken surfaces lacking cleavage planes)

FIGURE 3.9 Cross sections of mineral samples to illustrate degrees of development of cleavage—the tendency for a mineral to break along one or more sets of parallel planar surfaces called *cleavage planes*. The cleavage planes are surfaces of weak chemical bonding (attraction) between repeating, parallel layers of atoms in a crystal. Each different set of parallel cleavage planes is referred to as a *cleavage direction*. **A.** An *excellent cleavage* direction reflects light in one direction from a set of large, obvious, flat, parallel surfaces. **B.** A *good cleavage* direction reflects light in one direction from a set of small, flat, parallel, stair-like surfaces. A good example is galena (Figure 3.2). **C.** A *poor cleavage* direction reflects only a small amount of light from a set of very small, flat, parallel, surfaces that are not obvious. Remaining rays of light are reflected in all directions from fracture surfaces. **D.** *Fracture* refers to any break in a mineral that does not occur along a cleavage plane. Therefore, fracture surfaces are normally not flat and they never occur in parallel sets. Fracture can be characterized as conchoidal, uneven, hackly, splintery, or fibrous.

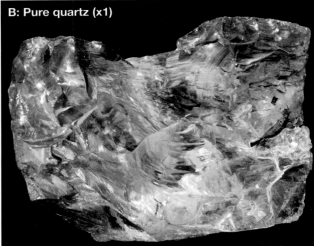

FIGURE 3.10 Quartz (silicon dioxide). These two samples are broken pieces of quartz mineral crystals that do not exhibit any crystal form. **A.** Milky quartz forms when the quartz has microscopic fluid inclusions, usually water. Note the nonmetallic (vitreous) luster and uneven fracture. **B.** Pure quartz (var. rock crystal) is colorless, transparent, nonmetallic and exhibits excellent conchoidal fracture.

Minerals of the pyroxene (e.g., augite) and amphibole (e.g., hornblende) groups generally are both dark-colored (dark green to black), opaque, nonmetallic minerals that have two good cleavage directions. The two groups of minerals are sometimes difficult to distinguish, so some people identify them collectively as *pyriboles*. However, pyroxenes can be distinguished from amphiboles on the basis of their cleavage. The two cleavages of pyroxenes intersect at 87° and 93°, or nearly right angles (Figure 3.14A). The two cleavages of amphiboles intersect at angles of 56° and 124° (Figure 3.14B). These angles can be measured in

Number of Cleavages and Their Directions	Name and Description of How the Mineral Breaks	Shape of Broken Pieces (Cleavage Directions are Numbered)	Illustration of Cleavage Directions
No cleavage (fractures only)	No parallel broken surfaces; May have conchoidal fracture (like glass)	Quartz	None (no cleavage)
1 cleavage	**Basal (book) cleavage** "Books" that split apart along flat sheets	Muscovite, biotite, chlorite (micas)	
2 cleavages intersect at or near 90°	**Prismatic cleavage** Elongated forms that fracture along short *rectangular* cross sections	Orthoclase 90° (K-spar) Plagioclase 86° & 94°, pyroxene (augite) 87° & 93°	
2 cleavages do not intersect at 90°	**Prismatic cleavage** Elongated forms that fracture along short *parallelogram* cross sections	Amphibole (hornblende) 56° & 124°	
3 cleavages intersect at 90°	**Cubic cleavage** Shapes made of cubes and parts of cubes	Halite, galena	
3 cleavages do not intersect at 90°	**Rhombohedral cleavage** Shapes made of rhombohedrons and parts of rhombohedrons	Calcite and dolomite 75° & 105°	
4 main cleavages intersect at 71° and 109° to form octahedrons, which split along hexagon-shaped surfaces; may have secondary cleavages at 60° and 120°	**Octahedral cleavage** Shapes made of octahedrons and parts of octahedrons	Fluorite	
6 cleavages intersect at 60° and 120°	**Dodecahedral cleavage** Shapes made of dodecahedrons and parts of dodecahedrons	Sphalerite	

FIGURE 3.11 Cleavage in minerals.

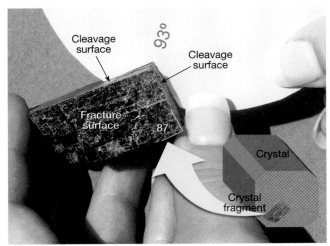

A. Cleavage in Pyroxenes (e.g., Augite)

FIGURE 3.12 Muscovite (white mica). Micas are aluminum silicate minerals that form stout crystals with *book* (basal) cleavage, because they split easily into paper-thin, transparent, flexible sheets, along planes of one excellent cleavage direction. Muscovite is a light-colored, colorless to brown mica, in contrast to biotite (black mica).

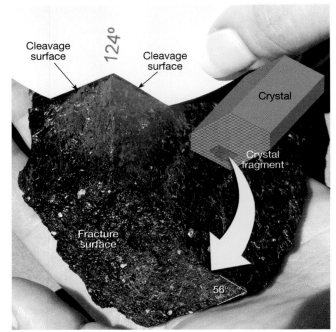

B. Cleavage in Amphiboles (e.g., Hornblende)

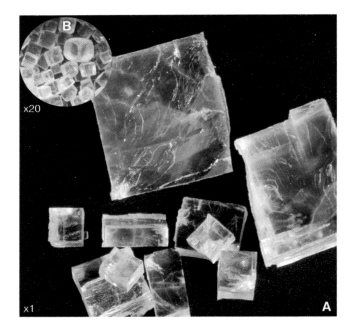

FIGURE 3.13 Halite, the mineral name for table salt and sodium chloride—NaCl. **A.** Fragmented hand sample showing excellent cubic cleavage. **B.** Crushed halite (table salt) from a salt shaker enlarged 20× to make the microscopic grains visible. Note cubic cleavage.

FIGURE 3.14 Pyroxenes and amphiboles are two groups of minerals with many similar properties. Minerals in both groups have similar hardness (5.5 – 6.0), color (usually green to black), luster (vitreous nonmetallic), and streak (white to pale gray). Minerals in both groups are also complex silicates containing some proportion of iron (Fe) and magnesium (Mg), so they are dark-colored *ferromagnesian silicates*. The main feature that distinguishes the two groups is their cleavage. Augite and other pyroxenes have two prominent cleavage directions that intersect at 87° and 93° (nearly right angles). Hornblende and other amphiboles have two prominent cleavage directions that intersect at 56° and 124°. These cleavage differences are the main way to distinguish pyroxenes (e.g., augite) from amphiboles (e.g., hornblende) in hand samples.

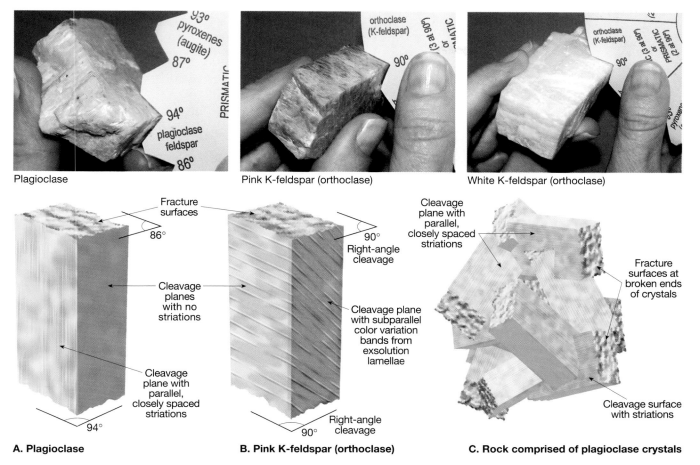

Plagioclase

Pink K-feldspar (orthoclase)

White K-feldspar (orthoclase)

A. Plagioclase   **B. Pink K-feldspar (orthoclase)**   **C. Rock comprised of plagioclase crystals**

FIGURE 3.15   Common feldspars and rock. Note how the cleavage goniometer can be used to distinguish potassium feldspar (K-feldspar, orthoclase) from plagioclase. The K-feldspar or orthoclase (Greek, *ortho*—right angle and *clase*—break) has perfect right-angle cleavage. Plagioclase (Greek, *plagio*—oblique angle and *clase*—break) does not. **A.** Illustration of a fragment of a plagioclase feldspar crystal showing two excellent cleavage directions, fractured ends, angles between cleavage directions, and *hairline striations* on a cleavage surface. The striations are caused by *twinning*: microscopic intergrowths between symmetrically-paired microcrystalline portions of the larger crystal. The striations occur only along one of the cleavage directions. They are not present in all samples of plagioclase. **B.** Broken piece of a K-feldspar (orthoclase) crystal showing two excellent cleavage directions, fractured ends, angles between cleavage directions, and intergrowths of thin, discontinuous, subparallel lamellae of plagioclase, called *exsolution lamellae.* The lamellae are actually microscopic layers of plagioclase that form as the mineral cools, like fat separates from soup when it is refrigerated. **C.** Hand sample of a rock that is an aggregate of intergrown plagioclase mineral crystals. Individual mineral crystals are discernible within the rock, particularly the cleavage surfaces that have characteristic hairline striations.

hand samples using the cleavage goniometer from GeoTools Sheet 1 at the back of this manual. Notice how a green cleavage goniometer was used to measure angles between cleavage directions in Figures 3.14 and 3.15.

Garnet (Figure 3.16) has no cleavage, but it fractures along very low angle, slightly curved *parting surfaces* that resemble cleavage. Parting surfaces develop along planes of chemical change or imperfection, so

they are not developed in all samples of garnet. Parting surfaces do not occur in parallel sets the way cleavage planes do.

Other Properties

Tenacity is the manner in which a substance resists breakage. Terms used to describe mineral tenacity include *brittle* (shatters like glass), *malleable* (like

FIGURE 3.16 Garnet, a hard (H = 7), complex silicate mineral used for sandpaper and as a gemstone. Garnet is brittle and has no cleavage. Some samples, such as this one, break along subparallel *parting* surfaces developed along planes of chemical change or imperfection. Parting does not occur along parallel even surfaces the way cleavage does.

FIGURE 3.17 Acid test (placing a drop of weak hydrochloric acid on the sample) was positive for this mineral (i.e., caused the mineral to effervesce), so this is a carbonate (CO_3-containing) mineral. Also note that the mineral occurs in several different colors and can be scratched by a wire (iron) nail. The yellow sample is a crystal of this mineral, but the other samples are fragments that reveal the mineral's characteristic cleavage angles.

modeling clay or gold; can be hammered or bent permanently into new shapes), *elastic* or *flexible* (like a plastic comb; bends but returns to its original shape), and *sectile* (can be carved with a knife).

Reaction to acid differs among minerals. Cool, dilute (1–3%) hydrochloric acid (HCl) applied from a dropper bottle is a common "acid test." All of the so-called *carbonate minerals* (minerals whose chemical composition includes carbonate, CO_3) will effervesce ("fizz") when a drop of such dilute HCl is applied to one of their freshly exposed surfaces (Figure 3.17). Calcite ($CaCO_3$) is the most commonly encountered carbonate mineral and effervesces in the acid test. Dolomite $[Ca,Mg(CO_3)_2]$ is another carbonate mineral that resembles calcite, but it will fizz in dilute HCl only if the mineral is first powdered. (It can be powdered for this test by simply scratching the mineral's surface with the tip of a rock pick, pocket knife, or nail.) If HCl is not available, undiluted vinegar can be used for the acid test, because it contains acetic acid (but the effervescence will be less violent).

Striations are straight "hairline" grooves on the cleavage surfaces of some minerals. This can be helpful in mineral identification. For example, you can use the striations of plagioclase feldspar (Figure 3.15A) to

distinguish it from potassium feldspar (K-feldspar, Figure 3.15B). *Plagioclase feldspars* have striations on surfaces of one of their two cleavage directions. *K-feldspars* may have lines that resemble striations at a glance. However, they are thin, discontinuous, subparallel exsolution lamellae (thin, discontinuous layers) of plagioclase within the K-feldspar.

Magnetism influences some minerals, such as magnetite. The test is simple: magnetite is attracted to a magnet. Lodestone is a variety of magnetite that is itself a natural magnet. It will attract steel paperclips. Some other minerals (e.g., hematite, bornite) may be weakly attracted to a magnet after they are heated.

Density is a measure of an object's mass (weighed in grams, g) divided by its volume (in cubic centimeters, cm^3). **Specific gravity (SG)** is the ratio of the density of a substance divided by the density of water. Since water has a density of $1\,g/cm^3$ and the units cancel out, specific gravity is the same number as density but without any units. For example, the mineral quartz has a density of $2.65\,g/cm^3$ so its specific gravity is 2.65 (i.e., SG = 2.65). *Hefting* is an easy way to judge the specific gravity of one mineral relative to another. This is done by holding a piece of the first mineral in one hand and holding an equal-sized piece of the second mineral in your other hand. Feel the difference in weight between the two samples (i.e., heft the samples). The sample that feels heavier has a higher specific gravity than the other. Most metallic minerals have higher specific gravities than nonmetallic minerals.

Questions

In Part B of this laboratory, you will be asked to analyze and identify mineral samples using mineral identification tables (Figures 3.18, 3.19, 3.20), a Mineral Database (Figure 3.21), and your skills of recognizing mineral properties. Complete the questions below to review mineral properties and develop some familiarity with the mineral identification tables and database before you proceed to Part B.

1. Indicate whether the luster of each of the following materials is metallic (M) or nonmetallic (NM):

   a. new coins:

   b. a brick:

   c. hard candy or a hard cough drop:

   d. ice:

   e. sharpened pencil lead:

   f. butter or margarine:

   g. a mirror:

2. What is the streak of each of the following substances?

   a. salt:

   b. wheat:

   c. pencil lead:

   d. charcoal:

3. What is the crystal form/habit of:

   a. quartz in Figure 3.1A?

   b. native copper in Figure 3.5?

4. Analyze the agate in Figure 3.1D, a massive form of quartz that is multicolored and banded. Also look up quartz in the Mineral Database (Figure 3.21) to find a list of some common varieties (var.) of quartz. Make a list of the names and descriptions of all of the varieties of quartz that are present in the agate in Figure 3.1D

5. A mineral can be scratched by a masonry nail or knife blade but not by a wire (iron) nail.

   a. Is this mineral hard or soft?

   b. What is the hardness number of this mineral on Mohs Scale?

   c. What mineral on Mohs Scale has this hardness?

6. A mineral can scratch calcite, and it can be scratched by a wire (iron) nail.

   a. What is the hardness number of this mineral on Mohs Scale?

   b. What mineral on Mohs Scale has this hardness?

7. The brassy, opaque, metallic mineral in Figure 3.6A is the same as the mineral in Figure 3.7.

   a. What is this mineral's hardness, and how can you tell?

   b. What is the crystal form (habit) of this mineral?

   c. Based on the mineral identification tables, what is the name of this mineral?

   d. Based on the Mineral Database, what is the chemical name and formula for this mineral?

   e. Based on the Mineral Database, how is this mineral used by society?

8. Analyze the mineral and figure caption in Figure 3.17.

   a. What is this mineral's hardness, and how can you tell?

   b. Very carefully cut out the cleavage goniometer from GeoTools Sheet 1 at the back of this manual. Be sure to cut the angles as exactly as possible. Use the cleavage goniometer to measure the angles between the flat cleavage surfaces of this mineral. What is the name of this kind of cleavage, and the broken shapes that it makes? (**Keep your cleavage goniometer for Part B.**)

   c. Based on the mineral identification tables, what is the name of this mineral?

   d. Based on the Mineral Database, what is the chemical name and formula for this mineral?

   e. Based on the Mineral Database, how is this mineral used by society?

9. A mineral sample weighs 27 grams and takes up 10.4 cubic centimeters of space. The SG (specific gravity) of this mineral is:

10. What products in your house or dormitory might be made from each of the minerals listed below? (Examine laboratory samples of them, if available. Also refer to the Mineral Database in Figure 3.21 as needed.)

    a. muscovite

    b. halite

    c. hematite

    d. feldspar

    e. galena

PART 3B: MINERAL IDENTIFICATION AND APPRECIATION

The ability to identify minerals is one of the most fundamental skills of an Earth scientist. It also is fundamental to identifying rocks, for you must first identify the minerals comprising them. Only after minerals and rocks have been identified can their origin, classification, and alteration be adequately understood. Mineral identification is based on your ability to describe mineral properties, as you did in Part 3A, and use identification charts (Figures 3.18–3.20). You will also need a Mineral Data Chart (Figure 3.22) to record your work.

Mineral Identification Procedures

To identify a mineral and its uses, use this step-by-step procedure:

A. Determine which minerals are metallic (M) and which ones are nonmetallic (NM). If you are uncertain about a mineral's luster, then it is probably nonmetallic. If you are looking at a rock, then consider each mineral separately.

B. For the *metallic* minerals (Figure 3.18):

1. Determine the mineral's hardness and record it in your Mineral Data Chart.

2. Determine the mineral's streak and record it in your Mineral Data Chart.

3. Determine and record other properties, such as color on fresh and tarnished surfaces, presence or absence of cleavage (and type of cleavage, if present), presence or absence of magnetism, tenacity, crystal form (if visible), or specific gravity.

4. Use the mineral properties to determine the probable name of the mineral in Figure 3.18. Check the Mineral Database (Figure 3.21) entry for that mineral name and compare its other properties against the properties you have recorded. Also note mineral uses. (If your mineral has different properties, then repeat your analysis to determine a more reasonable identification.)

5. Record the name and uses of the mineral in your Mineral Data Chart.

C. For the *nonmetallic* minerals (Figures 3.19 and 3.20):

1. Determine the mineral's hardness and record it in your Mineral Data Chart.

2. Determine the mineral's cleavage and record it (or the lack of it) in your Mineral Data Chart.

3. Determine and record the mineral's additional properties, such as color, tenacity, result of acid test, crystal form/habit (if visible), presence of striations or exsolution lamellae (if any), specific gravity, and specific kinds of nonmetallic luster.

4. Use the mineral properties to determine the probable name of the mineral in Figure 3.19 (light-colored nonmetallic minerals) or 3.20 (dark-colored nonmetallic minerals). Check the Mineral Database (Figure 3.21) entry for that mineral name and compare its other properties against those you have recorded. Note mineral uses. (If your mineral has different properties, then repeat your analysis to determine a more reasonable identification.)

5. Record the name and uses of the mineral in your Mineral Data Chart.

Question

11. Obtain a set of mineral samples according to your instructor's instructions. For each sample, fill in the Mineral Data Chart (Figure 3.22) using the procedures provided above.

PART 3C: MINERAL RESOURCES AND COMMODITIES

Use Internet resources to complete the items below. Refer to the *Laboratory Manual in Physical Geology* home page as needed: **http://www.prenhall.com/agi**

Questions

12. Choose a mineral resource that interests you—such as diamonds, gold, silver, emeralds, iron, copper, gypsum, or platinum.

a. What mineral did you pick?

b. What countries produce the majority of this resource?

c. What is the annual consumption (give amounts, years) of this resource by the United States?

d. What does the raw form (as extracted from the ground or a mine) of this resource look like, and how is it processed?

13. Write a short news story (one or several paragraphs) about a current event related to mining of a mineral resource that interests you.

METALLIC (M) MINERALS

Step 1: What is the mineral's hardness?	Step 2: What is the mineral's streak?	Step 3: Compare the mineral's physical properties to other characteristic properties below.	Step 4: Find mineral name(s) and check the mineral database for additional properties (Figure 3.21).
HARD (H > 5.5) Scratches glass Not scratched by masonry nail or knife blade	Dark gray	Color silvery gold; tarnishes brown; H 6–6.5; Cleavage absent to poor; Brittle; Crystals: cubes (often striated) or pyritohedrons	Pyrite
		Silvery dark gray to black; Tarnishes gray or rusty yellow brown; Strongly attracted to a magnet and may be magnetized; H 6–6.5; No cleavage; Crystals: octahedrons	Magnetite
HARD or **SOFT**	Yellow-brown	Color silvery brown to dark brown; Tarnishes dull yellow brown to brown; Amorphous; H 1–5.5; More common in softer (H 1–5) nonmetallic yellow brown forms	Limonite
	Brown	Color silvery black to black; Tarnishes gray to black; H 5.5–6; No cleavage; May be weakly attracted to a magnet; Crystals: octahedrons	Chromite
	Red to red-brown	Color steel gray to reddish silver; Tarnishes gray to dull red; May be attracted to a magnet; H 5–6.0; Also occurs in soft (H 1–5) nonmetallic earthy red forms	Hematite
SOFT (H ≤ 5.5) Does not scratch glass Scratched by masonry nail or knife blade	Dark gray	Color bright silvery gold; Tarnishes bronze brown brassy gold, or iridescent blue-green and red; H 3.5–4.0; Brittle; No cleavage; Crystals: tetrahedrons	Chalcopyrite
		Color brownish bronze; Tarnishes bright purple, blue, and/or red; May be weakly attracted to a magnet; H 3; Cleavage absent or poor; Rarely forms crystals	Bornite
		Color bright silvery gray; Tarnishes dull gray; Cleavage good to excellent; H 2.5; Crystals: cubes or octahedrons	Galena
		Color dark silvery gray to black; Can be scratched with your fingernail; Easily rubs off on your fingers and clothes, making them gray; H 1–2	Graphite
	Yellow-brown	Color dark brown to black; Forms layers of radiating microscopic crystals; H 5–5.5	Goethite
	White to pale yellow-brown	Color silvery yellow brown, silvery red, or black; Tarnishes brown or black. H 3.5–4.0; Cleavage excellent to good; Smells like rotten eggs when scratched/powdered	Sphalerite
	Copper	Color copper; Tarnishes dark brown or green; Malleable; No cleavage; H 2.5–3.0; Forms odd-shaped masses, nuggets, or dendritic forms	Copper (Native Copper)
	Gold	Color yellow gold; Does not tarnish; Malleable; No cleavage; H 2.5–3.0; Forms odd-shaped masses, nuggets, or dendritic forms	Gold (Native Gold)

FIGURE 3.18 Flowchart for identification of opaque minerals having a metallic (M) luster.

LIGHT-COLORED NONMETALLIC (NM) MINERALS

Step 1: What is the mineral's hardness?	Step 2: What is the mineral's cleavage?	Step 3: Compare the mineral's physical properties to other distinctive properties below.	Step 4: Find mineral name(s) and check the mineral database for additional properties (Figure 3.21).
HARD (H > 5.5) Scratches glass Not scratched by masonry nail or knife blade	Cleavage excellent or good	White or gray; 2 cleavages at nearly right angles and with striations; H 6	Plagioclase feldspar
		Orange, brown, white, gray, green, or pink; H 6; 2 cleavages at nearly right angles; exsolution lamellae	Potassium feldspar
		Pale brown, white, or gray; Long slender prisms; 1 excellent cleavage plus fracture surfaces; H 6–7	Sillimanite
		Blue, very pale green, white, or gray; Crystals are blades; H 4–7	Kyanite
	Cleavage poor or absent	Gray, white, or colored (dark red, blue, brown) hexagonal prisms with flat striated ends; H 9	Corundum Ruby (red var.), Sapphire (blue var.)
		Colorless, white, gray, or other colors; Greasy luster; Massive or hexagonal prisms and pyramids; Transparent or translucent; H 7	Quartz Milky Quartz (white var.), Citrine Quartz (yellow var.), Rose Quartz (pink var.)
		Opaque gray or white; Luster waxy; H 7	Chert (variety of quartz)
		Colorless, white, yellow, light brown, or pastel colors; Translucent or opaque; Laminated or massive; Cryptocrystalline; Luster waxy; H 7	Chalcedony (variety of quartz)
		Pale olive green to yellow; Conchoidal fracture; Transparent or translucent; Forms short stout prisms; H 7	Olivine
SOFT (H ≤ 5.5) Does not scratch glass Scratched by masonry nail or knife blade	Cleavage excellent or good	Colorless, white, yellow, green, pink, or brown; 3 excellent cleavages; Breaks into rhombohedrons; Effervesces in dilute HCl; H 3	Calcite
		Colorless, white, gray, creme, or pink; 3 excellent cleavages; Breaks into rhombohedrons; Effervesces in dilute HCl only if powdered; H 3.5–4	Dolomite
		Colorless or white with tints of brown, yellow, blue, black; Short tabular crystals and roses; Very heavy; H 3–3.5	Barite
		Colorless, white, or pastel colors; Massive or tabular crystals, blades, or needles; transparent to opaque. Can be scratched with your fingernail; H 2	Gypsum Selenite (colorless transparent var.) Alabaster (opaque white or pastel var.) Satin spar (fibrous silky translucent var.)
		Colorless, white, gray, or pale green, yellow, or red; Spheres of radiating needles; Luster silky; H 5–5.5	Natrolite
		Colorless, white, yellow, blue, brown, or red; Cubic crystals; Breaks into cubes; Salty taste; H 2.5	Halite
		Colorless, purple, blue, gray, green, yellow; Cubes with octahedral cleavage; H 4	Fluorite
		Colorless, yellow, brown, or red-brown; Short opaque prisms; Splits along 1 excellent cleavage into thin flexible transparent sheets; H 2–2.5	Muscovite (white mica)
	Cleavage poor or absent	Yellow crystals or earthy masses; Luster greasy; H 1.5–2.5; Smells like rotten eggs when powdered	Sulfur (Native sulfur)
		Opaque pale blue to blue-green; Amorphous crusts or massive; Very light blue streak; H 2–4	Chrysocolla
		Opaque green, yellow, or gray; Dull or silky masses or asbestos; White streak; H 2–5	Serpentine
		Opaque white, gray, green, or brown; Can be scratched with fingernail; Greasy or soapy feel; H 1	Talc
		Opaque earthy white to very light brown; Powdery, greasy feel; H 1–2	Kaolinite
		Colorless to white, orange, yellow, brown, blue, gray, green, or red; May have play of colors; Conchoidal fracture; H 5–5.5	Opal
		Colorless or pale green, brown, blue, white, or purple; Brittle hexagonal prisms; Conchoidal fracture; H 5	Apatite

FIGURE 3.19 Flowchart for identification of light-colored minerals with nonmetallic (NM) luster.

DARK-COLORED NONMETALLIC (NM) MINERALS

Step 1: What is the mineral's hardness?	Step 2: What is the mineral's cleavage?	Step 3: Compare the mineral's physical properties to other distinctive properties below.	Step 4: Find mineral name(s) and check the mineral database for additional properties (Figure 3.21).
HARD (H > 5.5) Scratches glass Not scratched by masonry nail or knife blade	Cleavage excellent or good	Translucent dark gray, blue-gray, or black; may have silvery iridescence; 2 cleavages at nearly 90° and with striations; H 6	Plagioclase feldspar
		Translucent brown, gray, green, or red; 2 cleavages at nearly right angles; exsolution lamellae; H 6	Potassium feldspar (K-spar)
		Dark green to black; 2 cleavages at nearly right angles (93° and 87°); H 5.5–6	Actinolite (Amphibole)
		Dark gray to black; 2 cleavages at about 56° and 124°; H 5.5–6	Hornblende (Amphibole)
		Dark green to black; 2 cleavages at nearly right angles (93° and 87°); H 5.5–6	Augite (Pyroxene)
	Cleavage poor or absent	Transparent or translucent gray, brown, or purple; Greasy luster; Massive or hexagonal prisms and pyramids; H 7	Quartz Smoky Quartz (black/brown var.), Amethyst (purple var.)
		Gray, black, or colored (dark red, blue, brown) hexagonal prisms with flat striated ends; H 9	Corundum Emery (black impure var.) Ruby (red var.), Sapphire (blue var.)
		Opaque red-brown or brown; Luster waxy; Cryptocrystalline; H 7	Jasper (variety of quartz)
		Transparent to translucent dark red to black; H 7	Garnet
		Opaque gray; Luster waxy; Cryptocrystalline; H 7	Chert (gray variety of quartz)
		Opaque black; Luster waxy; Cryptocrystalline; H 7	Flint (black variety of quartz)
		Black or dark green; Long striated prisms; H 7–7.5	Tourmaline
		Transparent or translucent olive green; Conchoidal fracture; Transparent or translucent; H 7	Olivine
		Opaque dark gray to black; Tarnishes gray to rusty yellow-brown; Cleavage absent; Strongly attracted to a magnet; May be magnetized; H 6–6.5	Magnetite
		Opaque green; Poor cleavage; H 6–7	Epidote
		Opaque brown prisms that interpenetrate to form crosses; H 7	Staurolite
SOFT (H ≤ 5.5) Does not scratch glass Scratched by masonry nail or knife blade	Cleavage excellent or good	Translucent to opaque yellow-brown to brown; may appear submetallic; Octahedral cleavage; H 3.5–4	Sphalerite
		Purple cubes or octahedrons; Octahedral cleavage; H 4	Fluorite
		Black short opaque prisms; Splits easily along 1 excellent cleavage into thin sheets; H 2.5–3	Biotite (black mica)
		Green short opaque prisms; Splits easily along 1 excellent cleavage into thin sheets; H 2–3	Chlorite
	Cleavage poor or absent	Opaque rusty brown or yellow-brown; Massive and amorphous; Yellow-brown streak; H 1–5.5	Limonite
		Opaque rusty brown to brown-gray rock with shades of gray, yellow, and white; Contains pea-sized spheres that are laminated internally; Pale brown streak; H 1–3	Bauxite
		Deep blue; Crusts, small crystals, or massive; Light blue streak; H 3.5–4	Azurite
		Opaque green or gray-green; Dull or silky masses or asbestos; White streak; H 2–5	Serpentine
		Opaque green in laminated crusts or massive; Streak pale green; Effervesces in dilute HCl; H 3.5–4	Malachite
		Translucent or opaque dark green; Can be scratched with your fingernail; Feels greasy or soapy; H 1	Talc
		Transparent or translucent green, brown, blue, or purple; Brittle hexagonal prisms; Conchoidal fracture; H 5	Apatite
		Opaque red or red-gray; H 1.5–5	Hematite

FIGURE 3.20 Flowchart for identification of dark- and medium-colored minerals with nonmetallic (NM) luster.

63

MINERAL DATABASE (Alphabetical Listing)

Mineral	Luster	Hardness	Streak	Distinctive Properties	Some Uses
ACTINOLITE (amphibole)	Nonmetallic (NM)	5.5–6	White	Color dark green or pale green; Forms needles, prisms, and asbestose fibers; Good cleavage at 56° and 124°; SG = 3.1	Gemstone (Nephrite), Asbestose products
AMPHIBOLE: See HORNEBLENDE and ACTINOLITE					
APATITE $Ca_5F(PO_4)_3$ calcium fluorophosphate	Nonmetallic (NM)	5	White	Color pale or dark green, brown, blue, white, or purple; Sometimes colorless; Transparent or opaque; Brittle; Conchoidal fracture; Forms hexagonal prisms; SG = 3.1–3.4	Used for pesticides and fertilizers
ASBESTOSE: fibrous varieties of AMPHIBOLE and SERPENTINE					
AUGITE (pyroxene) calcium ferromagnesian silicate	Nonmetallic (NM)	5.5–6	White to pale gray	Color dark green to gray; Forms short, 8-sided prisms; Two good cleavages that intersect at 87° and 93° (nearly right angles); SG = 3.2–3.5	Some pyroxene mined as an ore of lithium, for making steel
AZURITE $Cu_3(CO_3)_2(OH)_2$ hydrous copper carbonate	Nonmetallic (NM)	3.5–4	Light blue	Color a distinctive deep blue; Forms crusts of small crystals, opaque earthy masses, or short and long prisms; Brittle; Effervesces in dilute HCl; SG = 3.7–3.8	Ore of copper for pipes, electrical circuits, coins, ammunition, gemstone
BARITE $BaSO_4$ barium sulfate	Nonmetallic (NM)	3–3.5	White	Colorless to white, with tints of brown, yellow, blue, or red; Forms short tabular crystals and rose-shaped masses (Barite roses); Brittle; Cleavage good to excellent; Very heavy, SG = 4.3–4.6	Used in rubber, paint, glass, oil-well drilling fluids
BAUXITE Mixture of aluminum hydroxides	Nonmetallic (NM)	1–3	White	Brown earthy rock with shades of gray, white, and yellow; Amorphous; Often contains rounded pea-sized structures with laminations; SG = 2.0–3.0	Ore of Aluminum
BIOTITE MICA ferromagnesian potassium, hydrous aluminum silicate $K(Mg,Fe)_3 (Al,Si_3O_{10})(OH,F)_2$	Nonmetallic (NM)	2.5–3	Gray-brown	Color black, green-black, or brown-black; Cleavage excellent; Forms very short prisms that split easily into very thin, flexible sheets; SG = 2.7–3.1	Used for fire-resistant tiles, rubber, paint
BORNITE Cu_5FeS_4 copper-iron sulfide	Metallic (M)	3	Dark gray to black	Color brownish bronze; Tarnishes bright purple, blue, and/or red; May be weakly attracted to a magnet; H 3; Cleavage absent or poor; Forms dense brittle masses. Rarely forms crystals	Ore of copper for pipes, electrical circuits, coins, ammunition, brass, bronze
CALCITE $CaCO_3$ calcium carbonate	Nonmetallic (NM)	3	White	Usually colorless, white, or yellow, but may be green, brown, or pink; Opaque or transparent; Excellent cleavage in 3 directions not at 90°; Forms prisms, rhombohedrons, or scalenohedrons that break into rhombohedrons; Effervesces in dilute HCl; SG = 2.7	Used to make antacid tablets, fertilizer, cement; Ore of calcium
CHALCEDONY SiO_2 cryptocrystalline quartz	Nonmetallic (NM)	7	White*	Colorless, white, yellow, light brown, or other pastel colors in laminations; Often translucent; Conchoidal fracture; Luster waxy; Cryptocrystalline; SG = 2.5–2.8	Used as an abrasive; Used to make glass, gemstones (agate, chrysoprase)

*Streak cannot be determined with a streak plate for minerals harder than 6.5. They scratch the streak plate.

FIGURE 3.21 Mineral Database—alphabetical list of minerals and their properties and uses.

MINERAL DATABASE (Alphabetical Listing)

Mineral	Luster	Hardness	Streak	Distinctive Properties	Some Uses
CHALCOPYRITE $CuFeS_2$ copper-iron sulfide	Metallic (M)	3.5–4	Dark gray	Color bright silvery gold; Tarnishes bronze brown, brassy gold, or iridescent blue-green and red; Brittle; No cleavage; Forms dense masses or elongate tetrahedrons; SG = 4.1–4.3	Ore of copper for pipes, electrical circuits, coins, ammunition, brass, bronze
CHERT SiO_2 cryptocrystalline quartz	Nonmetallic (NM)	7	White*	Opaque gray or white; Luster waxy; Conchoidal fracture; SG = 2.5–2.8	Used as an abrasive; Used to make glass, gemstones
CHLORITE ferromagnesian aluminum silicate $(Mg,Fe,Al)_6(Si,Al)_4O_{10}(OH)_8$	Nonmetallic (NM)	2–2.5	White	Color dark green; Cleavage excellent; Forms short prisms that split easily into thin flexible sheets; Luster bright or dull; SG = 2–3	Used for fire-resistant tiles, rubber, paint, art sculpture medium
CHROMITE $FeCr_2O_4$ iron-chromium oxide	Metallic (M)	5.5–6	Dark brown	Color silvery black to black; Tarnishes gray to black; No cleavage; May be weakly attracted to a magnet; Forms dense masses or granular masses of small crystals (octahedrons).	Ore of chromium for making chrome, stainless steel, mirrors, paint and used in leather tanning
CHRYSOCOLLA $CuSiO_3 \cdot 2H_2O$ hydrated copper silicate	Nonmetallic (NM)	2–4	Very light blue	Color pale blue to blue-green; Opaque; Forms amorphous crusts or may be massive; Conchoidal fracture; Luster shiny or earthy; SG = 2.0–4.0	Ore of copper for pipes, electrical circuits, coins, ammunition; gemstone
COPPER (NATIVE COPPER) Cu copper	Metallic (M)	2.5–3	Copper	Color copper; Tarnishes brown or green; Malleable; No cleavage; Forms odd-shaped masses, nuggets, or dendritic forms; SG = 8.8–9.0	Ore of copper for pipes, electrical circuits, coins, ammunition, brass, bronze
CORUNDUM Al_2O_3 aluminum oxide	Nonmetallic (NM)	9	White*	Gray, white, black or colored (red, blue, brown, yellow) hexagonal prisms with flat striated ends; Opaque to transparent; Cleavage absent; SG = 3.9–4.1 H 9	Used for abrasive powders to polish lenses; gemstones (red ruby, blue sapphire); emery cloth
DOLOMITE $CaMg(CO_3)_2$ magnesian calcium carbonate	Nonmetallic (NM)	3.5–4	White	Color white, gray, creme, or pink; Usually opaque; Cleavage excellent in 3 directions; Breaks into rhombohedrons; Resembles calcite, but will effervesce in dilute HCl only if powdered; SG = 2.8–2.9	Ore of magnesium metal; soft abrasive; used to make paper
EPIDOTE complex silicate	Nonmetallic (NM)	6–7	White*	Color pale or dark green to yellow-green; Massive or forms striated prisms; Cleavage poor, SG = 3.3–3.5	Gemstone
FELDSPAR: See PLAGIOCLASE (Na-Ca Feldspars) and POTASSIUM FELDSPAR (K-Spar)					
FLINT SiO_2 cryptocrystalline quartz	Nonmetallic (NM)	7	White*	Color black to very dark gray; Opaque to translucent; Conchoidal fracture; Crypto-crystalline; SG = 2.5–2.8	Used as an abrasive; Used to make glass, gemstones
FLUORITE CaF_2 calcium fluoride	Nonmetallic (NM)	4	White	Colorless, purple, blue, gray, green, or yellow; Cleavage excellent; Crystals usually cubes; Transparent or opaque; Brittle; SG = 3.0–3.3	Source of fluorine for processing aluminum; flux in steel making

*Streak cannot be determined with a streak plate for minerals harder than 6.5. They scratch the streak plate.

FIGURE 3.21 (CONTINUED) Mineral Database—alphabetical list of minerals and their properties and uses.

MINERAL DATABASE (Alphabetical Listing)

Mineral	Luster	Hardness	Streak	Distinctive Properties	Some Uses
GALENA PbS lead sulfide	Metallic (M)	2.5	Gray to dark gray	Color bright silvery gray; Tarnishes dull gray; Forms cubes and octahedrons; Brittle; Cleavage good in three directions, so breaks into cubes; SG = 7.4–7.6	Ore of lead for TV glass, auto batteries, solder, ammunition, paint
GARNET complex silicate	Nonmetallic (NM)	7	White*	Color usually red, black, or brown; sometimes yellow, green, pink; Forms dodecahedrons; Cleavage absent but may have parting; Brittle; Translucent to opaque; SG = 3.5–4.3	Used as an abrasive; gemstone
GOETHITE $FeO(OH)$ iron oxide hydroxide	Metallic (M)	5–5.5	Yellow-brown	Color dark brown to black; Tarnishes yellow-brown; Forms layers of radiating microscopic crystals; SG = 3.3–4.3	Ore of iron for steel, brass, bronze, tools, vehicles, nails and bolts, bridges, etc.
GRAPHITE C carbon	Metallic (M)	1	Dark gray	Color dark silvery gray to black; Forms flakes, short hexagonal prisms, and earthy masses; Greasy feel; Very soft, Cleavage excellent in 1 direction; SG = 2.0–2.3	Used as a lubricant (as in graphite oil), pencil leads, fishing rods
GYPSUM $CaSO_4 \cdot 2H_2O$ calcium sulfate	Nonmetallic (NM)	2	White	Colorless, white, or gray; Forms tabular crystals, prisms, blades, or needles (satin spar variety); Transparent to translucent; Very soft; Cleavage good; SG = 2.3	Plaster-of-paris, wallboard, drywall, art sculpture medium (alabaster)
HALITE $NaCl$ sodium chloride	Nonmetallic (NM)	2.5	White	Colorless, white, yellow, blue, brown or red; Transparent to translucent; Brittle; Forms cubes; Cleavage excellent in 3 directions, so breaks into cubes; Salty taste; SG = 2.1–2.6	Table salt, road salt; Used in water softeners and as a preservative; Sodium ore
HEMATITE Fe_2O_3 iron oxide	Metallic (M) or Nonmetallic (NM)	1–6	Red to red-brown	Color silvery gray, reddish silver, black, or brick red; Tarnishes red; Opaque; Soft (earthy) and hard (metallic) varieties have same streak; Forms thin tabular crystals or massive; May be attracted to a magnet; SG = 4.9–5.3	Rouge makeup and polish: red pigment in paints, Ore of iron for steel tools, vehicles, nails and bolts, bridges, etc.
HORNBLENDE (amphibole) calcium ferromagnesian aluminum silicate	Nonmetallic (NM)	5.5	White to pale gray	Color dark gray to black; Forms prisms with good cleavage at 56° and 124°; Brittle; Splintery or asbestos forms; SG = 3.0–3.3	Fibrous varieties used for fire-resistant clothing, tiles, brake linings
JASPER SiO_2 cryptocrystalline quartz	Nonmetallic (NM)	7	White*	Color red-brown, or yellow; Opaque; Waxy luster; Conchoidal fracture; Cryptocrystalline; SG = 2.5–2.8	Used as an abrasive; Used to make glass, gemstones
KAOLINITE $Al_4(Si_4O_{10})(OH)_8$ hydrous aluminum silicate	Nonmetallic (NM)	1–2	White	Color white to very light brown; Commonly forms earthy, microcrystalline masses; Cleavage excellent but absent in hand samples; SG = 2.6	Used for pottery, clays, polishing compounds, pencil leads, paper
K-SPAR: See POTASSIUM FELSDPAR					
KYANITE $Al_2(SiO_4)O$ aluminum silicate	Nonmetallic (NM)	4–7	White*	Color blue, pale green, white, or gray; Translucent to transparent; Forms blades; SG = 3.6–3.7	High temperature ceramics, spark plugs

*Streak cannot be determined with a streak plate for minerals harder than 6.5. They scratch the streak plate.

FIGURE 3.21 (CONTINUED) Mineral Database—alphabetical list of minerals and their properties and uses.

MINERAL DATABASE (Alphabetical Listing)

Mineral	Luster	Hardness	Streak	Distinctive Properties	Some Uses
LIMONITE $Fe_2O_3 \cdot nH_2O$ hydrated iron oxide and/or $FeO(OH) \cdot nH_2O$ hydrated iron oxide hydroxide	Metallic (M) or Nonmetallic (NM)	1–5.5	Yellow-brown	Color yellow brown to dark brown; Tarnishes yellow to brown; Amorphous masses; Luster dull or earthy; Hard or soft; SG = 3.3–4.3	Yellow pigment; Ore of iron for steel tools, vehicles, nails and bolts, bridges, etc.
MAGNETITE Fe_3O_4 iron oxide	Metallic (M)	6–6.5	Dark gray	Color silvery gray to black; Opaque; Forms octahedrons; Tarnishes gray; No cleavage; Attracted to a magnet and can be magnetized; SG = 5.0–5.2	Ore of iron for steel, brass, bronze, tools, vehicles, nails and bolts, bridges, etc.
MALACHITE $Cu_2CO_3(OH)_2$ hydrous copper carbonate	Nonmetallic (NM)	3.5–4	Green	Color green, pale green, or gray green; Usually in crusts, laminated masses, or microcrystals; Effervesces in dilute HCl; SG = 3.6–4.0	Ore of copper for pipes, electrical circuits, coins, ammunition; gemstone

MICA: See BIOTITE and MUSCOVITE.

NATIVE COPPER: See COPPER.

NATIVE SULFUR: See SULFUR.

Mineral	Luster	Hardness	Streak	Distinctive Properties	Some Uses
NATROLITE (ZEOLITE) $Na_2(Al_2Si_3O_{10}) \cdot 2H_2O$ hydrous sodium aluminum silicate	Nonmetallic (NM)	5–5.5	White	Colorless, white, gray, or pale green, yellow, or red; Forms masses of radiating needles; Silky luster; SG = 2.2–2.4	Water softeners
MUSCOVITE MICA potassium hydrous aluminum silicate $KAl_2(Al,Si_3O_{10})(OH,F)_2$	Nonmetallic (NM)	2–2.5	White	Colorless, yellow, brown, or red-brown; Forms short opaque prisms; Cleavage excellent in 1 direction, can be split into thin flexible transparent sheets; SG = 2.7–3.0	Computer chip substrates, electrical insulation, roof shingles, facial makeup
OLIVINE $(Fe,Mg)_2SiO_4$ ferromagnesian silicate	Nonmetallic (NM)	7	White*	Color pale or dark olive-green to yellow, or brown; Forms short crystals that may resemble sand grains; Conchoidal fracture; Cleavage absent; Brittle; SG = 3.3–3.4	Gemstone (peridot); Ore of magnesium metal
OPAL $SiO_2 \cdot nH_2O$ hydrated silicon dioxide	Nonmetallic (NM)	5–5.5	White	Colorless to white, orange, yellow, brown, blue, gray, green, or red; may have play of colors (opalescence); Amorphous; Cleavage absent; Conchoidal fracture; SG = 1.9–2.3	Gemstone
PLAGIOCLASE FELDSPAR $NaAlSi_3O_8$ to $CaAl_2Si_2O_8$ calcium-sodium aluminum silicate	Nonmetallic (NM)	6	White	Colorless, white, gray, or black; may have iridescent play of color from within; Translucent; Forms striated tabular crystals or blades; Cleavage good in two directions at nearly 90°; SG = 2.6–2.8	Used to make ceramics, glass, enamel, soap, false teeth, scouring powders
POTASSIUM FELDSPAR $KAlSi_3O_8$ potassium aluminum silicate	Nonmetallic (NM)	6	White	Color orange, brown, white, green, or pink; Forms translucent prisms with subparallel exsolution lamellae; Cleavage excellent in two directions at nearly 90°; SG = 2.5–2.6	Used to make ceramics, glass, enamel, soap, false teeth, scouring powders
PYRITE ("fool's gold") FeS_2 iron sulfide	Metallic (M)	6–6.5	Dark gray	Color silvery gold; Tarnishes brown; H 6–6.5; Cleavage absent to poor; Brittle; Forms opaque masses, cubes (often striated) or pyritohedrons; SG = 4.9–5.2	Ore of sulfur, for sulfuric acid, explosives, fertilizers, pulp processing, insecticides

*Streak cannot be determined with a streak plate for minerals harder than 6.5. They scratch the streak plate.

FIGURE 3.21 (CONTINUED) Mineral Database—alphabetical list of minerals and their properties and uses.

MINERAL DATABASE (Alphabetical Listing)

Mineral	Luster	Hardness	Streak	Distinctive Properties	Some Uses
PYROXENE: See AUGITE.					
QUARTZ SiO_2 silicon dioxide	Nonmetallic (NM)	7	White*	Usually colorless, white, or gray but uncommon varieties occur in all colors; Transparent to translucent; Luster greasy; No cleavage; Forms hexagonal prism and pyramids; SG = 2.6–2.7 Some quartz varieties are: • var. flint (opaque black or dark gray) • var. smoky (transparent gray) • var. citrine (transparent yellow-brown) • var. amethyst (purple) • var. chert (opaque gray) • var. milky (white) • var. jasper (opaque red or yellow) • var. rock crystal (colorless) • var. rose (pink) • var. chalcedony (translucent, waxy luster)	Used as an abrasive; Used to make glass, gemstones
SERPENTINE $Mg_6Si_4O_{10}(OH)_8$ hydrous magnesian silicate	Nonmetallic (NM)	2–5	White	Color pale or dark green, yellow, gray; Forms dull or silky masses and asbestos forms; No cleavage; SG = 2.2–2.6	Fibrous varieties used for fire-resistant clothing, tiles, brake linings
SILLIMANITE $Al_2(SiO_4)O$ aluminum silicate	Nonmetallic (NM)	6–7	White	Color pale brown, white, or gray; One good cleavage plus fracture surfaces; Forms slender prisms; SG = 3.2	High-temperature ceramics
SPHALERITE ZnS zinc sulfide	Metallic (M) or Nonmetallic (NM)	3.5–4	White to pale yellow-brown	Color silvery yellow-brown, dark red, or black; Tarnishes brown or black; dodecahedral cleavage excellent to good; Smells like rotten eggs when scratched/ powdered; Forms misshapen tetrahedrons or dodecahedrons: SG = 3.9–4.1	Ore of zinc for die-cast automobile parts, brass, galvanizing, batteries
STAUROLITE iron magnesium zinc aluminum silicate	Nonmetallic (NM)	7	White to gray*	Color brown to gray-brown; Tarnishes dull brown; Forms prisms that interpenetrate to form natural crosses; Cleavage poor; SG = 3.7–3.8	Gemstone crosses called "fairy crosses"
SULFUR (NATIVE SULFUR) S sulfur	Nonmetallic (NM)	1.5–2.5	Pale yellow	Color bright yellow; Forms transparent to translucent crystals or earthy masses; Cleavage poor; Luster greasy to earthy; Brittle; SG = 2.1	Used for drugs, sulfuric acid, explosives, fertilizers, pulp processing, insecticides
TALC $Mg_3Si_4O_{10}(OH)_2$ hydrous magnesian silicate	Nonmetallic (NM)	1	White	Color white, gray, pale green, or brown; Forms cryptocrystalline masses that show no cleavage; Luster silky to greasy; Feels greasy or soapy (talcum powder); Very soft; SG = 2.7–2.8	Used for talcum powder, facial makeup, ceramics, paint, sculptures
TOURMALINE complex silicate	Nonmetallic (NM)	7–7.5	White*	Color usually opaque black or green, but may be transparent or translucent green, red, yellow, pink or blue; Forms long striated prisms with triangular cross sections; Cleavage absent; SG = 3.0–3.2	Crystals used in radio transmitters; gemstone
ZEOLITE: A group of calcium or sodium hydrous aluminum silicates. See NATROLITE.					

*Streak cannot be determined with a streak plate for minerals harder than 6.5. They scratch the streak plate.

FIGURE 3.21 (CONTINUED) Mineral Database—alphabetical list of minerals and their properties and uses.

FIGURE 3.22 Mineral Data Chart.

MINERAL DATA CHART

Sample Letter or Number	Luster*	Hardness	Cleavage	Color	Streak	Other Properties	Name (Fig. 3.18, 3.19, or 3.20)	Some Uses (Fig. 3.21)

*M = metallic, NM = nonmetallic

FIGURE 3.22 (CONTINUED) Mineral Data Chart.

MINERAL DATA CHART

Sample Letter or Number	Luster*	Hardness	Cleavage	Color	Streak	Other Properties	Name (Fig. 3.18, 3.19, or 3.20)	Some Uses (Fig. 3.21)

*M = metallic, NM = nonmetallic

FIGURE 3.22 (CONTINUED) Mineral Data Chart.

MINERAL DATA CHART

Sample Letter or Number	Luster*	Hardness	Cleavage	Color	Streak	Other Properties	Name (Fig. 3.18, 3.19, or 3.20)	Some Uses (Fig. 3.21)

*M = metallic, NM = nonmetallic

FIGURE 3.22 (CONTINUED) Mineral Data Chart.

MINERAL DATA CHART

Sample Letter or Number	Luster*	Hardness	Cleavage	Color	Streak	Other Properties	Name (Fig. 3.18, 3.19, or 3.20)	Some Uses (Fig. 3.21)

*M = metallic, NM = nonmetallic

FIGURE 3.22 (CONTINUED) Mineral Data Chart.

MINERAL DATA CHART

Sample Letter or Number	Luster*	Hardness	Cleavage	Color	Streak	Other Properties	Name (Fig. 3.18, 3.19, or 3.20)	Some Uses (Fig. 3.21)

*M = metallic, NM = nonmetallic

FIGURE 3.22 (CONTINUED) Mineral Data Chart.

MINERAL DATA CHART

Sample Letter or Number	Luster*	Hardness	Cleavage	Color	Streak	Other Properties	Name (Fig. 3.18, 3.19, or 3.20)	Some Uses (Fig. 3.21)

*M = metallic, NM = nonmetallic

74

Rock-Forming Processes and the Rock Cycle

OBJECTIVES

A. Know what rocks are made of and the names of some common rock textures.

B. Understand some processes that form igneous, sedimentary, and metamorphic rock groups and their characteristic textures.

C. Understand and be able to apply the "rock cycle" conceptual model.

D. Be able to observe and describe hand samples of rocks, then infer how each sample may have formed in relation to the rock cycle.

E. Be able to infer ways that rock cycle materials (elements, compounds, rocky materials) also cycle through and among Earth's other spheres.

MATERIALS

Pencil, colored pencils, eraser, laboratory notebook, magnifying lens (optional), and rock samples for Part 4B.

INTRODUCTION

Rocks are the solid materials that comprise most of Earth, our Moon, and the other rocky planets of our solar system. Nearly all rocks are solid aggregates of mineral grains (particles), either mineral crystals or clasts (broken pieces) of mineral crystals and rocks (e.g., pebbles, gravel, sand, and silt). There are, however, a few notable examples of rocks that are not comprised of mineral grains. For example, *obsidian* is a rock made of volcanic glass and *coal* is a rock made of plant fragments.

Rock-forming materials come from Earth's mantle (as molten rock called *magma* while underground and *lava* when it erupts to the surface), space (meteorites), organisms (parts of plants and animals), or the fragmentation and chemical decay of mineral crystals and other rocks. Environmental changes and processes affect these materials and existing rocks in ways that produce three main rock groups (Figure 4.1):

- **Igneous rocks** form when magma or lava cool to a solid form, either glass or masses of tightly intergrown mineral crystals. The crystals are large if they had a long time to grow in a slowly cooling magma, and they are small if they formed quickly in a rapidly cooling lava.

- **Sedimentary rocks** form mostly when mineral crystals and clasts (broken pieces, fragments) of plants, animals, mineral crystals, or rocks are compressed or naturally cemented together. They also form when mineral crystals precipitate from water to form a rocky mass such as *rock salt* or cave stalactites.

- **Metamorphic rocks** are rocks deformed or changed from one form to another (transformed) by intense heat, intense pressure, and/or the action of hot fluids. This causes the rock to recrystallize, fracture, change color, and/or flow. As the rock flows, the flat layers are folded and the mineral crystals are aligned like parallel needles or scales.

All rocks are part of a system of rock-forming processes, materials, and products that is often portrayed in a conceptual model called the **rock cycle** (Figure 4.2). The rock cycle model explains how all rocks can be formed, deformed, transformed, melted, and reformed as a result of environmental factors and natural processes that affect them.

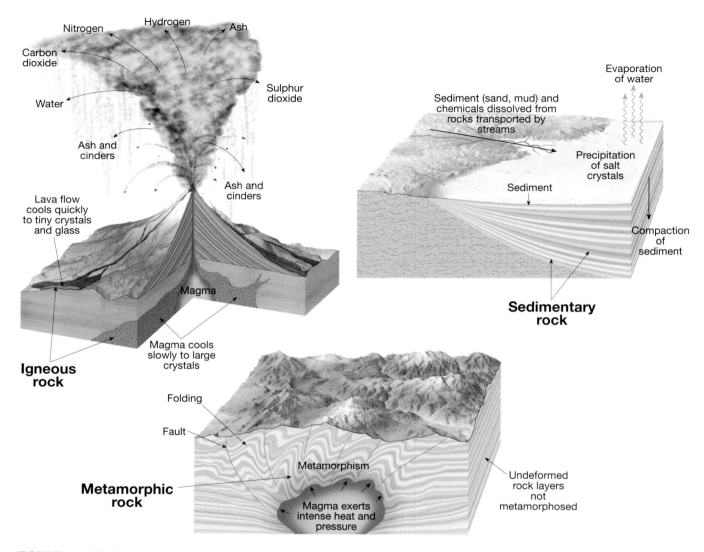

FIGURE 4.1 The formation of igneous, sedimentary, and metamorphic rock.

An idealized path (broad purple arrows) of rock cycling and redistribution of matter is illustrated in Figure 4.2, starting with igneous processes. If magma (from the mantle or lower crust) cools, then it solidifies into igneous rocks that are masses of glass or aggregates of intergrown mineral crystals. If these igneous rocks are uplifted, then sedimentary processes force other changes to occur. The igneous rocks are **weathered** (fragmented into grains, chemically decayed to residues, or even dissolved), **eroded** (worn away), and later deposited to form **sediment** (an accumulation of chemical residues and fragmented rocks, mineral crystals, plants or animals). Meteorites (dust and rocks from space) may be incorporated into the sediment. Sediment is **lithified** (hardened) into

sedimentary rock as it compacts under its own weight or gets naturally cemented with crystals precipitated from water. If the sedimentary rock is subjected to metamorphic processes (intense heat, intense pressure, or the chemical action of hot fluids), then it will *deform* (fold, fracture, or otherwise change its shape) and *transform* (change color, density, composition, and/or general form) to metamorphic rock. And if the heat is great enough, then the metamorphic rock will melt (an igneous process) to form another body of magma that will begin the cycle again.

Of course, not all rocks undergo change along such an idealistic path. There are *at least* three changes that each rock could undergo. The arrows in Figure 4.2 show that any rock from one group can

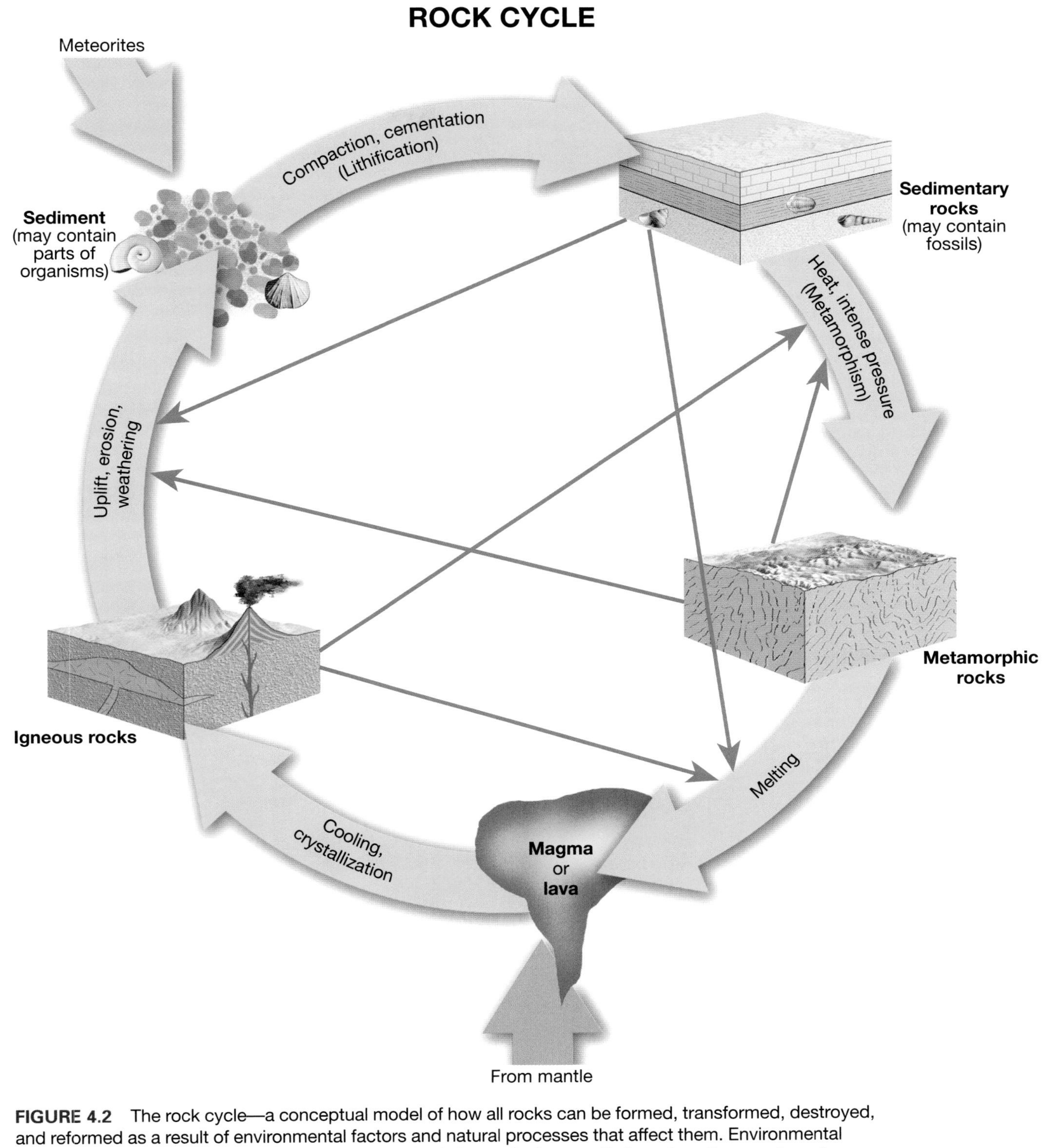

ROCK CYCLE

FIGURE 4.2 The rock cycle—a conceptual model of how all rocks can be formed, transformed, destroyed, and reformed as a result of environmental factors and natural processes that affect them. Environmental changes and processes affect these materials and existing rocks in ways that produce three main rock groups. Arrows show that a rock from one group can be transformed to either of the other two groups, or it can be recycled within its own group. An *idealized rock cycle path* is shown by the broad large arrows. But there are *at least* two other changes that each rock could undergo.

Igneous rocks form when any preexisting rocks melt to form magma/lava, which cools and solidifies (crystallizes or hardens into dense glass) into igneous rock. Wherever rocks are weathered (chemically decayed or broken apart) and eroded (worn away), they produce sediment that can be compacted and cemented into sedimentary rocks. Whenever any rocks are subjected to intense heating and pressure, they are deformed and transformed into metamorphic rocks.

COMMON IGNEOUS ROCK TEXTURES

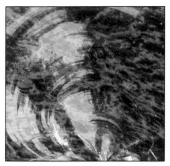

A   Dense hard glassy
texture ×1.0

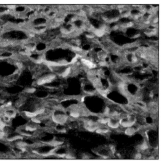

B   Vesicular (bubbly)
texture ×1.0

C   Randomly oriented small
crystals: fine-grained
crystalline texture ×1.0

D   Randomly oriented large
crystals: coarse-grained
crystalline texture ×1.0

COMMON METAMORPHIC ROCK TEXTURES

E   Equigranular
crystalline texture
×1.0

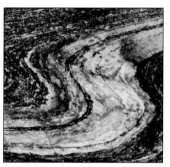

F   Layers are folded:
folded texture ×0.5

G   Crystalline texture. The
long crystals are foliated
(lay parallel to one
another) ×0.5

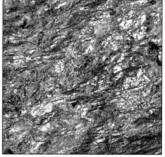

H   Crystalline texture. Flat,
scaly crystals are foliated
(lay parallel to one
another) ×1.0

COMMON SEDIMENTARY ROCK TEXTURES

I   No visible grains. Looks
like dry clay, silt, or mud
(a clastic texture). Note
fossil leaf. ×1.0

J   Grains are sand
(a clastic texture) ×2.0

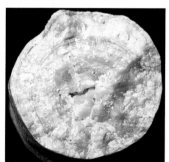

K   Crystalline texture. Crystals
are of different sizes in the
different layers ×0.5

L   Grains are sand, gravel,
and rounded pebbles
(a clastic texture) ×1.0

FIGURE 4.3 Some common textures of igneous, metamorphic, and sedimentary rocks. For use with
Figure 4.4.

FLOW CHART FOR CLASSIFICATION OF ROCKS AS IGNEOUS, SEDIMENTARY, OR METAMORPHIC

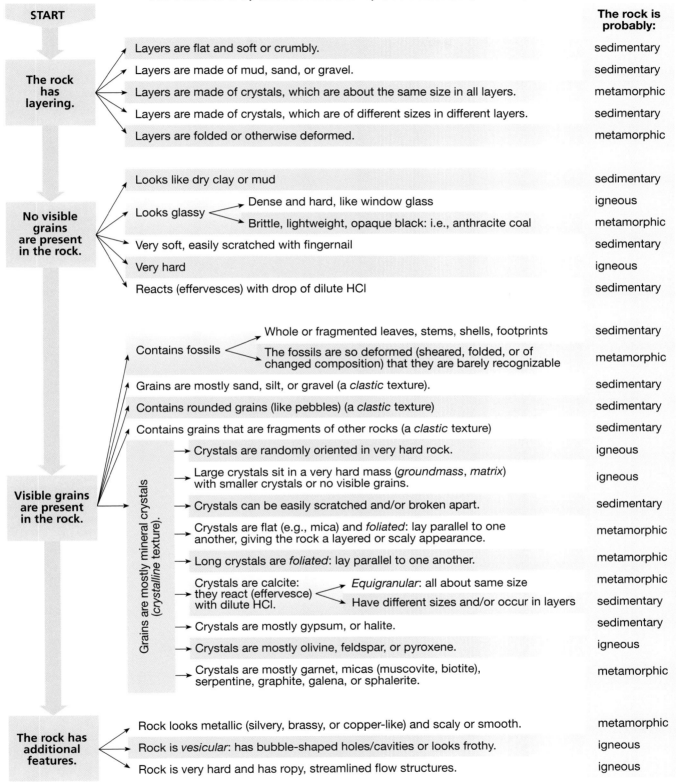

	The rock is probably:
START	
The rock has layering.	
Layers are flat and soft or crumbly.	sedimentary
Layers are made of mud, sand, or gravel.	sedimentary
Layers are made of crystals, which are about the same size in all layers.	metamorphic
Layers are made of crystals, which are of different sizes in different layers.	sedimentary
Layers are folded or otherwise deformed.	metamorphic
No visible grains are present in the rock.	
Looks like dry clay or mud	sedimentary
Looks glassy — Dense and hard, like window glass	igneous
— Brittle, lightweight, opaque black: i.e., anthracite coal	metamorphic
Very soft, easily scratched with fingernail	sedimentary
Very hard	igneous
Reacts (effervesces) with drop of dilute HCl	sedimentary
Visible grains are present in the rock.	
Contains fossils — Whole or fragmented leaves, stems, shells, footprints	sedimentary
— The fossils are so deformed (sheared, folded, or of changed composition) that they are barely recognizable	metamorphic
Grains are mostly sand, silt, or gravel (a *clastic* texture).	sedimentary
Contains rounded grains (like pebbles) (a *clastic* texture)	sedimentary
Contains grains that are fragments of other rocks (a *clastic* texture)	sedimentary
Grains are mostly mineral crystals (*crystalline* texture). Crystals are randomly oriented in very hard rock.	igneous
Large crystals sit in a very hard mass (*groundmass*, *matrix*) with smaller crystals or no visible grains.	igneous
Crystals can be easily scratched and/or broken apart.	sedimentary
Crystals are flat (e.g., mica) and *foliated*: lay parallel to one another, giving the rock a layered or scaly appearance.	metamorphic
Long crystals are *foliated*: lay parallel to one another.	metamorphic
Crystals are calcite: they react (effervesce) with dilute HCl. — *Equigranular*: all about same size	metamorphic
— Have different sizes and/or occur in layers	sedimentary
Crystals are mostly gypsum, or halite.	sedimentary
Crystals are mostly olivine, feldspar, or pyroxene.	igneous
Crystals are mostly garnet, micas (muscovite, biotite), serpentine, graphite, galena, or sphalerite.	metamorphic
The rock has additional features.	
Rock looks metallic (silvery, brassy, or copper-like) and scaly or smooth.	metamorphic
Rock is *vesicular*: has bubble-shaped holes/cavities or looks frothy.	igneous
Rock is very hard and has ropy, streamlined flow structures.	igneous

FIGURE 4.4 Flow chart for classification of rocks as igneous, sedimentary, or metamorphic.

be transformed to either of the other two groups *or* recycled within its own group. Igneous rock can be: (1) weathered and eroded to form sediment that is lithified to form sedimentary rock; (2) transformed to metamorphic rock by intense heat, intense pressure, and/or hot fluids; or (3) re-melted, cooled, and solidified back into another igneous rock. Sedimentary rock can be: (1) melted, cooled, and solidified into an igneous rock; (2) transformed to metamorphic rock by intense heat, intense pressure, and/or hot fluids; or (3) weathered and eroded back to sediment that is lithified back into another sedimentary rock. Metamorphic rock can be: (1) weathered and eroded to form sediment that is lithified into sedimentary rock; (2) melted, cooled, and solidified into igneous rock; or (3) re-metamorphosed into a different type of metamorphic rock by intense heat, intense pressure, or hot fluids.

Rock Properties

Rocks have many different properties. Among them are color, composition, and texture (Figure 4.3). The properties you observe in a rock are clues to its origin and are used to classify it (Figure 4.4).

Color of rocks is based on their **composition**—the chemicals or visible grains of Earth materials that comprise them. You have already learned that most rocks are comprised of mineral grains (Figures 4.3D, 4.3E, 4.3H, 4.3K), and that a few exceptions are comprised of glass (Figure 4.3A) or organic particles. Rocks vary more widely in their chemical composition, and some chemicals impart characteristic colors on rocks. For example, ferromagnesian-rich rocks (iron- and magnesium-rich rocks) generally have a dark color (Figures 4.3A, 4.3B, 4.3G) and ferromagnesian-poor rocks generally have a light color (Figures 4.3C, 4.3D, 4.3K). Multicolored crystalline rocks generally contain several kinds of mineral crystals (Figure 4.3D), and uniformly colored crystalline rocks generally contain a single kind of mineral crystal (Figures 4.3E, 4.3G, 4.3H, 4.3K).

Another very important property of rocks is **texture**—a description of the grains and other parts of a rock and their size, shape, and arrangement. Some rocks are dense, solid masses with no visible grains. An example is *obsidian*, a volcanic glass (with **glassy texture**; Figure 4.3A) that completely lacks visible grains. Some rocks have abundant visible grains, and geologists describe their size with the terms **fine-grained** (grains generally <1 mm and too small to identify without a hand lens) or **coarse-grained** (grains generally >1 mm and large enough to identify with your unaided eye). Some rocks are a mixture of

grains, dense parts with no grains, and even hollow spaces. The hollow spaces in **vesicular** rocks are bubbles (Figure 4.3B). But the hollow spaces can also be solution cavities (cave-like spaces), fractures, or irregular spaces between *poorly sorted* grains of diverse size (Figure 4.3L).

Rocks comprised of intergrown mineral crystals are said to have **crystalline texture,** and they are glittery when rotated in bright light. (The light reflects off of the flat crystal faces like tiny mirrors.) The mineral crystals may be **foliated**—oriented or lined up into patterns and/or layers that cause the rock to break or reflect light in a specific direction (Figures 4.3G, 4.3H). Or, the mineral crystals may be randomly arranged (Figure 4.3D), in which case the rock breaks randomly and irregularly. Mineral crystals in a rock can also be either a mixture of different sizes (Figure 4.3D) or else **equigranular** (all about the same size, Figure 4.3E).

Sedimentary rocks often contain grains that are clasts (fragments; broken pieces) of minerals, other rocks, plants, or animals and are said to have a **clastic texture** (Figures 4.3I, 4.3J, 4.3L). They may also include **fossils**—bones, impressions, tracks, or other evidence of ancient life (Figure 4.3I).

Metamorphic rocks have undergone *deformation* (a change in shape and/or texture caused by an applied stress or force) and transformation (change from one original texture, color, composition, and general form to a new and different one). The process of metamorphism produces fractures, faults, folds (Figure 4.3F), and foliated texture. **Foliated texture** in a rock occurs when the mineral crystals or other grains become foliated. There is a layered alignment of flat-sided or elongate mineral grains so they are parallel to one another, giving the appearance of layered scales (Figure 4.3H) or lined-up needles (Figure 4.3G). Foliated texture causes rocks to break and reflect light along the foliated layers of flat-sided minerals. Foliated metamorphic rocks that contain abundant mica minerals typically have a shiny, metallic-like, luster (Figure 4.3H).

Rock Classification

All rocks are classified as igneous, sedimentary, or metamorphic, based on their properties (Figures 4.3, 4.4). Some properties are characteristic of more than one rock type. For example, igneous, sedimentary, and metamorphic rocks all can be dark, light, or comprised of mineral particles. Therefore, it is essential to classify a rock based on more than one of its properties.

Recall that igneous rocks form when molten rock (rock liquefied by heat and pressure in the mantle) cools to a solid form. Molten rock exists both below Earth's surface (where it is called *magma*) and at Earth's surface (where it is called *lava*). Igneous rocks can have various textures, including crystalline, glassy, or vesicular (bubbly). They commonly contain mineral crystals of olivine, pyroxene, or feldspars. Igneous rocks from cooled lava flows may have ropy, streamlined shapes or layers (from repeated flows of lava). Igneous rocks usually lack fossils and organic grains.

Recall that sedimentary rocks form in two ways. **Lithification** is the hardening of sediment—masses of loose Earth materials such as pebbles, gravel, sand, silt, mud, shells, plant fragments, and products of chemical decay (clay, rust). **Precipitation** produces crystals that collect in aggregates, such as the rock salt that remains when ocean water evaporates. The lithification process occurs as layers of sediments are **compacted** (pressure-hardened) or **cemented** (glued together by tiny crystals precipitated from fluids in the pores of sediment).

Thus, most sedimentary rocks are layered and have a **clastic** texture (i.e., are made of grains called *clasts*—fragments of rocks, mineral crystals, shells, and plants—usually rounded into pebbles, gravel, sand, and mud). The sedimentary grains are arranged in layers due to sorting by wind or water. Fossils are also common in some sedimentary rocks.

The crystalline sedimentary rocks are layered aggregates of crystals precipitated from water. This includes the icicle-shaped stalactites that hang from the roofs of caves. Common minerals of these precipitated sedimentary rocks include calcite, dolomite, gypsum, or halite.

Recall that metamorphic rocks are rocks that have been deformed and transformed by intense heat, intense pressure, or the chemical action of hot fluids. Therefore, metamorphic rocks have textures indicating significant deformation (folds, extensive fractures, faults, and foliation). Fossils, if present, also are deformed (stretched or crushed). Metamorphic rocks often contain garnet, tourmaline, or foliated layers of mica. Serpentine, epidote, graphite, galena, and sphalerite occur only in metamorphic rocks. Metamorphism can occur over large regions, or in thin "contact" zones (like burnt crust on a loaf of bread) where the rock was in contact with magma or other hot fluids (Figure 4.1).

PART 4A: INTRODUCTION TO ROCKS AND THE ROCK CYCLE

Review the *Introduction, Rock Properties*, and *Rock Classification* sections of this laboratory and then proceed to Question 1 below. Your instructor may have you share with the class your observations and inferences from Question 1.

Questions

1. Figures 4.5–4.11 are photographs of rocks. For each photograph, record the following information:

   a. In column one of Figure 4.12, note the figure number of the rock sample photograph to be analyzed.

   b. In column two (blue) of Figure 4.12, list the rock properties that you can observe in the sample.

   c. In column three (pink) of Figure 4.12, classify the rock as igneous, sedimentary, or metamorphic.

   d. In column four (yellow) of Figure 4.12, describe, as best as you can, how the rock may have formed.

   e. On Figure 4.13 (on the back of Figure 4.12), write the figure number of the photograph/rock sample to show where it fits in the rock cycle model.

   f. In column five (green) of Figure 4.12, predict from the rock cycle (Figure 4.13) three different changes that the rock could undergo next if left in a natural setting.

2. Complete Figure 4.13. Color arrows in the rock cycle orange if they indicate a process leading to formation of igneous rocks, brown if they indicate a process leading to formation of sedimentary rocks, and green if they indicate a process leading to formation of metamorphic rocks. Place check marks in the table to indicate what rock group(s) is/are characterized by each of the processes and rock properties.

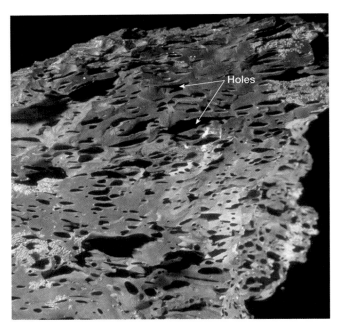

FIGURE 4.5 Photograph of a rock sample for analysis, classification, and evaluation. The black parts of the rock are sub-round holes. (×1.0)

FIGURE 4.6 Photograph of a rock sample for analysis, classification, and evaluation. (×1.0)

FIGURE 4.7 Photograph of a rock sample for analysis, classification, and evaluation. (×1.0)

FIGURE 4.8 Photograph of a rock sample for analysis, classification, and evaluation. (×0.5)

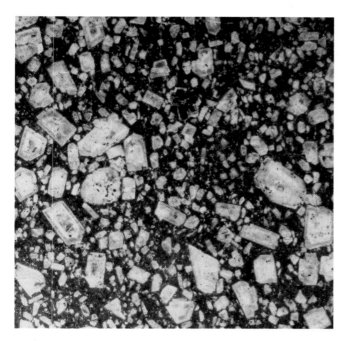

FIGURE 4.9 Photograph of a rock sample for analysis, classification, and evaluation. The white crystals are feldspar mineral crystals. (×1.0)

FIGURE 4.10 Photograph of a rock sample for analysis, classification, and evaluation. The rock effervesces in dilute hydrochloric acid (HCl). Notice that the weathered outer surface of this rock is brown. The rock has been broken open with a rock hammer to expose the fresh white crystals. (×1.0)

FIGURE 4.11 Photograph of a rock sample for analysis, classification, and evaluation. Notice how the rock breaks and its characteristic texture. (×1.0)

PART 4B: ROCK SAMPLES AND THE ROCK CYCLE

Obtain a set of numbered hand samples of rocks from your instructor or make a set yourself from local sources as directed by your instructor.

Questions

3. For each numbered rock sample, record the following information on Figure 4.14:

    a. In column one, note the figure number of the rock sample photograph to be analyzed.

    b. In column two, list the rock properties that you can observe in the sample.

    c. In column three, classify the rock as igneous, sedimentary, or metamorphic.

    d. In column four, describe, as best as you can, how the rock may have formed.

    e. In column five, predict from the rock cycle (Figure 4.2 or 4.13) three different changes that the rock could undergo next if left in a natural setting.

4. The scientific **Law of Conservation of Matter** states that matter (atoms and materials made from them) is neither created nor destroyed; it is simply recycled and redistributed in various forms and states (phases). The amount of matter comprising Earth has probably not changed significantly for millions of years, yet matter from the rock cycle can be exchanged and cycled through Earth's other spheres.

    a. Describe how matter from the rock cycle is exchanged (to and from) the atmosphere.

    b. Describe how matter from the rock cycle is exchanged (to and from) the hydrosphere.

    c. Describe how matter from the rock cycle is exchanged (to and from) the biosphere.

FIGURE 4.12 Rock analysis chart for Figures 4.5–4.11.

Sample Number	ROCK PROPERTIES (textures, minerals, fossils, etc.) Figures 4.3, 4.4	ROCK CLASSIFICATION (Igneous, Sedimentary, Metamorphic) Figure 4.4	HOW DID THE ROCK FORM?	WHAT ARE THREE CHANGES THE ROCK COULD UNDERGO? (according to the rock cycle model, Figure 4.2)
Figure 4.5				1. 2. 3.
Figure 4.6				1. 2. 3.
Figure 4.7				1. 2. 3.
Figure 4.8				1. 2. 3.
Figure 4.9				1. 2. 3.
Figure 4.10				1. 2. 3.
Figure 4.11				1. 2. 3.

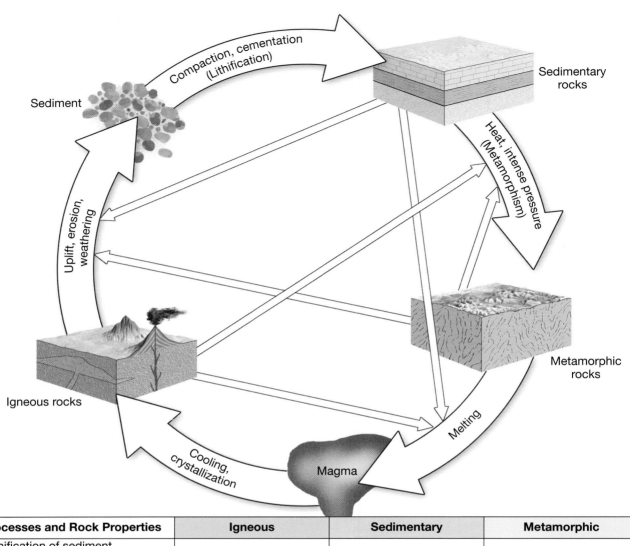

Processes and Rock Properties	Igneous	Sedimentary	Metamorphic
lithification of sediment			
intense heating (but no melting)			
crystals precipitate from water			
solidification of magma/lava			
melting of rock			
intense pressure			
compaction of sediment			
cementation of grains			
folding of rock			
pebbly, sandy, or muddy			
crystalline			
clastic			
foliated			
vesicular			
equigranular			
common fossils			
scaly and metallic-like luster			

FIGURE 4.13 Color arrows in the rock cycle orange if they indicate a process leading to formation of igneous rocks, brown if they indicate a process leading to formation of sedimentary rocks, and green if they indicate a process leading to formation of metamorphic rocks. Place check marks in the table to indicate what rock group(s) is/are characterized by each of the processes and rock properties.

FIGURE 4.14 Rock analysis chart for Part 4B.

Sample Number	ROCK PROPERTIES (textures, minerals, fossils, etc.) Figures 4.3, 4.4	ROCK CLASSIFICATION (Igneous, Sedimentary, Metamorphic) Figure 4.4	HOW DID THE ROCK FORM?	WHAT ARE THREE CHANGES THE ROCK COULD UNDERGO? (according to the rock cycle model, Figure 4.2)
				1. 2. 3.
				1. 2. 3.
				1. 2. 3.
				1. 2. 3.
				1. 2. 3.
				1. 2. 3.
				1. 2. 3.

FIGURE 4.14 (CONTINUED) Rock analysis chart for Part 4B

Sample Number	ROCK PROPERTIES (textures, minerals, fossils, etc.) Figures 4.3, 4.4	ROCK CLASSIFICATION (Igneous, Sedimentary, Metamorphic) Figure 4.4	HOW DID THE ROCK FORM?	WHAT ARE THREE CHANGES THE ROCK COULD UNDERGO? (according to the rock cycle model, Figure 4.2)
				1. 2. 3.
				1. 2. 3.
				1. 2. 3.
				1. 2. 3.
				1. 2. 3.
				1. 2. 3.
				1. 2. 3.

LABORATORY FIVE

Igneous Rocks and Volcanic Hazards

·CONTRIBUTING AUTHORS·

Harold E. Andrews • *Wellesley College*

James R. Besancon • *Wellesley College*

Claude E. Bolze • *Tulsa Junior College*

Margaret D. Thompson • *Wellesley College*

OBJECTIVES

A. Explore the geometry and origin of some intrusive and extrusive bodies of igneous rock.

B. Be able to describe and interpret textural features of igneous rocks.

C. Be able to describe compositional features of igneous rocks (mineralogy, color index).

D. Be able to classify common igneous rocks based on their texture, mineralogy, and color index.

E. Infer the origin of common igneous rocks.

F. Explore Internet resources to investigate hazards and human risks associated with active modern volcanoes.

MATERIALS

Pencil, eraser, laboratory notebook, hand magnifying lens (optional), mineral-identification tools of your choice, metric ruler and a chart for visual estimation of percent (from GeoTools Sheets 1 and 2 at the back of the manual), and samples of igneous rocks (obtain as directed by your instructor).

INTRODUCTION

Igneous rocks form when molten rock (rock liquefied by intense heat and pressure) cools to a solid state. When the molten rock cools, it always forms a mass of intergrown crystals and/or glass. Therefore, all igneous rocks and fragments of igneous rocks have crystalline or glassy textures. Even volcanic ash is microscopic fragments of igneous rock (mostly volcanic glass pulverized by an explosive volcanic eruption).

PART 5A: IGNEOUS PROCESSES AND ROCKS

Molten (heated until liquefied) rock exists in isolated bodies below Earth's surface, where it is called **magma.** In addition to its liquid molten rock portion, or melt, magma also contains dissolved gases (e.g., water, carbon dioxide, sulfur dioxide) and tiny crystals that may grow in size or abundance as the magma cools. Magma is under great pressure (like a bottled soft drink that has been shaken) and is less dense than the rocks that confine it. Like the blobs of heated "lava" in a lava lamp, the magma tends to rise and squeeze into Earth's cooler crust along any fractures or zones of weakness that it encounters. A body of magma that pushes its way into Earth's crust is called an **intrusion,** and it will eventually cool to form a coarse-grained **intrusive igneous rock** comprised of visible mineral crystals. If an intrusion of magma approaches Earth's surface, then the decrease in pressure allows its dissolved gases to separate from the magma as gas bubbles. This is like the bubbles of carbon dioxide that form when you open (release the pressure from) a pressurized bottle of beer or soft drink (containing dissolved carbon dioxide). When this happens to magma, the bubbly magma is called

lava, which may erupt (extrude) onto Earth's surface at volcanoes and cool to form a fine-grained **extrusive igneous rock** comprised of tiny crystals and/or glass.

Intrusions have different sizes and shapes. *Batholiths* (Figure 5.1) are massive intrusions (often covering regions of 100 km^2 or more in map view) that have no visible bottom. They form when small bodies of lava amalgamate (mix together) into one large body. To observe one model of this amalgamation process, watch the blobs of "lava" in a lighted lava lamp as they rise and merge into one large body (batholith) at the top of the lamp.

Smaller intrusions (see Figure 5.1) include *sills* (sheet-like intrusions that force their way between layers of bedrock), *laccoliths* (blister-like sills), *pipes* (vertical tubes or pipe-like intrusions that feed volcanoes), and *dikes* (sheet-like intrusions that cut across layers of bedrock). The dikes can occur as *sheet dikes* (nearly planar dikes that often occur in parallel pairs or groups), *ring dikes* (curved dikes that form circular patterns when viewed from above; they typically form under volcanoes), or *radial dikes* (dikes that develop from the pipe feeding a volcano; when viewed from above, they radiate away from the pipe). Study these three kinds of dikes in Figure 5.1.

Extrusive igneous processes produce *lava flows* and *pyroclastic deposits* (accumulations of rocky materials that have been fragmented and ejected by explosive volcanic eruptions). Extrusive (volcanic) igneous processes also present geologic hazards that place humans at risk.

Textures of Igneous Rocks

Texture of an igneous rock is a description of its constituent parts and their sizes, shapes, and arrangement. You should know the common textures of igneous rocks (highlighted in bold text below) and understand how they form (Figures 5.2 and 5.3). This will help you to classify *and* infer the origin of igneous rocks.

The size of mineral crystals in an igneous rock generally indicates the rate at which the lava or magma cooled to form a rock and the availability of the chemicals required to form the crystals. Large crystals require a long time to grow, so their presence generally means that a body of molten rock cooled slowly and contained ample atoms of the chemicals required to form the crystals. Tiny crystals generally indicate that the magma cooled more rapidly (there was not enough time for large crystals to form). Volcanic glass (no crystals) can indicate that a magma was quenched (cooled immediately), but most volcanic glass is the result of poor nucleation as described below.

The crystallization process depends on the ability of atoms in lava or magma to *nucleate*. *Nucleation* is the initial formation of a microscopic crystal, to which other atoms progressively bond. This is how a crystal grows. Atoms are mobile in a fluid magma, so they are free to nucleate. If such a fluid magma cools slowly, then crystals have time to grow—sometimes to many centimeters in length. However, if a magma is very viscous (thick and resistant to flow), then atoms cannot easily move to nucleation sites. Crystals may not form even by slow cooling. Rapid cooling of very viscous magma (with poor nucleation) can produce igneous rocks comprising volcanic glass, which are said to have a **glassy texture** (see Figure 5.2).

Several common terms are used to describe igneous rock texture on the basis of crystal size (Figure 5.2). Igneous rocks comprised of crystals too small to see without a hand lens (generally < 1 mm) have a fine-grained or **aphanitic texture** (from the Greek word for invisible). Those comprised of visible crystals have a **phaneritic texture** (coarse-grained; crystals 1–10 mm) or **pegmatitic texture** (very coarse-grained; > 1 cm).

Some igneous rocks have two distinct sizes of crystals. This is called **porphyritic texture** (see Figure 5.2). The large crystals are called *phenocrysts,* and the smaller, more numerous crystals form the *groundmass,* or *matrix.* Porphyritic textures may generally indicate that a body of magma cooled slowly at first (to form the large crystals) and more rapidly later (to form the small crystals). However, recall from above that crystal size can also be influenced by changes in magma composition or viscosity.

Combinations of igneous-rock textures also occur. For example, a *porphyritic-aphanitic* texture signifies that phenocrysts occur within an aphanitic matrix. A *porphyritic-phaneritic* texture signifies that phenocrysts occur within a phaneritic matrix.

When you examine an unopened pressurized bottle of soft drink or beer, no bubbles are present. But when you open the bottle (and hear a "swish" sound), you are releasing the pressure on the drink and allowing bubbles of carbon dioxide gas to escape from the liquid. Recall that magma behaves similarly. When its pressure is released near Earth's surface, it turns into bubbly lava that may erupt from a volcano. In fact, early stages of volcanic eruptions are eruptions of steam and other gases separated from magma. If the hot, bubbly lava cannot escape normally from the volcano, then the volcano may explode (like the top blowing off of a champagne bottle).

When gas bubbles get trapped in cooling lava they are called *vesicles,* and the rock is said to have a **vesicular texture.** Scoria is a textural name for a rock

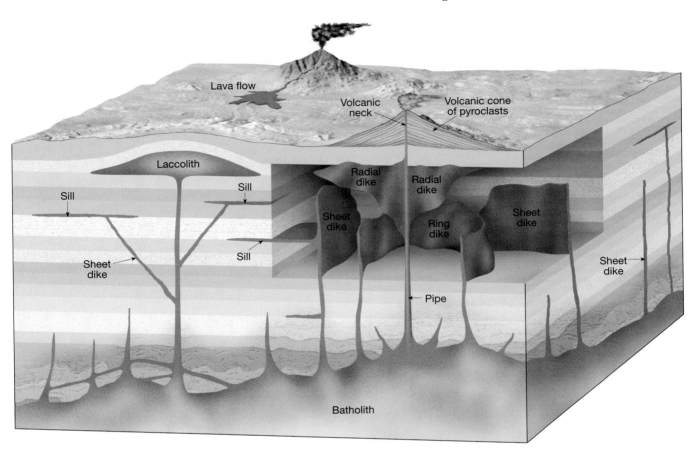

FIGURE 5.1 Illustration of the main types of intrusive and extrusive bodies of igneous rock.

having so many vesicles that it resembles a sponge. Pumice is glassy and has so many tiny vesicles that it resembles a frothy meringue and will float in water.

Pyroclasts (from Greek meaning "fire broken") are rocky materials that have been fragmented and/or ejected by explosive volcanic eruptions. They include *volcanic ash* fragments (pyroclasts <2 mm), *lapilli* or *cinders* (pyroclasts 2–64 mm), and *volcanic bombs* or *blocks* (pyroclasts >64 mm). Igneous rocks comprised of pyroclasts have a **pyroclastic texture** (see Figure 5.2). They include *tuff* (made of volcanic ash) and *volcanic breccia* (made chiefly of cinders and volcanic bombs).

Mineral Composition of Igneous Rocks

Mineral composition of an igneous rock is a description of the kinds and abundance of the mineral crystals that comprise the rock. You can estimate the abundance of any mineral in the rock using the charts for visual estimation of percent, cut from GeoTools Sheet 1 or Sheet 2 at the back of this manual.

Eight rock-forming minerals comprise most igneous rocks (Figures 5.2, 5.3): quartz, potassium feldspar (K-spar), plagioclase feldspar, muscovite mica, biotite mica, amphibole, pyroxene, and olivine. They are commonly divided into two groups on the basis of the darkness of their color (which is a function of their chemical compositions). *Quartz, plagioclase feldspar, potassium feldspar,* and *muscovite mica* are generally light-colored and form a group of **felsic minerals.** The name *felsic* refers to feldspars (*fel-*) and other silica-rich (*-sic*) minerals. *Biotite mica, amphibole, pyroxene,* and *olivine* are generally dark-colored and form a group of **mafic minerals.** The name *mafic* refers to the magnesium (*ma-*) and iron (*-fic*) in their chemical formulas, so they are also called *ferromagnesian* minerals.

Notice at the top of Figure 5.3 that the mineralogy of an igneous rock can be approximated based on a color index. **Color index (CI)** is the percentage (by volume) of the mafic (ferromagnesian, dark-colored) mineral crystals in the rock. Classification of an igneous rock depends on its mineralogy as estimated by color index or determined by direct identification of crystals of the eight rock-forming minerals

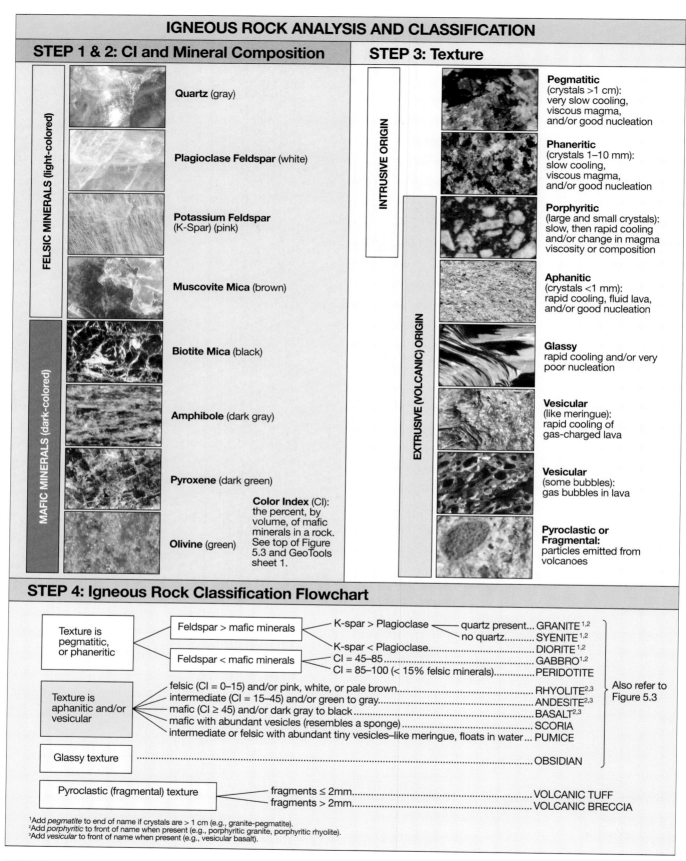

IGNEOUS ROCK ANALYSIS AND CLASSIFICATION

STEP 1 & 2: CI and Mineral Composition

FELSIC MINERALS (light-colored)

Quartz (gray)

Plagioclase Feldspar (white)

Potassium Feldspar (K-Spar) (pink)

Muscovite Mica (brown)

MAFIC MINERALS (dark-colored)

Biotite Mica (black)

Amphibole (dark gray)

Pyroxene (dark green)

Olivine (green)

Color Index (CI): the percent, by volume, of mafic minerals in a rock. See top of Figure 5.3 and GeoTools sheet 1.

STEP 3: Texture

INTRUSIVE ORIGIN

Pegmatitic (crystals >1 cm): very slow cooling, viscous magma, and/or good nucleation

Phaneritic (crystals 1–10 mm): slow cooling, viscous magma, and/or good nucleation

Porphyritic (large and small crystals): slow, then rapid cooling and/or change in magma viscosity or composition

EXTRUSIVE (VOLCANIC) ORIGIN

Aphanitic (crystals <1 mm): rapid cooling, fluid lava, and/or good nucleation

Glassy rapid cooling and/or very poor nucleation

Vesicular (like meringue): rapid cooling of gas-charged lava

Vesicular (some bubbles): gas bubbles in lava

Pyroclastic or Fragmental: particles emitted from volcanoes

STEP 4: Igneous Rock Classification Flowchart

Texture is pegmatitic, or phaneritic
- Feldspar > mafic minerals
  - K-spar > Plagioclase — quartz present... GRANITE[1,2]
  - — no quartz........... SYENITE[1,2]
  - K-spar < Plagioclase...................................... DIORITE[1,2]
- Feldspar < mafic minerals
  - CI = 45–85.. GABBRO[1,2]
  - CI = 85–100 (< 15% felsic minerals)...............PERIDOTITE

Texture is aphanitic and/or vesicular
- felsic (CI = 0–15) and/or pink, white, or pale brown..RHYOLITE[2,3]
- intermediate (CI = 15–45) and/or green to gray........................ANDESITE[2,3]
- mafic (CI ≥ 45) and/or dark gray to black ....BASALT[2,3]
- mafic with abundant vesicles (resembles a sponge)... SCORIA
- intermediate or felsic with abundant tiny vesicles–like meringue, floats in water ... PUMICE

Glassy texture .... OBSIDIAN

Pyroclastic (fragmental) texture
- fragments ≤ 2mm...VOLCANIC TUFF
- fragments > 2mm...VOLCANIC BRECCIA

Also refer to Figure 5.3

[1]Add *pegmatite* to end of name if crystals are > 1 cm (e.g., granite-pegmatite).
[2]Add *porphyritic* to front of name when present (e.g., porphyritic granite, porphyritic rhyolite).
[3]Add *vesicular* to front of name when present (e.g., vesicular basalt).

FIGURE 5.2 Igneous rock analysis and classification. **Step 1**—Estimate the rock's color index. **Step 2**— Identify the main rock-forming minerals if the mineral crystals are large enough to do so, and estimate the relative abundance of each mineral (using a Visual Estimation of Percent chart from GeoTools Sheet 1 or 2). **Step 3**—Identify the texture(s) of the rock. **Step 4**—Use the Igneous Rock Classification Flowchart to name the rock. Start on the left side of the flowchart, and work toward the right side to the rock name.

IGNEOUS ROCKS CLASSIFICATION

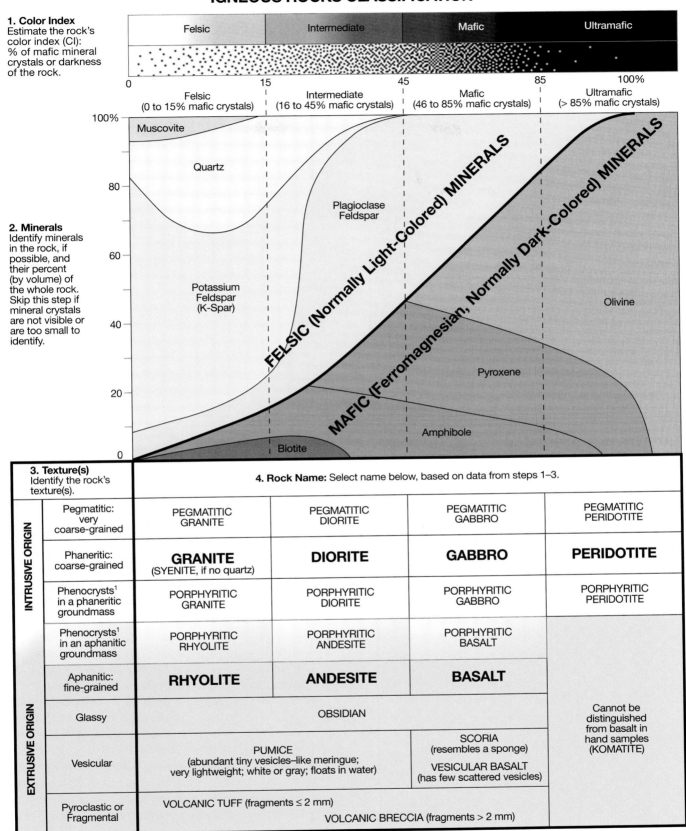

1. Color Index
Estimate the rock's color index (CI): % of mafic mineral crystals or darkness of the rock.

Felsic	Intermediate	Mafic	Ultramafic

0   15   45   85   100%

Felsic (0 to 15% mafic crystals)	Intermediate (16 to 45% mafic crystals)	Mafic (46 to 85% mafic crystals)	Ultramafic (> 85% mafic crystals)

2. Minerals
Identify minerals in the rock, if possible, and their percent (by volume) of the whole rock. Skip this step if mineral crystals are not visible or are too small to identify.

Muscovite, Quartz, Plagioclase Feldspar, Potassium Feldspar (K-Spar), Biotite — FELSIC (Normally Light-Colored) MINERALS

Olivine, Pyroxene, Amphibole — MAFIC (Ferromagnesian, Normally Dark-Colored) MINERALS

3. Texture(s)
Identify the rock's texture(s).

4. Rock Name: Select name below, based on data from steps 1–3.

INTRUSIVE ORIGIN					
	Pegmatitic: very coarse-grained	PEGMATITIC GRANITE	PEGMATITIC DIORITE	PEGMATITIC GABBRO	PEGMATITIC PERIDOTITE
	Phaneritic: coarse-grained	**GRANITE** (SYENITE, if no quartz)	**DIORITE**	**GABBRO**	**PERIDOTITE**
	Phenocrysts[1] in a phaneritic groundmass	PORPHYRITIC GRANITE	PORPHYRITIC DIORITE	PORPHYRITIC GABBRO	PORPHYRITIC PERIDOTITE
EXTRUSIVE ORIGIN	Phenocrysts[1] in an aphanitic groundmass	PORPHYRITIC RHYOLITE	PORPHYRITIC ANDESITE	PORPHYRITIC BASALT	
	Aphanitic: fine-grained	**RHYOLITE**	**ANDESITE**	**BASALT**	Cannot be distinguished from basalt in hand samples (KOMATITE)
	Glassy	OBSIDIAN			
	Vesicular	PUMICE (abundant tiny vesicles–like meringue; very lightweight; white or gray; floats in water)		SCORIA (resembles a sponge) VESICULAR BASALT (has few scattered vesicles)	
	Pyroclastic or Fragmental	VOLCANIC TUFF (fragments ≤ 2 mm) VOLCANIC BRECCIA (fragments > 2 mm)			

[1]Phenocrysts are crystals conspicuously larger than the finer grained groundmass (main mass, matrix) of the rock.

FIGURE 5.3 Igneous rock classification chart. Obtain data about the rock in steps 1–3, then use that data to select the name of the rock (step 4). Also refer to Figure 5.2 and the examples of classified igneous rocks in Figures 5.6–5.18.

(Figure 5.2). The large (coarse-grained) mineral crystals in phaneritic igneous rocks are generally easy to identify with the naked eye or a hand lens. But the tiny (fine-grained) mineral crystals in aphanitic igneous rocks are usually too small to identify with the naked eye, and one must rely on color index to estimate mineral composition.

Classifying Igneous Rocks

The complete classification of any igneous rock requires knowledge of its texture, color index, *and* the identity and abundance of specific minerals that comprise it (Figures 5.2, 5.3). Some igneous rocks are named on the basis of their texture, but most are named on the basis of their texture and mineral composition.

Color index (CI) is used to estimate the proportion of mafic and felsic mineral crystals in an igneous rock. If possible, the specific minerals and their abundance are then identified to complete the mineralogical analysis of the rock. Both color index and specific mineral identities are required for complete mineralogical analysis and classification of the rock.

Felsic igneous rocks have only 0–15% mafic mineral crystals (CI = 0−15), so they are generally very light-colored (Figure 5.3). Quartz and potassium feldspar are usually the most abundant mineral crystals in felsic rocks. **Intermediate igneous rocks** have 16–45% mafic mineral crystals (CI = 16−45), so they are more light-colored than dark (Figure 5.3). Plagioclase or potassium feldspar are generally the most abundant mineral(s) in intermediate rocks.

Mafic igneous rocks have 46–85% mafic mineral crystals (CI = 46−85), and they are dark-colored (Figure 5.3). Pyroxene and plagioclase are generally the most abundant mineral crystals in mafic igneous rocks. **Ultramafic igneous rocks** have 86–100% mafic minerals (CI = 86−100), so they are usually very dark-colored (Figure 5.3). Olivine and pyroxene are generally the most abundant mineral crystals in ultramafic igneous rocks.

The color index of an igneous rock is only an approximation of the rock's mineral composition. Whenever possible, the specific mafic or felsic minerals should also be identified. This is particularly important, because some mafic minerals (olivine) can be light-colored and some felsic minerals (labradorite feldspar) are dark-colored. Olivine is sometimes a pale yellow-green color (instead of dark green), and it could be mistaken for a mineral of the felsic group. Labradorite is a dark gray or black variety of plagioclase that could easily be mistaken for a mafic mineral.

Obsidian (volcanic glass) is also an exception to the color index rules. Its black color suggests that it is ultramafic when, in fact, most obsidian has less than 15% ferromagnesian constituents. (Ferromagnesian-rich obsidian does occur, but only rarely.)

The complete classification of any igneous rock requires classification charts like the ones in Figures 5.2 and 5.3, plus knowledge of the rock's color index, the identity and abundance of its specific minerals, and its texture(s).

Follow these steps to classify an igneous rock:

Steps 1 and 2: Identify the rock's color index (CI). Then, if possible, identify the minerals that make up the rock, and estimate the percentage of each.

- If the rock is fine-grained (aphanitic or porphyritic-aphanitic), or if you cannot identify its minerals, then you must estimate mineralogy based on the rock's color index. *Felsic* fine-grained rocks tend to be pink, white, or pale brown. *Intermediate* fine-grained rocks tend to be greenish gray. *Mafic* and *ultramafic* fine-grained rocks tend to be dark gray to black.

- If the rock is coarse-grained (phaneritic or pegmatitic), then estimate the color index and percentage abundance of quartz, feldspars, and mafic minerals. With this information, you can also characterize the rock as felsic, intermediate, mafic, or ultramafic.

Step 3: Identify the rock's texture(s) using Figure 5.2.

Step 4: Classify the rock using the flowchart in Figure 5.2 or the expanded classification chart in Figure 5.3.

- Use textural terms, such as porphyritic or vesicular, as adjectives. For example, you might identify a pinkish, aphanitic (fine-grained), igneous rock as a rhyolite. If it contains scattered phenocrysts, then you would call it a *porphyritic rhyolite*. Similarly, you should call a basalt with vesicles a *vesicular basalt*.

- The textural information can also be used to infer the origin of a volcanic rock. For example, vesicles (vesicular textures) imply that the rock formed by cooling of a gas-rich lava (vesicular and aphanitic). Pyroclastic texture implies violent volcanic eruption(s). Aphanitic texture implies more rapid cooling than phaneritic texture.

Bowen's Series of Mineral Crystallization and Reaction in Magma

When magma intrudes Earth's crust, it cools into a mass of mineral crystals and/or glass. Yet when geologists observe and analyze the igneous rocks in a single dike, sill, or batholith, they usually find that it contains more than just one kind of igneous rock. Apparently, more than one kind of igneous rock can *differentiate* (separate) from a single homogenous body of magma as it cools. American geologist, Norman L. Bowen made such observations in the early 1900s. He then devised and carried out laboratory experiments to study how magmas might evolve in ways that could explain the differentiation of multiple rock types from a single magma.

Other geologic investigations had already suggested that peridotite may comprise the top of Earth's mantle (Figure 1.18). So Bowen placed pieces of peridotite into *bombs*, strong pressurized ovens used to melt the rocks at high temperatures (1200–1400° C). Once melted, he would allow the molten rock (magma) to cool to a given temperature and remain at that temperature for a while in hopes of having it begin to crystallize. The rock was then quickly removed from the bomb and quenched (cooled by dunking it in water) to make any remaining molten rock (magma) form glass. Bowen then identified the mineral crystals that had formed at each temperature. His experiments showed that, as magmas cool, different silicate minerals crystalize in predictable series that are often summarized in a **Bowen's Reaction Series** diagram (Figure 5.4). One series is the continuous crystallization of plagioclase feldspar (on the right in the figure). Another series is the discontinuous crystallization of various mafic (ferromagnesian) silicate minerals (left).

Notice in Figure 5.4 how plagioclase feldspar (Figure 3.15) crystallizes continuously from high to low temperatures, yet the high-temperature plagioclase is calcium-rich (sodium-poor) and the low temperature plagioclase is sodium-rich (calcium-poor). As the magma begins to cool, calcium is depleted from it to form the calcium-rich plagioclase. But as the magma becomes calcium-poor, sodium takes its place in forming the plagioclase crystals. This is also observed in the field. Naturally formed plagioclase crystals are normally more calcium-rich (sodium-poor) at their centers and more sodium-rich (calcium-poor) in their outer edges.

Now notice the discontinuous series of crystallization in Figure 5.4. Bowen found that when one of these mafic minerals formed at high temperature it reacted with the magma at lower temperature to produce a different mineral. Olivine forms at the highest temperatures. If the olivine crystals remain in the magma as it cools (and do not settle out), then they react with the magma and are replaced by mineral crystals of pyroxene. If these pyroxene crystals remain in the magma as it cools, then they react with the magma and are replaced by mineral crystals of amphibole and biotite.

Finally, notice what happens at the bottom of Bowen's Reaction Series (Figure 5.4). Bowen found that at the lowest temperatures, where the last crystallization occurs, the remaining elements commonly form abundant potassium feldspar, muscovite, quartz, and rare gems such as emerald.

Bowen's laboratory investigations revealed one way that different kinds of igneous rocks can differentiate from a single, homogenous body of magma as it cools. If a mineral of the discontinuous series remains in the magma as it continues to cool, then it will react with the magma at a lower temperature and a different mineral will form. However, crystals that form continuously, and/or those that settle out of the magma as it cools, no longer react with the remaining magma. These crystals also take with them some of the chemicals that originally existed in the magma. In this way, crystallization and crystal settling remove chemical elements from the magma. These processes change the magma's composition and leave the body of cooling magma with a different combination of elements to form the next crystals. This is one way that intermediate and felsic magmas/rocks can differentiate from what started out as a mafic magma.

Bowen's Reaction Series clearly suggests a relationship between temperature, the composition of magmas, and the mineralogy and names of igneous rocks. It is laboratory-based evidence that ultramafic igneous rocks (peridotite, komatite) form at the highest temperatures, followed at lower temperatures by mafic rocks (gabbro, basalt), intermediate rocks (diorite, andesite), and felsic rocks (granite, rhyolite) (Figure 5.4). This makes Bowen's Reaction Series a useful conceptual model (like the rock cycle) for interpreting the origin of igneous rocks. For example, geologists have identified many natural examples where mafic magmas have differentiated into mafic and andesitic rocks and where andesitic magmas have differentiated into andesitic and felsic rocks. However, there is no known natural example of where a single ultramafic magma cooled and differentiated into all four main groups of igneous rocks (ultramafic, mafic, intermediate, felsic). This suggests that other factors are significant in changing the composition and temperature of a magma. For example, crystal settling may remove chemicals from a magma, while new chemicals are added to the magma as it melts and incorporates host bedrock—a process called *assimilation*. Assimilation may explain how mafic magmas at convergent plate boundaries evolve into larger bodies of intermediate and felsic magma by melting and incorporating crustal rocks that are rich in quartz and feldspar (Figure 5.5).

Bowen's Reaction Series is very generally reversed when rocks are heated. Earth materials react with their surroundings and melt at different temperatures as

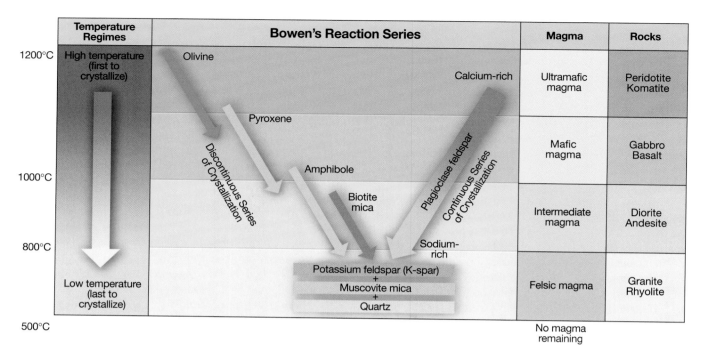

FIGURE 5.4 Bowen's Reaction Series—a laboratory-based conceptual model of one way that different kinds of igneous rocks can differentiate from a single, homogenous body of magma as it cools. See text for discussion.

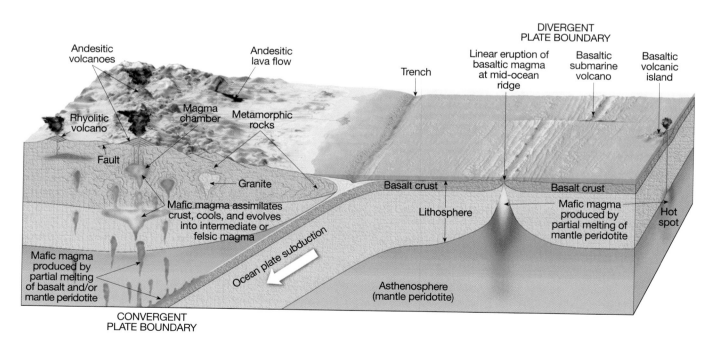

FIGURE 5.5 Formation of igneous rocks at a hot spot (such as the Hawaiian Islands), divergent plate boundary (mid-ocean ridge), and convergent plate boundary (subduction zone). See text for discussion.

they are heated. An analogy is a plastic tray of ice cubes, heated in an oven. The ice cubes would melt long before the plastic tray would melt (i.e., the ice cubes melt at a much lower temperature). As rocks are heated, their different mineral crystals melt at different temperatures. Therefore, at a given temperature, it is possible to have rocks that are partly molten and partly solid. This phenomenon is known as *partial melting* and Bowen's Reaction Series can be used to predict the sequence of melting for mineral crystals in a rock that is undergoing heating. Mineral crystals formed at low temperatures will melt at low temperatures, and mineral crystals formed at high temperatures will melt at high temperatures. However, the minerals in a particular group, say felsic or intermediate, do not all melt at once. Each mineral in the group has its own unique melting point at specific pressures. Thus, partial melting of mantle peridotite beneath hot spots and mid-ocean ridges produces mafic magma rather than ultramafic magma. When the mafic magma erupts as mafic lava along the mid-ocean ridges and hot spots (e.g., Hawaiian Islands), it cools to form basalt (Figure 5.5). A contributing factor in melting is water, which can lower the melting point of rocks. This may be how peridotite and/or basalt are partially melted in subduction zones at convergent plate boundaries (Figure 5.5).

Questions

1. Review the textures of rocks in Figures 5.6–5.18. Refer to Figure 5.19 and Step 3 of Figure 5.2. For each rock in Figure 5.19, identify the *texture(s)* present and infer the *origin* of the texture(s).

2. Review the textures, mineralogy, color indices, and classification of rocks in Figures 5.6–5.18. Then refer to the four rock samples in Figure 5.20. For each sample, identify the *texture(s), color index*, and *mineralogical composition* as requested. Use this data, Figures 5.2 and 5.3, and the steps listed on page 94 to classify and determine an igneous rock *name for each sample*.

3. Carefully study the various kinds of intrusive and extrusive igneous bodies in Figure 5.1 to understand their shapes and origins. Then analyze the stereogram (stereo pair of photographs) of Shiprock, New Mexico, in Figure 5.21. This "shiprock" feature is known to Navahos as *Tse Bi dahi*, or Rock with Wings. The light-colored portions of this stereogram are modern sand and gravel being transported by temporary streams after infrequent rains. The dark-colored features of

FIGURE 5.6 Granite—an intrusive, phaneritic igneous rock that has a low color index (light color) and is comprised chiefly of quartz and feldspar mineral crystals (see also Figures 1.18A and 3.1C). Ferromagnesian mineral crystals in granites generally include biotite and amphibole (hornblende). This sample contains pink potassium feldspar (K-spar), white plagioclase feldspar, gray quartz, and black biotite mica. Granites rich in pink potassium feldspar appear pink like this one, whereas those with white K-spar appear gray or white. Felsic rocks that resemble granite, but contain no quartz, are called *syenites*.

Quartz crystals

Mica crystals

Feldspar crystals

Photomicrograph (× 26.6)
Original sample width is 1.23 mm

FIGURE 5.7 Rhyolite—a felsic, aphanitic igneous rock that is the extrusive equivalent of a granite. It is usually light gray or pink. Some rhyolites resemble andesite (see Figure 5.9), so their exact identification must be finalized where possible by microscopic examination to verify the abundance of quartz and feldspar mineral crystals.

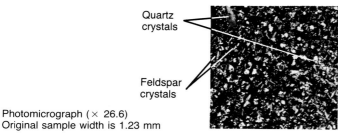

Quartz crystals

Feldspar crystals

Photomicrograph (× 26.6)
Original sample width is 1.23 mm

FIGURE 5.8 Diorite—an intrusive, phaneritic igneous rock that has an intermediate color index and is comprised chiefly of plagioclase feldspar and ferromagnesian mineral crystals. The ferromagnesian mineral crystals are chiefly amphibole (hornblende). Quartz is only rarely present and only in small amounts (<5%).

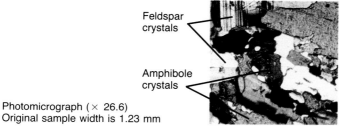

Feldspar crystals

Amphibole crystals

Photomicrograph (× 26.6)
Original sample width is 1.23 mm

FIGURE 5.9 Andesite—an intermediate, aphanitic igneous rock that is the extrusive equivalent of diorite. It is usually medium-to-dark gray. Some andesites resemble rhyolite (Figure 5.7), so their identification must be finalized by microscopic examination to verify the abundance of plagioclase feldspar and ferromagnesian mineral crystals. This sample has a porphyritic-aphanitic texture, because it contains phenocrysts of black amphibole (hornblende) set in the aphanitic groundmass.

Amphibole phenocryst

Groundmass of feldspar and ferromagnesian mineral crystals

Feldspar phenocrysts

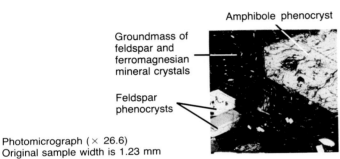

Photomicrograph (× 26.6)
Original sample width is 1.23 mm

FIGURE 5.10 Gabbro—a mafic, phaneritic igneous rock comprised chiefly of ferromagnesian and plagioclase mineral crystals. The ferromagnesian mineral crystals usually are pyroxene (augite). Quartz is absent.

Plagioclase feldspar crystals

Pyroxene crystals

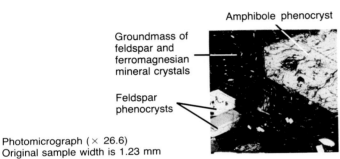

Photomicrograph (× 26.6)
Original sample width is 1.23 mm

FIGURE 5.11 Basalt—a mafic, aphanitic igneous rock that is the extrusive equivalent of gabbro, so it is dark gray. This sample has a vesicular (bubbly) texture. Microscopic examination of basalts reveals that they are comprised chiefly of plagioclase and ferromagnesian mineral crystals. The ferromagnesian mineral crystals generally are pyroxene, but they also may include olivine or magnetite. Glass also may be visible between mineral crystals. Basalt forms the floors of all modern oceans (beneath the mud and sand) and is the most abundant aphanitic igneous rock on Earth.

Ferromagnesian
mineral
crystals

Plagioclase
feldspar
crystals

Glass

Photomicrograph (× 26.6)
Original sample width is 1.23 mm

10x close-up of peridotite

FIGURE 5.12 Peridotite—an intrusive, phaneritic igneous rock having a very high color index (>95%) and comprised essentially of ferromagnesian mineral crystals. This sample is a peridotite composed of olivine mineral crystals; such a peridotite also is called *dunite*. Similarly, a peridotite composed of pyroxene mineral crystals is called *pyroxenite*. Also refer to Figure 1.18B (peridotite xenoliths).

FIGURE 5.13 Obsidian—an extrusive igneous rock comprised of dark glass (volcanic glass). It forms when very viscous lava is cooled very suddenly, or *quenched*. Such a glassy texture is also called *hyaline* (the Greek word for glass) texture. Some obsidian contains phenocrysts of feldspar that are visible in hand samples; such texture is called *porphyritic-glass* or *porphyritic-hyaline*. Some obsidian also contains microscopic plagioclase feldspar crystals, which impart a glittery reflectiveness; gemstone manufacturers call this "golden-sheen" obsidian.

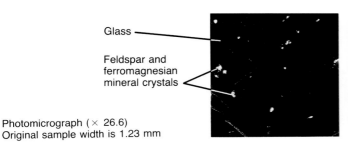

Glass

Feldspar and ferromagnesian mineral crystals

Photomicrograph ($\times$ 26.6)
Original sample width is 1.23 mm

FIGURE 5.14 Scoria—an extrusive igneous rock with a mafic color index and such abundant adjacent vesicles that it resembles the texture of a sponge. Scoria can form from the cooling of lava flows that are dense and frothy (bubbly, like whipped egg whites). Scoria also can develop from the cooling of gas-charged lava that is explosively ejected from volcanoes, forming scoria cinders.

FIGURE 5.15 Pumice—a glassy extrusive igneous rock, generally white to dark gray, having very abundant adjacent vesicles. In these properties, pumice is similar to scoria (Figure 5.14). However, pumice is less dense than scoria. Its density is so low that it floats on water. In 1992 a submarine volcanic eruption near Fiji produced a floating boulder field of pumice in the Pacific Ocean.

FIGURE 5.16 Volcanic breccia—an extrusive igneous rock comprised chiefly of pyroclasts more than 2 mm in diameter. Recall that *pyroclasts* (from the Greek, "fire broken") are rocky materials that have been fragmented and ejected by explosive volcanic eruptions. They include *volcanic ash* fragments (pyroclasts <2 mm), *lapilli* and *cinders* (pyroclasts 2–64 mm), and *volcanic bombs* or *blocks* (pyroclasts >64 mm, Figure 5.17). The pyroclasts in this sample are angular pieces of obsidian (volcanic glass, Figure 5.13).

|←——————— 10 cm ———————→|

FIGURE 5.17 A volcanic bomb. Volcanic bombs are aphanitic masses of cooled lava that were violently ejected from volcanoes and then solidified while in the air. As such, many volcanic bombs have the shapes of teardrops, ribbons, and raindrops (this example). Actually, they are cooled "lava drops"!

FIGURE 5.18 Tuff—an extrusive, pyroclastic igneous rock comprised chiefly of volcanic ash (pyroclasts <2 mm). Tuff has a dull, earthy appearance, as in this example. This sample also includes tiny shards of volcanic glass and brown *lapilli* (pyroclasts 2–64 mm) in the tuff.

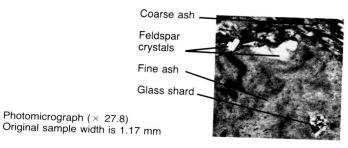

Coarse ash

Feldspar crystals

Fine ash

Glass shard

Photomicrograph (× 27.8)
Original sample width is 1.17 mm

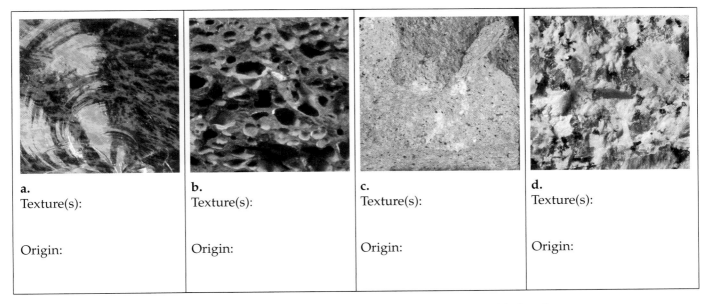

a.
Texture(s):

Origin:

b.
Texture(s):

Origin:

c.
Texture(s):

Origin:

d.
Texture(s):

Origin:

FIGURE 5.19 Four samples (actual size) of igneous rocks for analysis. Identify the texture(s) of each sample. Then use the textural information to infer the rock's origin (how it formed).

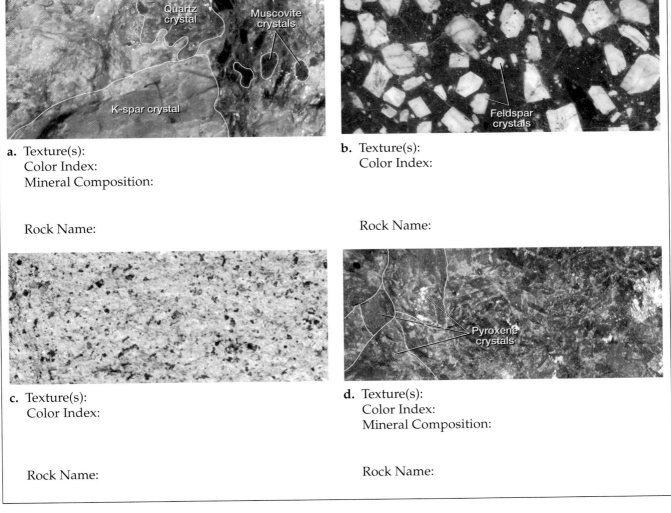

a. Texture(s):
   Color Index:
   Mineral Composition:

   Rock Name:

b. Texture(s):
   Color Index:

   Rock Name:

c. Texture(s):
   Color Index:

   Rock Name:

d. Texture(s):
   Color Index:
   Mineral Composition:

   Rock Name:

FIGURE 5.20 Four samples (actual size) of igneous rocks for analysis.

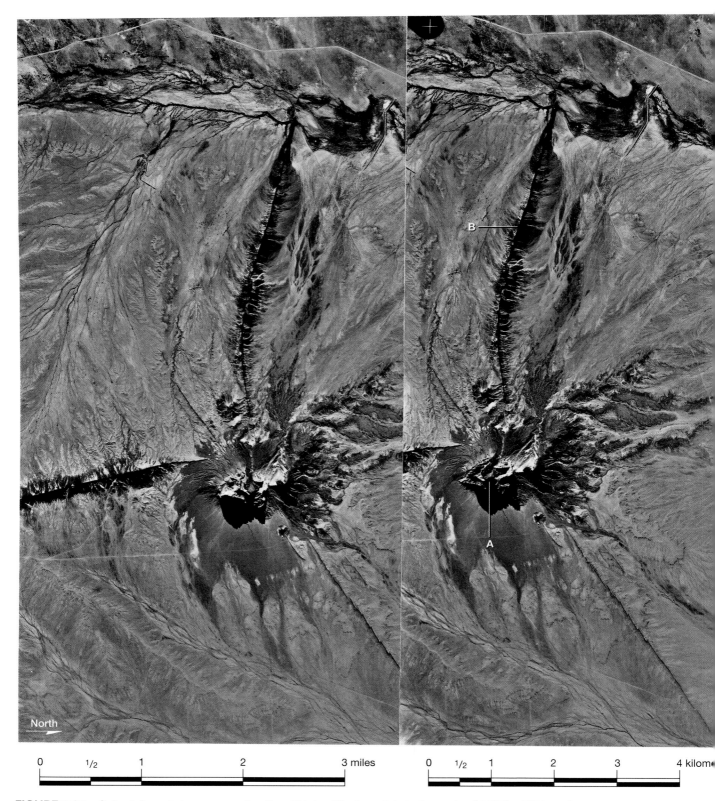

FIGURE 5.21 Color-infrared stereogram of national high-altitude aerial photographs (NHAP) of Shiprock, New Mexico, 1991. Scale 1:58,000. To view in stereo: (a) note that the figure is two images, (b) hold figure at arm's length, (c) cross your eyes until the two images become four images, (d) slightly relax your eyes so the two center images merge in stereo. To view using a pocket stereoscope, refer to Figure 9.23. (Courtesy of U.S. Geological Survey)

Mount Rushmore National Monument carved from igneous rock.

FIGURE 5.22 Mount Rushmore is carved from a body of igneous rock that crops out in the Black Hills of South Dakota.

this Rock (labeled A) with Wings (labeled B) are igneous rocks about 27 million years old.

a. What kind of igneous body is labeled A?

b. What kind of igneous body is labeled B?

c. What was the main feature of the landscape that existed here about 27 million years ago (when these bodies of igneous rock were lava)? Explain your reasoning.

4. Review Figure 5.1. Then analyze the map and photograph in Figure 5.22. Notice that Mount Rushmore is carved from a large body of igneous rock (red) that crops out in the center of the Black Hills. This body of igneous rock formed in Precambrian time (about 1.8 Ga) and was pushed up through layers of Paleozoic and Mesozoic sedimentary rocks during the Cenozoic Era to produce a bulls-eye pattern on geologic maps (i.e., Figure 10.16B).

a. What is the approximate area of the Black Hills igneous rock (colored red) in this map view?

b. What kind of igneous body is this? Explain your reasoning.

5. Review Figure 5.1. Then study the portion of a geologic map of Pennsylvania (Figure 5.23). The green-colored areas are exposures of 200–220 million-year-old Mesozoic (Triassic) sand and mud that were deposited in lakes, streams, and fields of a long, narrow valley. The red-colored areas are bodies of igneous rock about 190 million years old (Jurassic).

a. Notice the igneous bodies labeled A. Based on their geometries (as viewed from above, in map view), what kind of igneous bodies could they

be? Explain your reasoning, because more than one answer is possible.

b. Notice the igneous bodies labeled B. Based on their geometries (as viewed from above, in map view), these are probably what kind of igneous bodies? Explain your reasoning.

c. If you could have seen the landscape that existed in this part of Pennsylvania about 190 million years ago (when the bodies of igneous rock were lava), then what else would you have seen on the landscape besides valleys, streams, lakes, and fields? Explain your reasoning.

6. In your notebook, *make a list of the eight rock-forming minerals* common to igneous rocks (Figure 5.2) *and note the distinctive properties of each one* (from the Mineral Database in Figure 3.21).

7. Refer back to the four rock samples in Figure 5.20 that you analyzed and classified in Question 2. Write a reasonable explanation of how each of these rock samples may have formed on the basis of its texture(s) and mineralogy.

PART 5B: DESCRIPTION AND INTERPRETATION OF IGNEOUS ROCK SAMPLES

Question

8. Obtain a set of igneous rocks as directed by your instructor. Then fill in the information below on an Igneous Rocks Worksheet such as Figure 5.24. For each rock in the set:

a. Record the rock's sample identification number or letter.

105

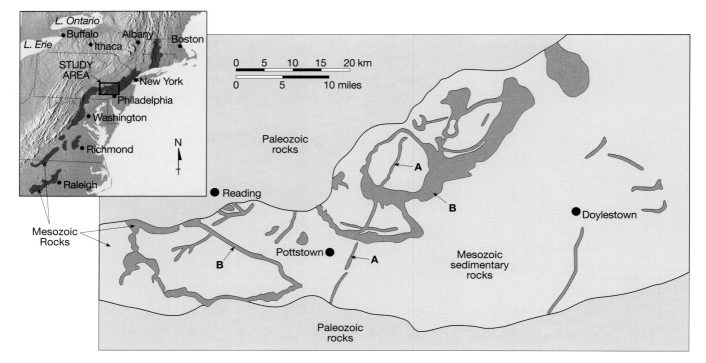

FIGURE 5.23 Geologic map of a portion of southeastern Pennsylvania. Red areas are bodies of Mesozoic igneous rock (basalt, Figure 5.10) about 190 million years old (Jurassic). Green areas are Mesozoic (Triassic) sands and muds (hardened into sandstones and mudstones) that are about 200–220 million years old. Older Paleozoic and Precambrian rocks are colored pale brown.

b. Estimate the rock's color index using the visual estimation bars at the top of Figure 5.3 and/or the Visual Estimation of Percent cut-out in GeoTools Sheet 1 or 2.

c. List the minerals present (Figure 5.2, Step 2) and visually estimate their percent abundance (for phaneritic or pegmatitic rocks).

d. Describe the rock's texture(s). Refer to Figure 5.2, Step 3.

e. Determine the rock's name (Step 4 in Figures 5.2 or 5.3).

f. Describe how the rock may have formed (Figures 5.4, 5.5).

PART 5C: VOLCANIC HAZARDS AND HUMAN RISKS

There are hundreds of active volcanoes on Earth today, and any one of them can cause hazards that place humans at risk. Geologists not only identify hazards, they also develop or recommend ways to *mitigate* them (by reducing the extent of the hazards or taking steps to reduce the risk associated with them). Internet servers contain ample information about where modern volcanoes are located, where volcanoes are now erupting, volcanic hazards, and the risks being taken by people who live near volcanoes. Search for realtime information about Earth's active volcanoes by browsing servers identified on the laboratory manual Web site or searching for others.

Questions

9. Name and briefly describe as many different kinds of volcanic hazards as you can.

10. Find information about an active volcano that is erupting somewhere in the world at this time.

    a. What is the name of the volcano?

    b. Exactly where is the volcano located?

    c. What volcanic hazards presently exist at or near the volcano?

11. What active volcano is located nearest to your home? What hazard(s), if any, do you think the volcano poses to you?

12. Find an example of volcanic hazards *mitigation*. Write a short description of what hazard was mitigated, how it was mitigated, and the results (or anticipated results) of the mitigation efforts.

FIGURE 5.24 Igneous Rocks Worksheet.

IGNEOUS ROCKS WORKSHEET

Sample Number or Letter	Texture(s) Present (Figure 5.2)	Minerals Present and Their % Abundance (Figure 5.2)	Color Index (Figure 5.3)	Rock Name from Figure 5.2 or 5.3	How did the rock form relative to Bowen's Reaction Series (Figure 5.4) and Intrusive/Extrusive Processes?

FIGURE 5.24 (CONTINUED) Igneous Rocks Worksheet.

IGNEOUS ROCKS WORKSHEET

Sample Number or Letter	Texture(s) Present (Figure 5.2)	Minerals Present and Their % Abundance (Figure 5.2)	Color Index (Figure 5.3)	Rock Name from Figure 5.2 or 5.3	How did the rock form relative to Bowen's Reaction Series (Figure 5.4) and Intrusive/Extrusive Processes?

Sedimentary Rocks, Processes, and Environments

·CONTRIBUTING AUTHORS·

Harold Andrews • *Wellesley College*

James R. Besancon • *Wellesley College*

Pamela J.W. Gore • *Dekalb College*

Margaret D. Thompson • *Wellesley College*

OBJECTIVES

A. Be able to describe and interpret textural features of sedimentary rocks.

B. Be able to describe compositional features of sedimentary rocks so you can classify the rocks as detrital (siliciclastic), biochemical (bioclastic), or chemical.

C. Be able to determine names of common sedimentary rocks based on their textures and compositions.

D. Be able to identify and interpret common kinds of sedimentary structures.

E. Infer the origin of sedimentary rock units based on their textures, compositions, and sedimentary structures.

F. Infer Earth history by interpreting a sequence of strata, from bottom to top.

MATERIALS

Pencil, eraser, laboratory notebook, hand magnifying lens (optional), metric ruler and sediment grain-size scale (from GeoTools Sheets 1 and 2 at the end of the manual), and samples of sedimentary rocks (obtain as directed by your instructor).

INTRODUCTION

Sediments are loose grains and chemical residues of Earth materials, including rock fragments, mineral grains, parts of plants or animals, and rust (hydrated iron oxide residue). All sediments have a *source* (place of origin) where they were produced by the biochemical processes of plants or animals, or by chemical and physical weathering of parts of organisms (shells, leaves, logs) and inorganic materials (rocks, minerals).

Chemical weathering is the decomposition or dissolution of Earth materials (see Figure 1.4). For example, feldspar and mica mineral grains decompose to clay minerals (such as kaolinite). Calcite decomposes to calcium and bicarbonate ions in aqueous solution. Olivine decomposes to iron oxide residues and magnesium oxide residues. Halite (sodium chloride) dissolves to form salt water (sodium chloride in aqueous solution).

Physical (mechanical) weathering is the cracking, scratching, crushing, abrasion, or other physical disintegration of Earth materials. This process causes big rocks to be disintegrated into *clasts* (broken pieces), including *rock fragments* and *mineral grains* (whole crystals or fragments of crystals). It causes logs and animal shells to be reduced to peat and shell gravel.

The products of weathering processes are **sediments** (loose grains and chemical residues of rocks, minerals, plants, or animals), and **aqueous**

solutions (mixtures of water and other chemicals formed by the dissolution and chemical decay of rocky or organic materials). Gravel, sand, mud, peat, and accumulations of seashells on a beach are all examples of sediment. Rust (hydrated iron oxide) on a piece of iron is an example of a chemical residue type of sediment. Ocean water is an aqueous solution containing many dissolved materials, including several kinds of salt.

Sedimentary rocks form when sediments are compressed together or otherwise hardened (like mud hardened by the Sun forms *adobe*), or when masses of intergrown mineral crystals precipitate from aqueous solutions (like the rock salt that remains when ocean water is evaporated).

PART 6A: SEDIMENTARY PROCESSES AND ROCKS

Wherever sediments accumulate, or crystals precipitate from aqueous solutions, they form a deposit or *sedimentary unit*. Some sedimentary units are unconsolidated (unhardened), and some are consolidated (hardened) into sedimentary rock. Therefore, every sedimentary unit is comprised of sediment or sedimentary rock and can be described, identified, and interpreted on the basis of its texture and composition.

Textures of Sediments and Sedimentary Rocks

The same processes of weathering, transportation, precipitation, and *deposition* (settling to rest and forming a deposit) that contribute to the formation of a sediment or sedimentary rock also contribute to forming its texture. The **texture** of a sediment or sedimentary rock is a description of its constituent parts and their sizes, shapes, and arrangement.

Sediments can be transported over great distances by wind, water, or ice. During this process, the sedimentary grains are dragged, bounced, rolled, and carried. This can cause the grains to be scratched, broken, or abraded (sharp edges worn away), all of which affect *grain shape*. A fresh rock fragment, mineral grain, or seashell has sharp edges and is described as *angular*. The more the grain is transported, the more *rounded* and smaller it generally becomes (Figure 6.1). Angular grains get rounded, and rounded grains get *well-rounded*. Rock fragments and mineral crystals get broken and abraded down to gravel, gravel gets broken and abraded down to sand, and sand gets broken and abraded into silt and clay.

Different velocities of wind and water currents are capable of transporting and naturally separating different densities and sizes of sediments from one another. This process is called *sorting* (see Figure 6.1). *Poorly sorted* sediments are composed of many different sizes and/or densities of grains mixed together. *Well-sorted* sediments, however, are comprised of grains that are of similar size and/or density.

Well-sorted sediments usually are comprised of *well-rounded* grains, because the grains have been abraded and rounded during transportation (see Figure 6.1). Poorly sorted sediments usually are *angular* (have sharp corners), because of the lack of abrasion during transportation.

Grain size (Figure 6.1) usually is expressed in *Wentworth classes*, named after C. K. Wentworth, an American geologist who devised the scale in 1922. Here are the Wentworth grain-size classes commonly used by sedimentologists in describing sediments:

- **gravel** includes grains larger than 2 mm in diameter (granules, pebbles, cobbles, and boulders).

- **sand** includes grains from 1/16 mm to 2 mm in diameter (in decimal form, 0.0625 mm to 2.000 mm). This is the size range of grains in a sandbox. The grains are visible and feel very gritty when rubbed between your fingers.

- **silt** includes grains from 1/256 mm to 1/16 mm in diameter (in decimal form, 0.0039 mm to 0.0625 mm). Grains of silt are usually too small to see, but you can still feel them as very tiny gritty grains when you rub them between your fingers or teeth.

- **clay** includes grains less than 1/256 mm diameter (in decimal form, 0.0039 mm). Clay-sized grains are too small to see, and they feel smooth (like chalk dust) when rubbed between your fingers or teeth. Note that the word *clay* is used not only to denote a grain size, but also a clay mineral. However, clay mineral crystals are usually clay-sized.

Sedimentary rocks that form when crystals precipitate from aqueous solutions have a **crystalline texture** (clearly visible crystals; see Figure 6.1) or **microcrystalline texture** (crystals too small to identify; see Figure 6.1). As the crystals grow, they interfere with each other and form an intergrown and interlocking texture that also holds the rock together.

Composition of Sediments and Sedimentary Rocks

The **composition** of a sediment or sedimentary rock is a description of the kinds and abundances of grains that comprise it. Sediments and sedimentary rocks are

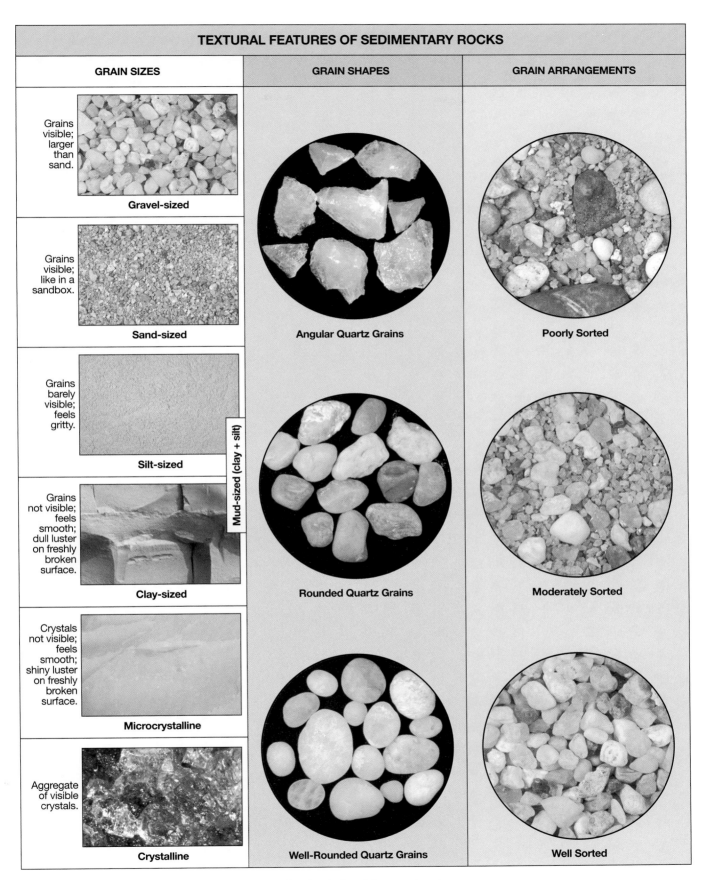

FIGURE 6.1 Textural features of sedimentary rocks. Scale for all images is ×1

COMPOSITIONAL CLASSIFICATION OF SEDIMENTARY ROCKS

A. DETRITAL (SILICICLASTIC) — made mostly of rock fragments, quartz grains, feldspar grains, or clay minerals

Breccia:
made mostly of angular gravel
(usually rock fragments)

Mudstone and **Shale:**
made mostly of clay
minerals

Conglomerate:
made mostly of rounded gravel
and sand grains
(usually quartz grains)

Arkose:
made mostly of
feldspar grains

B. BIOCHEMICAL (BIOCLASTIC) — made mostly of grains that are fragments or shells of organisms (plants or animals)

**Biochemical/Bioclastic
Limestone:**
made mostly of shells
and shell fragments

Peat:
made mostly of plant
fragments

Coal:
made of carbon/charcoal
from plants

C. CHEMICAL — made mostly of mineral crystals precipitated from aqueous solutions and/or chemical residues (e.g., rust)

Rock Gypsum:
made mostly of gypsum
mineral crystals

Rock Salt:
made mostly of halite
mineral crystals

Ironstone:
made mostly of iron-bearing
mineral crystals like this hematite

Ironstone:
made mostly of iron-bearing
residues like this limonite

Chemical Limestone:
made mostly of calcite
(or aragonite) mineral crystals

Dolostone:
made mostly of dolomite
mineral crystals

Chert
made of microcrystalline
quartz varieties

FIGURE 6.2 Compositional classification of sedimentary rocks. Scale for all images is ×1

classified as biochemical (bioclastic), chemical, or detrital (siliciclastic) on the basis of their composition (Figure 6.2). **Biochemical (bioclastic)** sediments and rocks are comprised mostly of the remains of organisms, such as shells, plant fragments, and carbon. **Chemical** sediments and rocks are comprised mostly of intergrown mineral crystals precipitated from aqueous solutions and chemical residues. The precipitated minerals commonly include gypsum, halite, hematite, limonite, calcite, dolomite, and chert (microcrystalline variety of quartz). **Detrital** (Latin, "from rubbing or wearing away") sediments and rocks are composed mostly of *detrital* grains—worn rock fragments and mineral grains that were weathered and transported from their source. Because detrital sedimentary grains are mostly *clasts* (broken pieces) of silicate minerals such as quartz, feldspars, micas, and clay minerals, detrital sediments and rocks are also referred to as **siliciclastic**.

Formation of Sedimentary Rocks

Lithification is the hardening of sediment (masses of loose Earth materials such as pebbles, gravel, sand, silt, mud, shells, plant fragments, mineral crystals, and products of chemical decay) to produce rock. The lithification process usually occurs as layers of sediment are **compacted** (pressure-hardened, Figure 6.3) or **cemented** together (glued together by tiny crystals or chemical residues precipitated from fluids in the pores of sediment, Figures 6.4, 6.5). However, it is also possible to form a dense hard mass of intergrown crystals directly, as they precipitate from aqueous solutions (Figures 6.6 and 6.7).

Sand (a sediment) can be *compacted* until it is pressure-hardened into sandstone (a sedimentary rock). Alternatively, sandstone can form when sand grains are *cemented* together by chemical residues or the growth of interlocking microscopic crystals in pore spaces of the rock (void spaces among the grains). Rock salt and rock gypsum are examples of sedimentary rocks that form by the *precipitation* of aggregates of intergrown and interlocking crystals during the evaporation of salt water or brine.

Ocean water is the most common aqueous solution and variety of salt water on Earth. As it evaporates, a

A. Start with a handful of mud.

B. Compact the mud by squeezing it in your fist.

C. Release your grip to observe a piece of mudstone.

FIGURE 6.3 Compaction of a handful of mud to form a lump of mudstone. The more the mud is compacted, the harder it will become.

variety of minerals precipitate in a particular sequence. The first mineral to form in this sequence is aragonite (calcium carbonate). Gypsum forms when about 50–75% of the ocean water has evaporated, and halite (table salt) forms when 90% has evaporated. Ancient rock salt units buried under modern Lake Erie probably formed from evaporation of an ancient ocean. The salt units were then buried under superjacent layers of mud and sand, long before Lake Erie formed on top of them (see Figure 6.6).

Quartz sand (sediment)

×2

CEMENTATION

SEDIMENTARY ROCK:

1. Sandstone with white calcite or quartz cement.

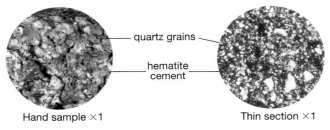

quartz grains

calcite cement

Hand sample ×1   Thin section ×1

2. Sandstone with reddish hematite cement.

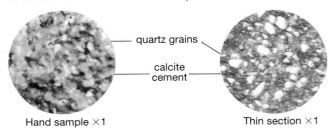

quartz grains

hematite cement

Hand sample ×1   Thin section ×1

3. Sandstone with brown, black, or yellow limonite cement.

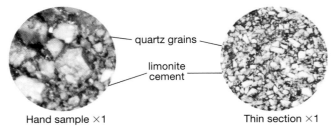

quartz grains

limonite cement

Hand sample ×1   Thin section ×1

FIGURE 6.4   Cementation of quartz sand to form sandstone.

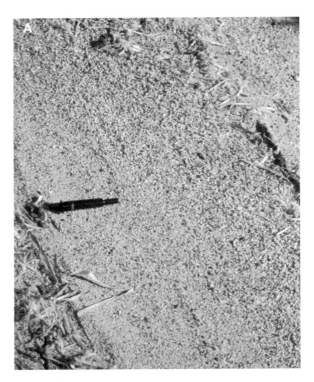

A

B

×1

C

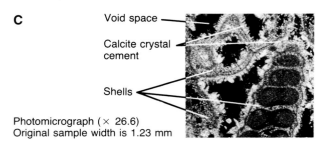

Void space

Calcite crystal cement

Shells

Photomicrograph (× 26.6)
Original sample width is 1.23 mm

FIGURE 6.5   Formation of the biochemical (bioclastic) limestone. **A.** Shell gravel and blades of the sea grass *Thalassia* have accumulated on a modern beach of Crane Key, Florida. Note pen (12 cm long) for scale. **B.** Sample of gravel like that shown in part A, but it is somewhat older and has been cemented together with calcite to form lime-stone. **C.** Photomicrograph of a thin section of the sample shown in B. Note that the rock is very porous and that it is cemented with films of microscopic calcite crystals that have essentially glued the shells together.

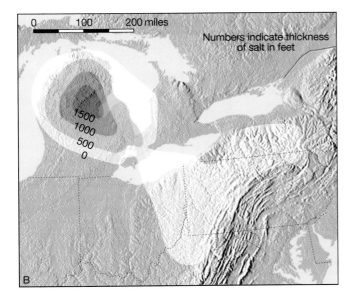

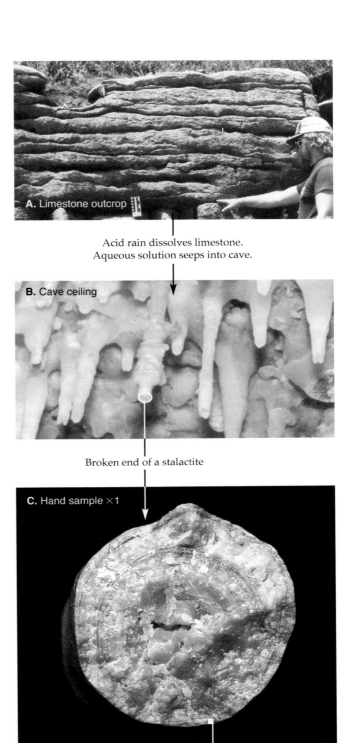
A. Limestone outcrop

Acid rain dissolves limestone.
Aqueous solution seeps into cave.

B. Cave ceiling

Broken end of a stalactite

C. Hand sample ×1

Photomicrograph of laminations

D. Thin section

Microcrystalline calcite

Microcrystalline calcite with iron impurity

Pore spaces

Photomicrograph (× 70.1)
Original sample width is 0.47 mm

FIGURE 6.6 Rock salt, a chemical sedimentary rock with crystalline texture. **A.** Hand sample from salt mines deep below Lake Erie reveals that rock salt is an aggregate of intergrown halite mineral crystals. **B.** Map showing the thickness and distribution of rock salt deposits formed about 400 million years ago, when a portion of the ocean was trapped and evaporated in what is now the Great Lakes region, millions of years before any lakes existed.

FIGURE 6.7 Formation of the chemical sedimentary rock, travertine. **A.** Limestone bedrock is dissolved by acidic rain near the Earth's surface. **B.** The resulting aqueous solution of water, calcium ions, and bicarbonate ions seeps into caves. As the solution drips from the roof of a cave, it forms icicle-shaped stalactites. **C.** Broken end of a stalactite reveals that it is actually an aggregate of chemically precipitated calcite crystals. **D.** Thin section photomicrograph reveals that the concentric laminations of the stalactite are caused by variations in iron impurity and porosity of the calcite layers.

115

SEDIMENTARY ROCK ANALYSIS AND CLASSIFICATION

STEP 1: Composition. What materials comprise most of the rock?		STEP 2: What are the rock's texture and other distinctive properties?			STEP 3: Name the rock based on your analysis in steps 1 and 2.		
Detrital (Siliciclastic) sediment grains: fragmented rocks and/or silicate mineral crystals	Rock fragments and/or quartz grains and/or feldspar grains and/or clay minerals (e.g., kaolinite) Detrital sediment is derived from the mechanical and chemical weathering of continental (land) rocks, which are comprised mostly of silicate minerals. Detrital sediment is also called terrigenous (land derived) sediment.	Angular gravel, poorly sorted grains larger than 2 mm			BRECCIA*		Detrital (Siliciclastic) sedimentary rocks
		Rounded gravel, poorly sorted grains larger than 2 mm			CONGLOMERATE*		
		Mostly sand (1/16 – 2 mm grains). May contain fossils	Mostly quartz		QUARTZ SANDSTONE	SANDSTONE	
			Mostly feldspar		ARKOSE		
			Mostly rock fragments		LITHIC SANDSTONE		
			Sand is mixed with much mud		WACKE (GRAYWACKE)		
		No visible grains	Mud (< 1/16 mm)	Mostly silt. May contain fossils	Breaks into blocks or layers	SILTSTONE	MUDSTONE
				Mostly clay. May contain fossils	Fissile (splits easily into layers)	SHALE	
					Crumbles into blocks	CLAYSTONE	
Biochemical (Bioclastic) sediment grains: fragments/shells of organisms	Plant fragments and/or charcoal	Brown porous rock with visible plant fragments that are easily broken apart from one another			PEAT		Biochemical (Bioclastic) sedimentary rocks
		Dull, dark brown, brittle rock; fossil plant fragments may be visible			LIGNITE		
		Black, layered, brittle rock; may be sooty or bright			BITUMINOUS COAL		
	Shells and shell/coral fragments, and/or calcareous microfossils	Mostly gravel-sized shells and shell or coral fragments; (Figure 6.5)			COQUINA	LIMESTONE	
		Mostly sand-sized shell fragments; often contains a few larger whole fossil shells			CALCARENITE (FOSSILIFEROUS LIMESTONE)		
		Silty, earthy rock comprised of the microscopic shells of calcareous phytoplankton (microfossils); may contain a few visible fossils			CHALK		
		No visible grains	No visible grains in most of the rock. May break with conchoidal fracture. May contain a few visible fossils in the micrite		MICRITE		
Mineral crystals (inorganic) or chemical residues (e.g., rust)	Calcite crystals and/or calcite spheres and/or microcrystalline calcite/aragonite	Comprised of spherical grains that resemble miniature pearls (< 2 mm), called ooliths or ooids			OOLITIC LIMESTONE	LIMESTONE	Chemical sedimentary rocks
		Microcrystalline masses or masses of visible crystals (Figure 6.7); may have cavities, pores, and/or faint layering; usually light colored			TRAVERTINE		
	Microcrystalline dolomite	Effervesces in dilute HCl only if powdered. Usually light colored. (Commonly forms from alteration of limestone)			DOLOSTONE		
	Halite mineral crystals	Salty taste, visible crystals, brittle. (Figure 6.6)			ROCK SALT		
	Gypsum mineral crystals	Gray, white, or colorless. Visible crystals or microcrystalline. Can be scratched with your fingernail			ROCK GYPSUM		
	Iron-bearing minerals crystals or residues	Dark-colored, heavy, amorphous chemical residues (limonite) or microcrystalline nodules (e.g., hematite, goethite)			IRONSTONE		
	Microcrystalline varieties of quartz (flint, chalcedony, chert, jasper)	Microcrystalline, may break with a conchoidal fracture. Hard (scratches glass). Usually gray, brown, black or mottled mixture of those colors. May contain fossils, as the silica in most chert is derived from dissolution of siliceous phytoplankton ooze (diatoms, radiolaria)			CHERT (a siliceous rock)		

*Modify name as quartz breccia/conglomerate, arkose breccia/conglomerate, lithic breccia/conglomerate or wacke breccia/conglomerate as done for sandstones.

FIGURE 6.8 Sedimentary rock analysis and classification. See text for steps to analyze and name a sedimentary rock.

Hand sample

**5X close-up of
hand sample**

FIGURE 6.9 Photograph of hand sample X and close-up photomicrograph of same (magnified 5×) for example on Sedimentary Rocks Worksheet. See Row 1 of the worksheet (Figure 6.10) to see how this rock's composition, texture, and origin were described.

Classifying Sedimentary Rocks

The complete classification of a sedimentary rock requires knowledge of its composition, texture(s), and other distinctive properties (Figure 6.8). The same information used to name the rock can also be used to infer its origin. Refer to the example for sample X (Figures 6.9 and 6.10).

Follow these steps to classify a sedimentary rock:

Step 1: Determine and record the rock's general composition as *biochemical (bioclastic), chemical,* or *detrital (siliciclastic)* with reference to Figure 6.2, and record a description of the specific kinds and abundances of grains that comprise the rock. Refer to the categories for composition in the left-hand column of Figure 6.8.

Step 2: Record a description of the rock's texture(s) with reference to Figure 6.1. Also record any other of the rock's distinctive properties as categorized in the center columns of Figure 6.8.

Step 3: Determine the name of the sedimentary rock by categorizing the rock from left to right across Figure 6.8. Use the compositional, textural, and special properties data from Steps 1 and 2 (left side of Figure 6.8) to deduce the rock name (right side of Figure 6.8). For assistance, refer to the discussion below.

Step 4: After you have named the rock, then you can use information from Steps 1 and 2 to infer the origin of the rock. See the example for sample X (Figures 6.9 and 6.10).

The main kinds of biochemical (bioclastic) sedimentary rocks are limestone, peat, lignite, and coal. Biochemical limestone is comprised of animal skeletons (usually seashells, coral, or microscopic shells), as in Figures 6.2 and 6.5. Differences in the density and size of the constituent grains of a biochemical (bioclastic) limestone can also be used to call it a **coquina, calcarenite (fossiliferous limestone), micrite,** or **chalk** (Figure 6.8). **Peat** (Figure 6.2) is a very porous brown rock with visible plant fragments (like peat moss). **Lignite** is brown but more dense than peat. **Bituminous coal** (coal in Figure 6.2) is a black rock made of charcoal or brittle shiny layers of carbon from plants.

There are seven main kinds of chemical (inorganic) sedimentary rocks in the classification in Figure 6.8. **Travertine** is a mass of intergrown calcite crystals that may have faint layering, cavities, or pores (Figure 6.7C). **Oolitic limestone** is composed mostly of tiny spherical grains that resemble beads or miniature pearls and are made of concentric layers of microcrystalline aragonite or calcite. **Dolostone** (Figure 6.2) is an aggregate of dolomite mineral crystals that are usually microcrystalline. Because calcite and dolomite closely resemble one another, the best way to tell them apart is with the "acid test" that you learned in Laboratory 3. Calcite will effervesce (fizz) in dilute HCl, but dolomite will fizz *only* if it is powdered first. **Rock gypsum** is an aggregate of gypsum crystals, and **rock salt** is an aggregate of halite crystals (Figures 6.2 and 6.6). Two other chemical sedimentary rocks are **chert** (comprised of microcrystalline quartz) and **ironstone** (comprised of hematite, limonite, or other iron-bearing minerals or chemical residues).

The main kinds of detrital (siliciclastic) sedimentary rocks are mudstone, sandstone, breccia, and conglomerate. It is very difficult to tell the percentage of clay or silt in a sedimentary rock with the naked eye, so sedimentary rocks comprised of clay and/or silt are commonly called **mudstone.** Mudstone that is *fissile* (splits apart easily into layers) can be called **shale.** Mudstone can also be called siltstone or claystone, depending upon whether silt or clay is the most abundant grain size. Any detrital rock comprised mostly of sand-sized grains is simply called **sandstone** (see Figures 6.4 and 6.8); although you can distinguish among *quartz sandstone* (comprised mostly of quartz grains), *arkose* (comprised mostly of feldspar grains), *lithic sandstone* (comprised mostly of rock fragments), or *wacke* (comprised of a mixture of sand-sized and mud-sized grains). **Breccia** and **conglomerate** are both comprised of gravel-sized grains and are often poorly sorted or moderately sorted. But the grains in breccia are angular, and the grains in conglomerate are rounded or well rounded.

Questions

1. What specific kind of biochemical limestone is shown in Figure 6.5B? Explain how you determined this name.

2. What simple chemical test could you use to distinguish chalk from white claystone? Explain.

3. How would you distinguish limestone from dolostone using dilute HCl?

4. How would you distinguish a sample of calcarenite (fossiliferous limestone) from a sample of mudstone with fossil shells?

5. How would you distinguish a sample of peat from a sample of mudstone with fossil plant fragments?

PART 6B: HAND SAMPLE ANALYSIS AND INTERPRETATION

Question

6. Obtain a set of sedimentary rocks (as directed by your instructor) and analyze the rocks one at a time. For each sample, complete a line on the Sedimentary Rocks Worksheet (Figure 6.10) using the steps to classify a sedimentary rock that you learned in Part 6A.

FIGURE 6.10 Sedimentary Rocks Worksheet.

SEDIMENTARY ROCKS WORKSHEET

Sample Number or Letter	Composition (Figures 6.2 and 6.8)	Textural and Other Distinctive Properties (Figure 6.1 and 6.8)	Rock Name (Figure 6.8)	How did the rock form? (See Figures 6.3–6.7)
Fig. 6.9	Clastic rock composed mostly of orange feldspar and about 10% quartz grains	Mostly gravel-sized, angular, poorly sorted grains	Arkose breccia (or arkose sandstone)	Preexisting rock (probably granite) was weathered. Grains were not rounded or sorted much, so they were not transported very far from their source. Grains were mixed with some green silt, deposited, and hardened (compaction?) into rock.

FIGURE 6.10 (CONTINUED) Sedimentary Rocks Worksheet.

SEDIMENTARY ROCKS WORKSHEET

Sample Number or Letter	Composition (Figures 6.2 and 6.8)	Textural and Other Distinctive Properties (Figure 6.1 and 6.8)	Rock Name (Figure 6.8)	How did the rock form? (See Figures 6.3–6.7)

FIGURE 6.10 (CONTINUED) Sedimentary Rocks Worksheet.

SEDIMENTARY ROCKS WORKSHEET

Sample Number or Letter	Composition (Figures 6.2 and 6.8)	Textural and Other Distinctive Properties (Figure 6.1 and 6.8)	Rock Name (Figure 6.8)	How did the rock form? (See Figures 6.3–6.7)

FIGURE 6.10 (CONTINUED)   Sedimentary Rocks Worksheet.

SEDIMENTARY ROCKS WORKSHEET

Sample Number or Letter	Composition (Figures 6.2 and 6.8)	Textural and Other Distinctive Properties (Figure 6.1 and 6.8)	Rock Name (Figure 6.8)	How did the rock form? (See Figures 6.3–6.7)

FIGURE 6.10 (CONTINUED)   Sedimentary Rocks Worksheet.

SEDIMENTARY ROCKS WORKSHEET

Sample Number or Letter	Composition (Figures 6.2 and 6.8)	Textural and Other Distinctive Properties (Figure 6.1 and 6.8)	Rock Name (Figure 6.8)	How did the rock form? (See Figures 6.3–6.7)

FIGURE 6.10 (CONTINUED) Sedimentary Rocks Worksheet.

SEDIMENTARY ROCKS WORKSHEET

Sample Number or Letter	Composition (Figures 6.2 and 6.8)	Textural and Other Distinctive Properties (Figure 6.1 and 6.8)	Rock Name (Figure 6.8)	How did the rock form? (See Figures 6.3–6.7)

PART 6C: SEDIMENTARY STRUCTURES AND ENVIRONMENTS

A variety of structures occur in sedimentary rocks (Figure 6.11). Some form by purely physical processes, and others form as a result of the activities of plants or animals. Therefore, the specific kinds of sedimentary structures can be used as indicators of environments where they normally form today.

Sedimentary Structures

One of the most obvious sedimentary structures is layering of sediments. Most layers of sediment, or **strata** (plural of *stratum,* a single layer), accumulate in nearly horizontal sheets. Strata less than 1 cm thick are called *laminations*; strata 1 cm or more thick are called *beds* (see Figure 6.11).

Surfaces between strata are called **bedding planes.** These represent surfaces of exposure that occurred between sedimentary depositional events. To illustrate, imagine a series of storms, each of which causes sediment to be deposited in puddles. Each storm is a sedimentary depositional event. Between storms, deposition stops, and the surface of the sediment in the puddles (bedding plane surface) becomes exposed to the sorting action of water in the puddles or to the processes of weathering as dry surfaces after the puddles evaporate.

Most strata are deposited in nearly horizontal sheets. However, some stratification is inclined and is referred to as **cross-stratification** or **cross-bedding** (see Figure 6.11). Sediment transported in a single direction by water or air currents commonly forms **current ripple marks** or sand dunes. Sediment transported by back-and-forth water motions or very gentle waves skimming the bottom of a lake or ocean commonly forms **oscillatory ripple marks** (Figure 6.11). Both types of ripple marks are internally cross-stratified, and the cross-strata are inclined in the direction of water/air flow. This information is useful for interpreting the kinds of environments in which the strata formed. For example, cross-strata inclined in just one general direction indicate flow of air or water in just one direction (downstream or downwind). If a sequence of cross-strata is inclined in opposite directions (**bimodal cross-bedding** in Figure 6.11), then the environment in which the sequence formed must have water/wind that changed direction back and forth. An example would be water currents associated with tides.

Individual strata also may be **graded** (Figure 6.11). Normally, graded beds are sorted from coarse at the bottom to fine at the top. This feature is caused when sediment-laden currents suddenly slow as they enter a standing body of water, or as current flow terminates abruptly.

Flutes (Figure 6.11) are scoop-shaped or V-shaped depressions scoured into a sediment surface by the erosional, winnowing action of currents. Natural casts of flutes are called **flute casts.** Flutes and flute casts indicate current direction, because they flare out (widen) in the down-current direction.

Many sedimentary rocks also contain structures that formed shortly after deposition of the sediments that compose them. For example, **mudcracks** often form while moist deposits of mud dry and shrink, and **raindrop impressions** may form on terrestrial (land) surfaces (Figure 6.11). Animals make tracks, trails, and burrows (Figure 6.11) that can be preserved in sedimentary rocks. Such traces of former life are called **trace fossils.**

Sedimentary Environments

Sediments are deposited in many different environments. Some of these environments are illustrated in Figure 6.12. Each environment has characteristic sediments, sedimentary structures, and organisms that can become **fossils** (any evidence of prehistoric life). The information gained from grain characteristics, sedimentary structures, and fossils can be used to infer what ancient environments (**paleoenvironments**) were like in comparison to modern ones.

Questions

7. Complete the questions in Figure 6.13.

8. Complete the questions in Figure 6.14.

9. Complete the questions in Figure 6.15.

PART 6D: INTERPRETATION OF A STRATIGRAPHIC SEQUENCE

As sediments accumulate, they cover up the sediments that were already deposited at an earlier (older) time. Environments also change through time, as layers of sediment accumulate. Therefore at any particular location, bodies of sediment have accumulated in different times and environments. These bodies of sediment then changed into rock units, which have different textures, compositions, and sedimentary structures.

SEDIMENTARY STRUCTURES

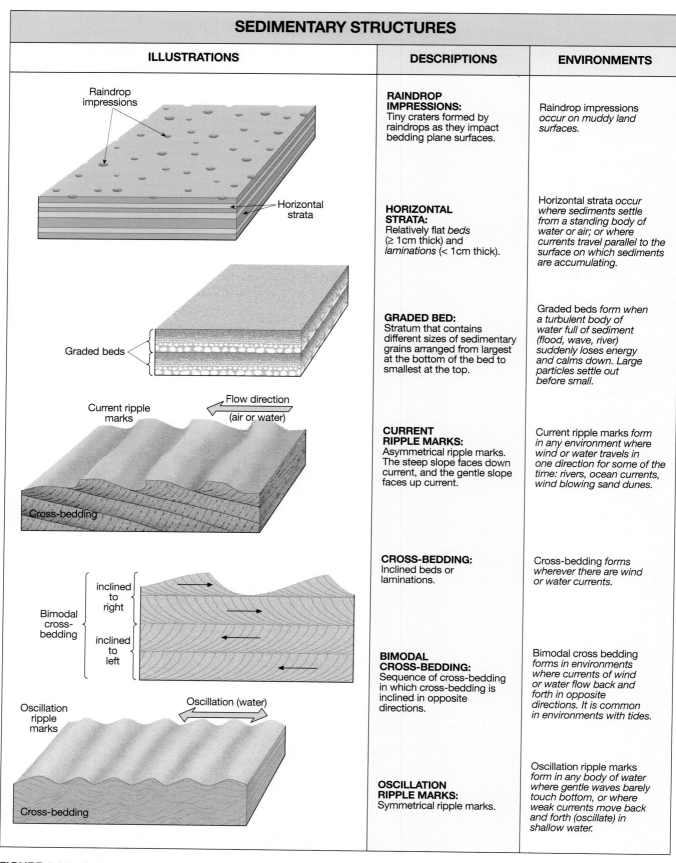

ILLUSTRATIONS	DESCRIPTIONS	ENVIRONMENTS
Raindrop impressions / Horizontal strata	**RAINDROP IMPRESSIONS:** Tiny craters formed by raindrops as they impact bedding plane surfaces.	Raindrop impressions *occur on muddy land surfaces.*
Graded beds	**HORIZONTAL STRATA:** Relatively flat *beds* (≥ 1cm thick) and *laminations* (< 1cm thick).	Horizontal strata *occur where sediments settle from a standing body of water or air; or where currents travel parallel to the surface on which sediments are accumulating.*
	GRADED BED: Stratum that contains different sizes of sedimentary grains arranged from largest at the bottom of the bed to smallest at the top.	Graded beds *form when a turbulent body of water full of sediment (flood, wave, river) suddenly loses energy and calms down. Large particles settle out before small.*
Current ripple marks / Flow direction (air or water) / Cross-bedding	**CURRENT RIPPLE MARKS:** Asymmetrical ripple marks. The steep slope faces down current, and the gentle slope faces up current.	Current ripple marks *form in any environment where wind or water travels in one direction for some of the time: rivers, ocean currents, wind blowing sand dunes.*
inclined to right / Bimodal cross-bedding / inclined to left	**CROSS-BEDDING:** Inclined beds or laminations.	Cross-bedding *forms wherever there are wind or water currents.*
	BIMODAL CROSS-BEDDING: Sequence of cross-bedding in which cross-bedding is inclined in opposite directions.	Bimodal cross bedding *forms in environments where currents of wind or water flow back and forth in opposite directions. It is common in environments with tides.*
Oscillation ripple marks / Oscillation (water) / Cross-bedding	**OSCILLATION RIPPLE MARKS:** Symmetrical ripple marks.	Oscillation ripple marks *form in any body of water where gentle waves barely touch bottom, or where weak currents move back and forth (oscillate) in shallow water.*

FIGURE 6.11 Sedimentary structures.

126

SEDIMENTARY STRUCTURES

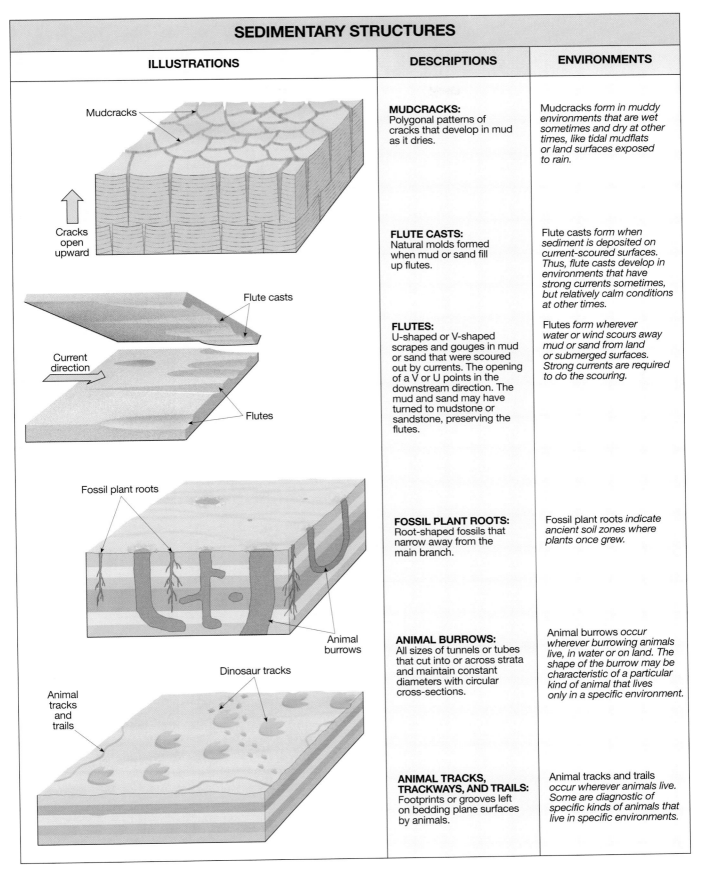

ILLUSTRATIONS	DESCRIPTIONS	ENVIRONMENTS
Mudcracks / Cracks open upward	**MUDCRACKS:** Polygonal patterns of cracks that develop in mud as it dries.	Mudcracks *form in muddy environments that are wet sometimes and dry at other times, like tidal mudflats or land surfaces exposed to rain.*
Flute casts / Current direction / Flutes	**FLUTE CASTS:** Natural molds formed when mud or sand fill up flutes.	Flute casts *form when sediment is deposited on current-scoured surfaces. Thus, flute casts develop in environments that have strong currents sometimes, but relatively calm conditions at other times.*
	FLUTES: U-shaped or V-shaped scrapes and gouges in mud or sand that were scoured out by currents. The opening of a V or U points in the downstream direction. The mud and sand may have turned to mudstone or sandstone, preserving the flutes.	Flutes *form wherever water or wind scours away mud or sand from land or submerged surfaces. Strong currents are required to do the scouring.*
Fossil plant roots / Animal burrows	**FOSSIL PLANT ROOTS:** Root-shaped fossils that narrow away from the main branch.	Fossil plant roots *indicate ancient soil zones where plants once grew.*
Animal tracks and trails / Dinosaur tracks	**ANIMAL BURROWS:** All sizes of tunnels or tubes that cut into or across strata and maintain constant diameters with circular cross-sections.	Animal burrows *occur wherever burrowing animals live, in water or on land. The shape of the burrow may be characteristic of a particular kind of animal that lives only in a specific environment.*
	ANIMAL TRACKS, TRACKWAYS, AND TRAILS: Footprints or grooves left on bedding plane surfaces by animals.	Animal tracks and trails *occur wherever animals live. Some are diagnostic of specific kinds of animals that live in specific environments.*

FIGURE 6.11 (CONTINUED) Sedimentary structures.

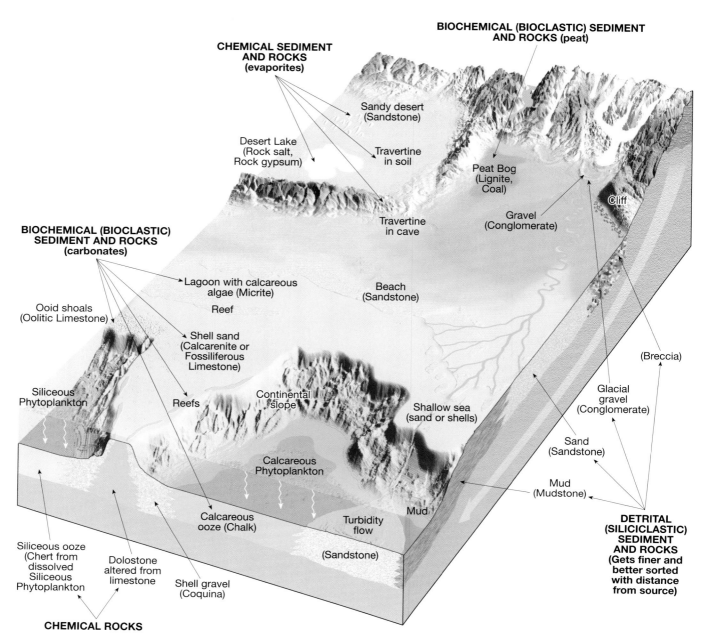

FIGURE 6.12 Some named modern environments where sediments and sedimentary rocks are forming now.

A succession of rock strata or units, one on top of the other, is called a *stratigraphic sequence.* If you interpret each rock unit of the stratigraphic sequence in order, from oldest (at the base) to youngest (at the top), then you will know what happened over a given portion of geologic history for the site where the stratigraphic sequence is located.

Question

10. A stratigraphic sequence of Permian rocks (approximately 270 million years old) from northeast

Kansas is pictured in Figure 6.16. Close-up pictures of hand samples from the rock units and field descriptions of the rock units are also provided. Use all of the information that is provided in the figure to fill in the paleoenvironment represented by each rock unit in the sequence. Then work from bottom to top, and shade in the narrow righthand columns to indicate the "record of change." When you are done, you will see how environments changed in Kansas over about 400,000 years of the Permian Period.

The rock exposed in this central Pennsylvania outcrop is calcarenite (fossiliferous limestone). The fossils in the rock are sand-sized fragments of seashells. The rock is named the Loyalhanna Member of the Mauch Chunk Formation. It is of Mississippian age (about 340 million years old).

a. Notice that the strata (sedimentary layers) are not horizontal. Draw arrows on the picture to show the flow direction(s) of water currents that formed these inclined strata. (Refer to Figure 6.11 as needed.)

b. What is the name of this kind of stratification? (Refer to Figure 6.11 and be as specific as you can.)

c. What do you think caused the water to flow as it did to make this kind of stratification?

d. Describe (as best as you can) what the environment was like here about 340 million years ago when these strata were formed.

FIGURE 6.13 See Question 7. Photograph of an outcrop to analyze and evaluate.

A. Modern dog tracks in mud with mudcracks on a tidal flat, St Catherines Island, Georgia

B. Triassic rock (about 215 m.y. old) from southeast Pennsylvania with the track of a three-toed *Coelophysis* dinosaur

C. Venn Diagram Comparison

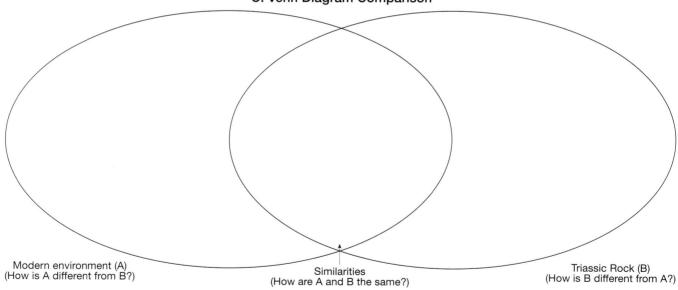

Modern environment (A)
(How is A different from B?)

Similarities
(How are A and B the same?)

Triassic Rock (B)
(How is B different from A?)

D. What was the Pennsylvania ecosystem (environment + organisms) like where *Coelophysis* walked in southeast Pennsylvania about 215 m.y. ago?

FIGURE 6.14   See Question 8. Photographs of rock samples to analyze and evaluate.

A. Pennsylvanian-age rock from Kansas (290 m.y. old)

Sand-sized fragments of fossil shells comprise the rock

10× close-up of thin section

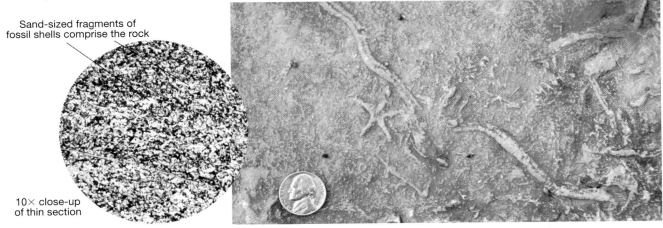

B. Modern sea-floor environment, 40 m (130 ft) deep, near Massachusetts (10 miles north of Cape Cod). Detrital (siliciclastic) sediment:
- 1% gravel
- 90% sand
- 9% mud

0   10 cm

C. Venn Diagram Comparison

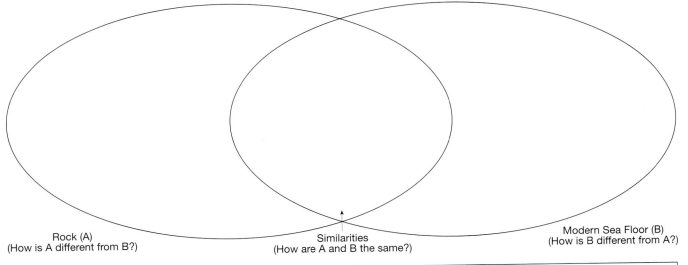

Rock (A)
(How is A different from B?)

Similarities
(How are A and B the same?)

Modern Sea Floor (B)
(How is B different from A?)

D. What was the Kansas ecosystem (environment + organisms) like 290 m.y. ago?

FIGURE 6.15 See Question 9. Photographs and photomicrograph to analyze and evaluate.
(Photograph B provided by U.S. Geological Survey (Open File Report OFR 00-427))

OUTCROP	HAND SAMPLE Bedding-plane surface	DESCRIPTION OF ROCK UNIT	DESCRIPTION OF PALEOENVIRONMENT REPRESENTED BY THE ROCK UNIT	RECORD OF CHANGE				
				ocean (marine)	muddy bay/estuary	evaporating bay	peat bog or swamp	land
		Tan skeletal limestone with shells of many kinds of marine organisms, bimodal cross-bedding, oscillation ripple marks, animal burrows, flutes, flute casts, and chert.						
	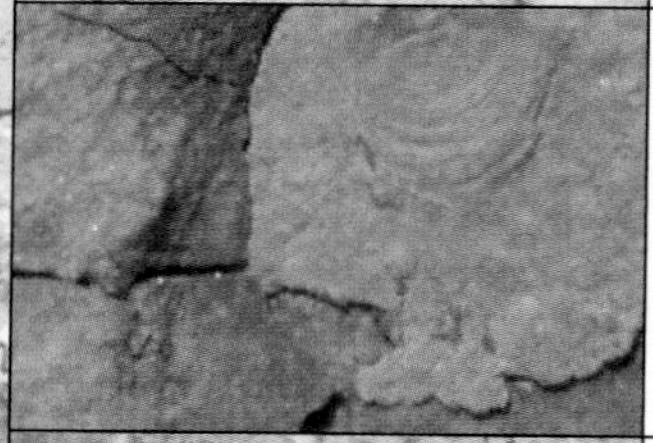	Gray silty mudstone (shale) with animal burrows, fossil clams, fossil plant fragments, and current ripple marks.						
		Red and gray silty mudstone with raindrop impressions, fossil roots, and mudcracks.						
		Gray silty mudstone with abundant gypsum layers and crystals.						
	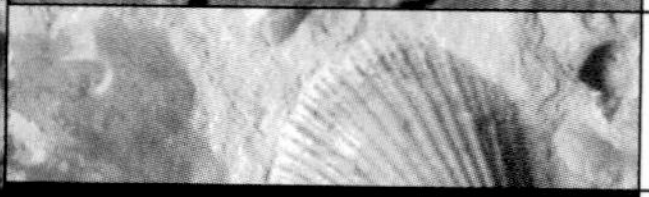	Tan skeletal limestone with bimodal cross-bedding.						
		Coal	peat bog or swamp				▓	
	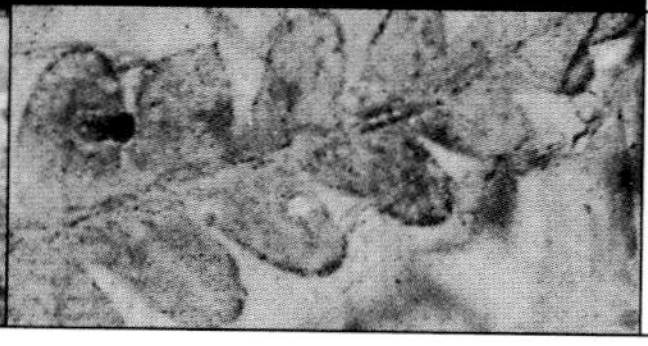	Gray silty mudstone with mudcracks and fossil ferns.	Probably moist muddy land where ferns grew; mudcracks formed in dry periods.					▓

1 METER

FIGURE 6.16 See Question 10. Permian stratigraphic sequence (approximately 270 million years old, exposed along Interstate Route 70 in eastern Kansas) to analyze and evaluate. Write a concise description of the paleoenvironment represented by each rock unit (pink column). Then shade in the narrow Record of Change columns to infer how the environments changed over the time that these sediments were deposited.

Metamorphic Rocks, Processes, and Resources

•CONTRIBUTING AUTHORS•

Harold E. Andrews • *Wellesley College*

James R. Besancon • *Wellesley College*

Margaret D. Thompson • *Wellesley College*

OBJECTIVES

A. Be able to describe and interpret textural and compositional features of metamorphic rocks.

B. Be able to determine the names, parent rocks (protoliths), and uses of common metamorphic rocks, based on their textures and mineralogical compositions.

C. Infer the relative grades of metamorphism that common metamorphic rocks have undergone.

MATERIALS

Pencil, eraser, laboratory notebook, hand magnifying lens (optional), metric ruler, mineral identification materials of your choice, and samples of metamorphic rocks (obtain as directed by your instructor).

INTRODUCTION

The word *metamorphic* is derived from Greek and means "of changed form." **Metamorphic rocks** are rocks changed from one form to another (metamorphosed) by intense heat, intense pressure, or the action of watery hot fluids (Figures 7.1, 7.2, 7.3). Think of metamorphism as it occurs in your home. *Heat* can be used to metamorphose bread into toast, *pressure* can be used to compact an aluminum can into a flatter and more compact form, and the chemical action of

watery hot fluids (boiling water, steam) can be used to change raw vegetables into cooked forms. Inside Earth, all of these metamorphic processes are more intense and capable of changing a rock from one form (size, shape, texture, color, and/or mineralogy) to another. Therefore, every metamorphic rock has a **parent rock** (or *protolith*), the rock type that was metamorphosed. Parent rocks can be any of the three main rock types: igneous rock, sedimentary rock, or even metamorphic rock (i.e., metamorphic rock can be metamorphosed again).

Figure 7.1 illustrates how a regional intrusion of magma (that cooled to form granite) has metamorphosed parent rocks to new metamorphic forms of rock. Mafic and ultramafic igneous rocks were metamorphosed to serpentinite. Sedimentary conglomerate, sandstone, and limestone parent rocks were metamorphosed to *metaconglomerate, quartzite,* and *marble.* Shale was metamorphosed to *slate, phyllite, schist,* and *gneiss,* depending on the grade (intensity) of metamorphism from low-grade (slate) to medium-grade (phyllite, schist), to high-grade (gneiss). *Hornfels* formed only in a narrow zone of contact metamorphism next to the intrusion of magma.

Different grades of metamorphism produce characteristic changes in the texture and mineralogy of the rock, which you will study below. Some common metamorphic rock-forming minerals include quartz, feldspars, muscovite, biotite, chlorite, garnet, tourmaline, calcite, dolomite, serpentine, talc, kyanite, sillimanite, and amphibole (hornblende). You should

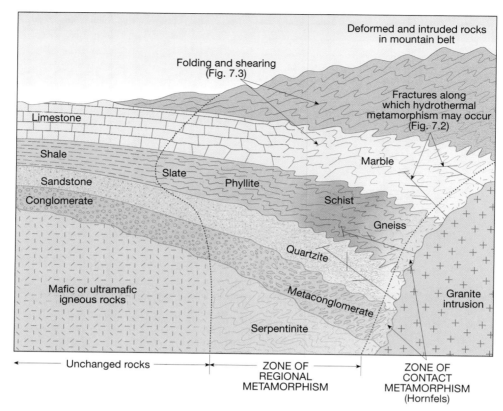

FIGURE 7.1 Metamorphism of a region by the heat, pressure, and chemical action of watery hot (hydrothermal) fluids associated with a large magma intrusion that cooled to form granite (granite intrusion). Some of the preexisting *parent rocks* are far removed from the intrusion and remain unchanged. Closer to the intrusion, the parent rocks were changed in form within a zone of regional metamorphism. Mafic igneous rocks were metamorphosed to serpentinite. Sedimentary conglomerate, sandstone, and limestone parent rocks were metamorphosed to *metaconglomerate*, *quartzite*, and *marble*. Shale was metamorphosed to *slate*, *phyllite*, *schist*, and *gneiss* depending on the grade (intensity) of metamorphism from low-grade (slate) to medium-grade (phyllite, schist), to high-grade (gneiss).

Notice two scales of metamorphism in the figure. **Contact metamorphism** occurred in narrow zones next to the contact between parent rock and intrusive magma and along fractures in the parent rock that were intruded by hydrothermal fluids. The zone of contact metamorphism next to the intrusive magma was changed to *hornfels* by intense heat and chemical reaction with the magma. Zones of contact metamorphism have widths on the order of millimeters to tens-of-meters. **Regional metamorphism** occurred over a larger region, throughout the mountain belt, and was accompanied by folding and shearing of rock layers.

familiarize yourself with all of these minerals by reviewing their distinctive properties in the Mineral Database (see Figure 3.28).

PART 7A: METAMORPHIC PROCESSES AND ROCKS

There are two main scales at which metamorphic processes occur: contact and regional (see Figure 7.1). **Contact metamorphism** occurs locally, adjacent to

igneous intrusions and along fractures that are in contact with watery hot (hydrothermal) fluids. The latter process is called *hydrothermal metamorphism*, and it involves condensation of gases to form liquids, which may precipitate mineral crystals along the fractures such as in Figure 7.2. Contact metamorphism is caused by conditions of low to moderate pressure, intense heating, and reaction with the metamorphosing magma or hydrothermal fluids over days to thousands of years. The intensity of contact metamorphism is greatest at the contact between parent rock

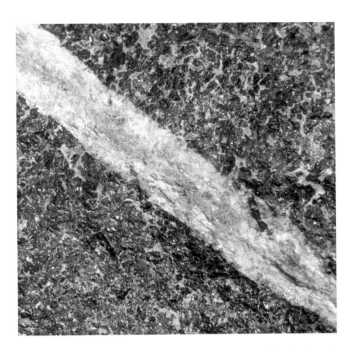

FIGURE 7.2 Hydrothermal mineral deposits. The dark part of this rock is chromite (chromium ore) that was precipitated from *hydrothermal fluids* (watery hot fluids). The light-colored minerals form a *vein* of zeolites (a group of light-colored hydrous aluminum silicates formed by low-grade metamorphism). The vein formed when directed pressure fractured the chromite deposit, hydrothermal fluids intruded the fracture, and the zeolites precipitated from the hydrothermal fluids as they cooled (making a *healed* fracture and a *vein* of zeolites).

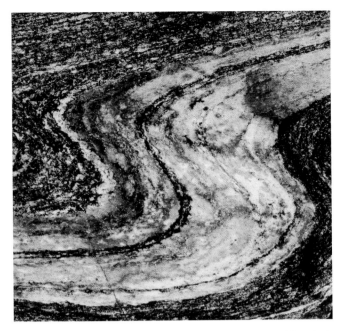

FIGURE 7.3 Folded and foliated (layered) gneiss. The dark minerals are muscovite, and the white minerals are quartz. Some of the quartz has been stained brown by iron. Regional metamorphism caused this normally rigid and brittle rock to be bent into *folds* without breaking. The flat mica mineral grains have been sheared (smeared) into layers called *foliations*. Metamorphic rocks with a layered appearance or texture are *foliated* metamorphic rocks. Figure 7.2 is a *nonfoliated* metamorphic rock because it lacks layering.

and intrusive magma or hydrothermal fluids. The intensity then decreases rapidly over a short distance from the magma or hydrothermal fluids. Thus, zones of contact metamorphism are narrow, on the order of millimeters to tens-of-meters thick.

Regional metamorphism occurs over very large areas (regions), such as deep within the cores of rising mountain ranges (see Figure 7.1), and generally is accompanied by folding of rock layers (see Figure 7.3). Regional metamorphism is caused by large igneous intrusions that form and cool over long periods (thousands to tens-of-millions of years), the moderate to extreme pressure and heat associated with deep burial or tectonic movements of rock, and/or the very widespread migration of hot fluids from one region to another along rock fractures and pore spaces.

The distinction between contact and regional metamorphism often is blurred. Contact metamorphism may be caused by small igneous intrusions, or by the local effects of hydrothermal fluids from some distance away that are traveling along fractures or

other voids. Regional metamorphism may be caused by large intrusions, tectonism, and/or the action of abundant and widespread hydrothermal fluids associated with large intrusions. One kind of metamorphism replaces another, so that rocks undergo both regional and contact metamorphism. Most major intrusions are preceded by contact metamorphism and followed by regional metamorphism.

The **mineralogical composition** of a metamorphic rock is a description of the kinds and *relative* abundances of mineral crystals that comprise the rock. Information about the relative abundances of the minerals is important for constructing a complete name for the rock and understanding metamorphic changes that formed the mineralogy of the rock. Mineralogical composition of a parent rock may change during metamorphism as a result of changing pressure, changing temperature, and/or the chemical action of hydrothermal fluids. Mineralogical composition may also stay the same, whereas the texture of the rock changes. **Recrystallization** is a process whereby small

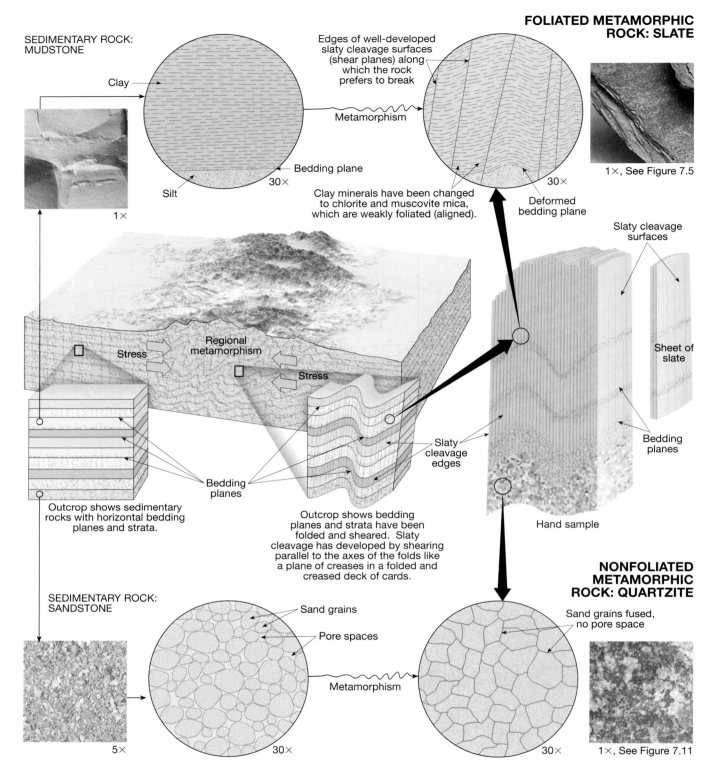

SEDIMENTARY ROCK: MUDSTONE

Clay

Silt

1×

Edges of well-developed slaty cleavage surfaces (shear planes) along which the rock prefers to break

Metamorphism

Bedding plane

30×

FOLIATED METAMORPHIC ROCK: SLATE

1×, See Figure 7.5

Clay minerals have been changed to chlorite and muscovite mica, which are weakly foliated (aligned).

30×

Deformed bedding plane

Slaty cleavage surfaces

Sheet of slate

Bedding planes

Stress

Regional metamorphism

Stress

Slaty cleavage edges

Bedding planes

Outcrop shows sedimentary rocks with horizontal bedding planes and strata.

Outcrop shows bedding planes and strata have been folded and sheared. Slaty cleavage has developed by shearing parallel to the axes of the folds like a plane of creases in a folded and creased deck of cards.

Hand sample

NONFOLIATED METAMORPHIC ROCK: QUARTZITE

SEDIMENTARY ROCK: SANDSTONE

Sand grains

Pore spaces

Metamorphism

Sand grains fused, no pore space

5×

30×

30×

1×, See Figure 7.11

FIGURE 7.4 Foliated and nonfoliated metamorphic rock formed by regional metamorphism. The mudstone and sandstone (sedimentary rocks) occur in layers separated by relatively flat, horizontal, bedding planes. Regional metamorphism compresses the sedimentary rock layers and bedding planes until they are folded (bent) and sheared across the layering into flat, parallel sheets of slate that slide past one another. The flat parallel surfaces between the layers of slate are called *slaty cleavage* surfaces (because they resemble cleavage in minerals).

Photomicrograph illustrations (in circles) show microscopic effects of the metamorphism. The layers of mudstone are metamorphosed to *slate* (Figure 7.5), in which the chlorite and muscovite mineral crystals are also *foliated* (aligned and layered subparallel to the shear planes). The sandstone (comprised of quartz sand grains and pore spaces) is metamorphosed to the harder, more dense, nonfoliated metamorphic rock *quartzite*, which is comprised of fused quartz sand grains. Notice that the shear planes are not obvious in the quartzite, because the sand grains roll and move about easily as the rock deforms.

Slaty cleavage surfaces

Clay minerals have been changed to chlorite and muscovite mica, which are weakly foliated (aligned).

Edges of well-developed slaty cleavage (shear planes) along which the rock prefers to break

A. Hand samples, 1×

B. Side view, 30×

FIGURE 7.5 Slate—a foliated metamorphic rock with dull luster, excellent slaty cleavage, and no visible grains. Slate forms from low-grade metamorphism of mudstone (shale, claystone). Clay minerals of the mudstone parent rock change to foliated chlorite and muscovite mineral crystals. Slate splits into hard, flat sheets (usually less than 1 cm thick) along its well-developed *slaty cleavage* (Figure 7.4). It is used to make roofing shingles and classroom blackboards.

crystals of one mineral will slowly convert to fewer, larger crystals of the same mineral, without melting of the rock. For example, microscopic calcite crystals in seashells that comprise a limestone can recrystallize to form a mass of visible calcite crystals in metamorphic marble.

Neomorphism is one way that mineralogical composition actually changes during metamorphism. In this process, minerals not only recrystallize but also form different minerals from the same chemical elements. For example, shales comprised mainly of clay minerals, quartz grains, and feldspar grains may change to a metamorphic rock comprised mainly of muscovite and garnet.

The most significant mineralogical changes occur during **metasomatism.** In this process, chemicals are added or lost. For example, anthracite coal is a relatively pure aggregate of carbon that forms when the volatile chemicals like nitrogen, oxygen, and methane are driven off from peat or bituminous coal by pressure and heating. Hornfels sometimes has a spotted appearance caused by the partial decomposition of just some of its minerals. In still other cases, one mineral may decompose (leaving only cavities or molds where its crystals formerly existed) and be simultaneously replaced by a new mineral of slightly or wholly different composition.

Textures of Metamorphic Rocks

Texture of a metamorphic rock is a description of its constituent parts and their sizes, shapes, and arrangements. Two main groups of metamorphic rocks are distinguished on the basis of their characteristic textures, *foliated* and *nonfoliated.*

Foliated metamorphic rocks (foliated textures) exhibit **foliations**—*layering* and parallel alignment of platy (flat) mineral crystals, such as micas. All metamorphic rocks with a layered appearance are foliated. This usually forms as a result of pressure (shearing and smearing of crystals) and recrystallization. Crystals of minerals such as tourmaline, hornblende, and kyanite can also be foliated because their crystalline growth occurred during metamorphism and had a preferred orientation in relation to the directed pressure. Specific kinds of foliated textures are described below:

• **Slaty rock cleavage**—*a very flat foliation* (resembling mineral cleavage) developed along flat, parallel, closely spaced shear planes (microscopic faults) in tightly folded clay- or mica-rich rocks (Figure 7.4). Rocks with excellent slaty cleavage are called *slate* (Figure 7.5), which is used to make roofing shingles and classroom blackboards. The flat surface of a blackboard or sheet of roofing slate is a slaty cleavage surface.

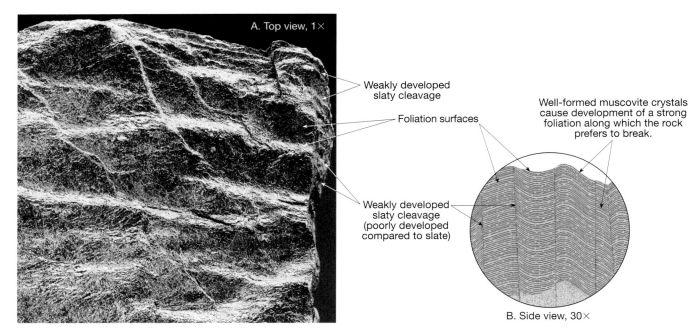

FIGURE 7.6 Phyllite—a foliated, fine-grained metamorphic rock, with a satiny, green, silver, or brassy metallic luster and a wavy foliation with a wrinkled appearance (*phyllite texture*). Phyllite forms from low-grade metamorphism of mudstone (shale, claystone), slate, or other rocks rich in clay, chlorite, or mica. When the very fine-grained mineral crystals of clay, chlorite, or muscovite in dull mudstone or slate are metamorphosed to form the phyllite, they become recrystallized to larger sizes and are aligned into a wavy and/or wrinkled foliation (*phyllite texture*) that is satiny or metallic. This is the wavy foliation along which phyllite breaks. Slaty cleavage may be poorly developed. It is not as obvious as the wavy and/or wrinkled foliation surfaces. The phyllite grade of metamorphism is between the low grade that produces slate (Figure 7.5) and the intermediate grade that produces schist (Figure 7.7).

- **Phyllite texture**—*a wavy and/or wrinkled foliation* of fine-grained *platy minerals* (mainly muscovite or chlorite crystals) that gives the rock a satiny or metallic luster. Rocks with phyllite texture are called *phyllite* (Figure 7.6). The phyllite texture is normally developed oblique or perpendicular to a weak slaty cleavage, and it is a product of intermediate-grade metamorphism.

- **Schistosity**—*a scaly glittery layering* of visible (medium- to coarse-grained) *platy minerals* (mainly micas and chlorite) *and/or linear alignment of long prismatic crystals* (tourmaline, hornblende, kyanite). Rocks with schistosity break along scaly, glittery foliations and are called *schist* (Figure 7.7). Schists are a product of intermediate-to-high grades of metamorphism.

- **Gneissic banding**—*alternating layers or lenses of light and dark medium- to coarse-grained minerals.* Rock with gneissic banding is called *gneiss* (Figures 7.3 and 7.8). Ferromagnesian minerals usually form the dark bands. Quartz or feldspars usually form the light bands. Most gneisses form by high-grade metamorphism (including recrystallization) of clay- or mica-rich rocks such as shale (see Figure 7.1), but they can also form by metamorphism of igneous rocks such as granite and diorite.

Nonfoliated metamorphic rocks have no obvious layering (i.e., no foliations), although they may exhibit stretched fossils or long, prismatic crystals (tourmaline, amphibole) that have grown parallel to the pressure field. Nonfoliated metamorphic rocks are mainly characterized by the following textures:

- **Crystalline texture (nonfoliated)**—a medium- to coarse-grained aggregate of intergrown, usually equal-sized (equigranular), visible crystals. *Marble* is a nonfoliated metamorphic rock that typically exhibits an obvious crystalline texture (Figure 7.9).

Visible well-formed
muscovite crystals

Edges of weak slaty
cleavage surfaces

Top view of foliation surfaces, 1×

FIGURE 7.7 Schist—a medium- to coarse-grained, scaly (like fish scales), foliated metamorphic rock formed by intermediate-grade metamorphism of mudstone, shale, slate, phyllite, or other rocks rich in clay, chlorite, or mica. Schist forms when clay, chlorite, and mica mineral crystals are foliated as they recrystallize to larger, more visible crystals of chlorite, muscovite, or biotite. This gives schist its scaly foliated appearance called *schistosity*. Slaty cleavage or *crenulations* (sets of tiny folds) may be present, but schist breaks along its scaly, glittery schistosity. It often contains porphyroblasts of garnet, kyanite, sillimanite, or tourmaline mineral crystals. The schist grade of metamorphism is intermediate between the lower grade that produces phyllite (see Figure 7.6) and the higher grade that produces gneiss (see Figure 7.8). Also see chlorite schist in Figure 7.14.

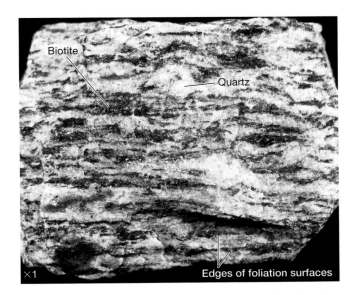

Biotite

Quartz

×1   Edges of foliation surfaces

FIGURE 7.8 Gneiss—a medium- to coarse-grained metamorphic rock with *gneissic banding* (alternating layers or lenses of light and dark minerals). Generally, light-colored layers are rich in quartz or feldspars and alternate with dark layers rich in biotite mica, hornblende, or tourmaline. Most gneisses form by high-grade metamorphism (including recrystallization) of clay or mica-rich rocks such as shale (see Figure 7.1), mudstone, slate, phyllite, or schist. However, they can also form by metamorphism of igneous rocks such as granite and diorite. The compositional name of the rock in this picture is biotite quartz gneiss.

FIGURE 7.9 Marble—a fine- to coarse-grained, nonfoliated metamorphic rock with a crystalline texture formed by tightly interlocking grains of calcite or dolomite. Marble forms by intermediate- to high-grade metamorphism of limestone or dolostone. Marble is a dense aggregate of nearly equal-sized crystals (see photograph), in contrast to the porous and random-sized crystal aggregate of its parent rock, limestone (see Figure 6.5).

Calcite crystals

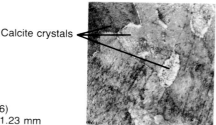

Photomicrograph (× 26.6)
Original sample width is 1.23 mm

Enlarged 5×

- **Microcrystalline texture**—a fine-grained aggregate of intergrown microscopic crystals (as in a sugar cube). *Hornfels* (Figure 7.10) is a nonfoliated metamorphic rock that has a microcrystalline texture.

- **Sandy texture**—a medium- to coarse-grained aggregate of fused, sand-sized grains that resembles sandstone. *Quartzite* is a nonfoliated metamorphic rock with a sandy texture (Figure 7.11) remaining from its sandstone parent rock.

- **Glassy texture**—a homogeneous texture with no visible grains or other structures and breaks along glossy surfaces; said of materials that resemble glass, such as *anthracite coal* (Figure 7.12).

Besides the main features that distinguish foliated and nonfoliated metamorphic rocks, there are some features that can occur in any metamorphic rock. They include:

- **Stretched or sheared grains**—deformed pebbles, fossils, or mineral crystals that have been stretched out (Figure 7.13), shortened, or sheared.

- **Porphyroblastic texture**—an arrangement of large crystals, called *porphyroblasts*, set in a finer-grained groundmass (Figure 7.14). It is analogous to porphyritic texture in igneous rocks.

- **Hydrothermal veins**—fractures "healed" (filled) by minerals that precipitated from hydrothermal fluids (see Figure 7.2).

FIGURE 7.10 Hornfels—a fine-grained, nonfoliated metamorphic rock having a dull luster and a microcrystalline texture (that may appear smooth or sugary). It is usually very hard and dark in color, but it sometimes has a spotted appearance caused by patchy chemical reactions with the metamorphosing magma or hydrothermal fluid. Hornfels forms by contact metamorphism of any rock type.

FIGURE 7.11 Quartzite—a medium- to coarse-grained, nonfoliated metamorphic rock comprised chiefly of fused quartz grains that give the rock its *sandy texture*. Compare the fused quartz grains of this quartzite sample (see photomicrograph) with the porous sedimentary fabric of quartz sandstones in Figure 6.4.

×1

Quartz sand grains

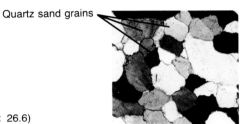

Photomicrograph (× 26.6)
Original sample width is 1.23 mm

- **Folds**—bends in rock layers that were initially flat, like a folded stack of paper (see Figure 7.3).

- **Lineations**—lines on rocks at the edges of foliations, shear planes, slaty cleavage, folds, or aligned crystals.

Classification of Metamorphic Rocks

Metamorphic rocks are mainly classified according to their texture and mineralogical composition. This information is valuable for naming the rock and determining how it formed from a parent rock (protolith). It is also useful for inferring how the metamorphic rock could be used as a commodity for domestic or industrial purposes. You can analyze and classify metamorphic rocks with the aid of Figure 7.15, which also provides information about parent rocks and how the metamorphic rocks are commonly used. Record all of your work on a Metamorphic Rocks Worksheet (Figure 7.19).

Follow these steps to analyze and classify a metamorphic rock:

Step 1: *Determine and record the rock's textural features.* Determine and record if the rock is foliated or nonfoliated (see Figure 7.4), and what other specific kinds of textural features are present. Use this information to work from left to right across the three columns of Step 1 in Figure 7.15, and match the rock texture to one of the specific categories there.

×1

FIGURE 7.12 Anthracite coal—a fine-grained, nonfoliated metamorphic rock, also known as *hard coal* (because it cannot easily be broken apart like its parent rock, bituminous or soft coal). Anthracite has a smooth, homogeneous, glassy texture and breaks along glossy, curved (conchoidal) fractures. It is formed by low- to intermediate-grade metamorphism of bituminous coal, lignite, or peat.

Sedimentary Rock

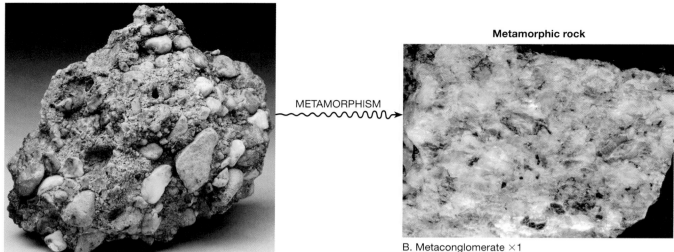

A. Conglomerate parent rock (protolith) ×1

METAMORPHISM

Metamorphic rock

B. Metaconglomerate ×1

FIGURE 7.13 Metaconglomerate—metamorphosed conglomerate. **A.** Conglomerate is a detrital (clastic) sedimentary rock comprised chiefly of rounded, odd-sized grains of gravel (grains larger than 2 mm, coarser than sand). Conglomerate breaks around the grains (pebbles, granules, sand) that stand out in this image. **B.** Metaconglomerate forms when conglomerate is heated (and softened) under directed pressure or tension. The stout, rounded grains of conglomerate are compressed and fused together to form a denser mass of metaconglomerate. Metaconglomerate breaks through the fused grains rather than around them.

Step 2: *Determine and record the rock's mineralogical composition and/or other distinctive features.* List the minerals in order of increasing abundance, and distinguish between porphyroblasts and mineralogy of the groundmass making up most of the rock. Use this information and any other distinctive features to match the rock to one of the categories in Step 2 of Figure 7.15.

Step 3: Recall how you categorized the rock in Steps 1 and 2. Use this information to work from left to right across Figure 7.15 and determine the name of the rock. You can also modify the rock name by adding the names of minerals present in the rock in order of their increasing abundance. If the rock is porphyroblastic, then you can add this to the name as well (e.g., Figure 7.14).

Step 4: After you have determined the metamorphic rock name in Step 3, look to the right along the same row of Figure 7.15 and find the name of a parent rock (protolith) for that kind of metamophic rock.

Step 5: After you have determined the parent rock in Step 4, look to the right along the same row of Figure 7.15 and find out what the rock is commonly used for.

Chlorite

Pyrite porphyroblast (brassy cube)

×1

FIGURE 7.14 Porphyroblastic texture—large, visible crystals of one mineral occur in a fine-grained groundmass of one or more other minerals. This medium-grained chlorite schist contains porphyroblasts of pyrite (brassy metallic cubes) in a groundmass of chlorite. The rock can be called porphyroblastic chlorite schist or pyrite chlorite schist.

METAMORPHIC ROCK ANALYSIS AND CLASSIFICATION

	STEP 1: What are the rock's textural features?		STEP 2: What are the rock's mineralogical composition and/or other distinctive features?	STEP 3: Metamorphic rock name		STEP 4: What was the parent rock?	STEP 5: What is the rock used for?
FOLIATED	Fine grained or no visible grains	Flat slaty cleavage is well developed	Dull luster; breaks into hard flat sheets along the slaty cleavage	SLATE[1]		Mudstone or shale	Roofing slate, table tops, floor tile, and blackboards
		Phyllite texture well developed more than slaty cleavage	Breaks along wrinkled or wavy foliation surfaces with shiny metallic luster	PHYLLITE[1]	INCREASING METAMORPHIC GRADE	Mudstone, shale, or slate	Construction stone, decorative stone, sources of gemstones
	Medium to coarse grained	Schistosity: foliation formed by alignment of visible crystals; rock breaks along scaly foliation surfaces; crystalline texture	Visible sparkling crystals of platy minerals (chlorite, biotite, muscovite), bladed crystals (kyanite), or prismatic crystals (amphiboles, tourmaline, sillimanite); breaks along scaly foliated surfaces	SCHIST[1] Chlorite schist Muscovite schist Biotite schist Kyanite schist Amphibole schist Tourmaline schist Sillimanite schist		Mudstone, shale, slate, or phyllite	
		Gneissic banding: minerals segregated into alternating layers gives the rock a banded texture in side view; crystalline texture	Visible crystals of two or more minerals in alternating light and dark foliated layers	GNEISS[1]		Mudstone, shale, slate, phyllite, schist, granite, or diorite	Construction stone, decorative stone, sources of gemstones
FOLIATED OR NONFOLIATED	Medium to coarse grained		Mostly visible crystals of amphibole (usually glossy black hornblende)	AMPHIBOLITE		Basalt, Gabbro, or Ultramafic igneous rocks	Construction stone
NONFOLIATED	Fine grained or no visible grains	Glassy texture; slaty cleavage may barely be visible	Black glossy rock that breaks along uneven or conchoidal fractures (Figure 7.12)	ANTHRACITE COAL		Peat, Lignite, Bituminous coal	Highest grade coal for clean burning fossil fuel
		Microcrystalline texture	Usually a dull dark color; very hard	HORNFELS		Any rock type	
		Microcrystalline texture or no visible grains. May have fibrous asbestos form	Serpentine; dull or glossy; color usually shades of green	SERPENTINITE		Basalt, Gabbro, or Ultramafic igneous rocks	Decorative stone
		Microcrystalline or no visible grains	Talc; can be scratched with your fingernail; shades of green, gray, brown, white	SOAPSTONE		Basalt, Gabbro, or Ultramafic igneous rocks	Art carvings, electrical insulators, talcum powder
	Fine to coarse grained	Sandy texture	Quartz sand grains fused together; grains will not rub off like sandstone, usually light colored	QUARTZITE[1]		Sandstone	Construction stone, decorative stone
		Microcrystalline (resembling a sugar cube) or medium to coarse crystalline texture	Calcite (or dolomite) crystals of nearly equal size and tightly fused together; calcite effervesces in dilute HCl; dolomite effervesces only if powdered	MARBLE[1]		Limestone	Art carvings, construction stone, decorative stone, source of lime for agriculture
		Conglomeratic texture, but breaks across grains	Pebbles may be stretched or cut by rock cleavage	META-CONGLOMERATE		Conglomerate	Construction stone, decorative stone

[1]Modify rock name by adding names of minerals in order of increasing abundance. For example, garnet muscovite schist is a muscovite schist with a small amount of garnet.

FIGURE 7.15 Five-step chart for metamorphic rock analysis and classification. See text for description of steps.

FIGURE 7.16 Serpentinite—a nonfoliated metamorphic rock comprised chiefly of serpentine, which gives the rock its green color. Serpentinite forms by the low- to intermediate-grade metamorphism of mafic or ultramafic igneous rocks (basalt, gabbro, peridotite). Serpentinite often contains minor amounts of talc, magnetite, and chlorite. This sample contains some fibrous serpentine called *asbestos*.

FIGURE 7.17 A metamorphic rock sample to analyze in Question 3.

Example:

Study the metamorphic rock in Figure 7.16. Step 1—Notice that the rock has no obvious layering, so it is nonfoliated. It also has a microcrystalline form with some asbestos. Step 2—Notice that the rock is comprised of the mineral serpentine and is many shades of green color. Step 3—By moving from left to right across the chart in Figure 7.15, you should be able to find that the rock name is serpentinite. Step 4—Notice that the parent rock for serpentinite was basalt, gabbro, or an ultramafic igneous rock. Step 5—Notice that serpentinite is commonly used as a green decorative stone.

Questions

1. Analyze the rock sample in Figure 7.3. The parent rock for this metamorphic rock had flat layers that were folded during metamorphism. Describe a process that could account for how this rigid gneiss was folded without breaking during regional metamorphism. (*Hint*: How could you bend a brittle candlestick without breaking it?)

2. Common metamorphic rock-forming minerals include quartz, plagioclase feldspar, potassium feldspar, muscovite, biotite, chlorite, garnet, tourmaline, calcite, dolomite, serpentine, talc, kyanite, sillimanite, and amphibole (hornblende). You must be able to identify these minerals in hand samples of metamorphic rocks. Make a list of these minerals in your laboratory notebook and use the Mineral Database (see Figure 3.21) to list the distinctive properties of each. Analyze samples of each mineral (if available) as you work, so you develop an ability to recognize each mineral based on its distinctive properties. (If you have trouble distinguishing one mineral from another, then you should use your list of mineral names and distinctive properties to compose a flowchart for the identification of these minerals.)

3. Analyze the metamorphic rock sample pictured in Figure 7.17 using the analysis and classification chart in Figure 7.15.
   a. What are the rock's textural features?
   b. What are the rock's mineralogical composition and/or other distinctive features?
   c. What is the metamorphic rock name?
   d. What was the parent rock?
   e. What is the rock used for?

4. Analyze the metamorphic rock sample pictured in Figure 7.18 using the analysis and classification chart in Figure 7.15.

a. What are the rock's textural features?

b. What are the rock's mineralogical composition and/or other distinctive features?

c. What is the metamorphic rock name?

d. What was the parent rock?

e. What is the rock used for?

FIGURE 7.18 A metamorphic rock sample to analyze in Question 4.

FIGURE 7.19 Metamorphic Rocks Worksheet.

METAMORPHIC ROCKS WORKSHEET

Sample Letter or Number	Texture(s) (Figures 7.4, 7.15—Step 1)	Mineral Composition and Other Distinctive Properties (Figure 7.15, Step 2)	Rock Name (Figure 7.15, Step 3)	Parent Rock (Figure 7.15, Step 4)	Uses (Figure 7.15, Step 5)
	☐ foliated ☐ nonfoliated				
	☐ foliated ☐ nonfoliated				
	☐ foliated ☐ nonfoliated				
	☐ foliated ☐ nonfoliated				
	☐ foliated ☐ nonfoliated				

FIGURE 7.19 (CONTINUED) Metamorphic Rocks Worksheet.

METAMORPHIC ROCKS WORKSHEET

Sample Letter or Number	Texture(s) (Figures 7.4, 7.15—Step 1)	Mineral Composition and Other Distinctive Properties (Figure 7.15, Step 2)	Rock Name (Figure 7.15, Step 3)	Parent Rock (Figure 7.15, Step 4)	Uses (Figure 7.15, Step 5)
	☐ foliated ☐ nonfoliated				
	☐ foliated ☐ nonfoliated				
	☐ foliated ☐ nonfoliated				
	☐ foliated ☐ nonfoliated				
	☐ foliated ☐ nonfoliated				

FIGURE 7.19 (CONTINUED) Metamorphic Rocks Worksheet.

METAMORPHIC ROCKS WORKSHEET

Sample Letter or Number	Texture(s) (Figures 7.4, 7.15—Step 1)	Mineral Composition and Other Distinctive Properties (Figure 7.15, Step 2)	Rock Name (Figure 7.15, Step 3)	Parent Rock (Figure 7.15, Step 4)	Uses (Figure 7.15, Step 5)
	☐ foliated ☐ nonfoliated				
	☐ foliated ☐ nonfoliated				
	☐ foliated ☐ nonfoliated				
	☐ foliated ☐ nonfoliated				
	☐ foliated ☐ nonfoliated				

FIGURE 7.19 **(CONTINUED)** Metamorphic Rocks Worksheet.

METAMORPHIC ROCKS WORKSHEET

Sample Letter or Number	Texture(s) (Figures 7.4, 7.15—Step 1)	Mineral Composition and Other Distinctive Properties (Figure 7.15, Step 2)	Rock Name (Figure 7.15, Step 3)	Parent Rock (Figure 7.15, Step 4)	Uses (Figure 7.15, Step 5)
	☐ foliated ☐ nonfoliated				
	☐ foliated ☐ nonfoliated				
	☐ foliated ☐ nonfoliated				
	☐ foliated ☐ nonfoliated				
	☐ foliated ☐ nonfoliated				

Dating of Rocks, Fossils, and Geologic Events

•CONTRIBUTING AUTHORS•

Jonathan Bushee • *Northern Kentucky University*

John K. Osmond • *Florida State University*

Raman J. Singh • *Northern Kentucky University*

OBJECTIVES

A. Learn and be able to apply techniques for relative age dating of Earth materials and events.

B. Use fossils to date some rock bodies and infer some of Earth's history.

C. Learn and be able to apply techniques for absolute age dating of Earth materials and events.

D. Be able to apply relative and absolute dating techniques to analyze two field sites and infer their geologic history.

E. Be able to apply relative dating techniques to analyze logs of five wells and correlate among them.

MATERIALS

Pencil, eraser, laboratory notebook, calculator, and colored pencils (optional) plus a ruler and protractor cut from GeoTools Sheets at the back of the manual.

INTRODUCTION

If you could dig a hole deep into Earth's crust, then you would encounter the **geologic record,** layers of rock stacked one atop the other like pages in a book. As each new layer of sediment or rock forms today, it covers the older layers of the geologic record beneath

it and becomes the youngest layer of the geologic record. Thus, rock layers form a *sequence* from oldest at the bottom to youngest at the top. They also have different colors, textures, chemical compositions, and **fossils** (any evidence of ancient life) depending on the environmental conditions under which they were formed. Geologists have studied sequences of rock layers wherever they are exposed in mines, quarries, river beds, road cuts, wells, and mountain sides throughout the world. They have also *correlated* the layers (traced them from one place to another) across regions and continents. Thus, the geologic record of rock layers is essentially a stack of stone pages in a giant natural book of Earth history. And like the pages in any old book, the rock layers have been folded, fractured (cracked), torn (faulted), and even removed by geologic events.

Geologists tell time based on relative and absolute dating techniques. **Relative age dating** is the process of determining when something formed or happened in relation to other things. For example, if you have a younger brother and an older sister, then you could describe your relative age by saying that you are younger than your sister and older than your brother. **Absolute age dating** is the process of determining when something formed or happened in exact units of time such as days, months, or years. Using the example above, you could describe your absolute age just by saying how old you are in years.

Geologists "read" and infer Earth history from rocky outcrops and geologic cross sections by observing rock layers, recognizing geologic structures, and evaluating age relationships among the layers and structures. The so-called *geologic time scale* (Figure 1.3, p. 4) is a chart of named intervals of the geologic record and their ages in both relative and absolute time. It has taken thousands of geoscientists, from all parts of the world, more than a century to construct the present form of the geologic time scale.

Just as authors organize books according to sections, chapters, and pages, geologists have subdivided the rock layers of the geologic record into named eonothems (the largest units), erathems, systems, series, stages, and zones of rock on the basis of fossils, minerals, and other historical features they contained. These physical divisions of rock also represent specific intervals of geologic time. An *eonothem* of rock represents an eon of time, an *erathem* of rock represents an era of time, a *system* of rock represents a period of time, and so on in the table below.

ROCK UNITS (Division of the Geologic Record)	CORRESPONDING GEOLOGIC TIME UNITS
Eonothem (largest)	Eon of time (longest unit)
Erathem	Era of time
System	Period of time
Series	Epoch of time
Stage	Age of time
Zone	Chron of time

PART 8A: DETERMINING RELATIVE AGES OF ROCKS BASED ON THEIR PHYSICAL RELATIONSHIPS

A geologist's initial challenge in the field is to subdivide the local sequence of sediments and bodies of rock into mappable units that can be correlated from one site to the next. Subdivision is based on color, texture, rock type, or other physical features of the rocks, and the mappable units are called **formations.** Formations can be subdivided into *members,* or even individual strata. Surfaces between any of these kinds of units are *contacts.*

Geologists use six basic laws for determining relative age relationships among bodies of rock based on their physical relationships. They are:

- **Law of Original Horizontality**—*Sedimentary layers (strata) and lava flows were originally deposited as relatively horizontal sheets, like a layer cake. If they are no longer horizontal or flat, it is because they*

have been displaced by subsequent movements of the Earth's crust.

- **Law of Lateral Continuity**—*Lava flows and strata extend laterally in all directions until they thin to nothing (pinch out) or reach the edge of their basin of deposition.*

- **Law of Superposition**—*In an undisturbed sequence of strata or lava flows, the oldest layer is at the bottom of the sequence and the youngest is at the top.*

- **Law of Inclusions**—*Any piece of rock (clast) that has become included in another rock or body of sediment must be older than the rock or sediment into which it has been incorporated.* Such a clast (usually a rock fragment, crystal, or fossil) is called an **inclusion.** The surrounding body of rock is called the **matrix** (or groundmass). Thus, an inclusion is older than its surrounding matrix.

- **Law of Cross-Cutting**—*Any feature that cuts across a rock or body of sediment must be younger than the rock or sediment that it cuts across.* Such crosscutting features include fractures (cracks in rock), faults (fractures along which movement has occurred), or masses of magma (*igneous intrusions*) that cut across preexisting rocks before they cooled. When a body of magma intrudes preexisting rocks, a narrow *zone of contact metamorphism* usually forms in the preexisting rocks adjacent to the intrusion.

- **Law of Unconformities**—*Surfaces called unconformities represent gaps in the geologic record that formed wherever layers were not deposited for a time or else layers were removed by erosion.* Most contacts between adjacent strata or formations are conformities, meaning that rocks on both sides of them formed at about the same time. An unconformity is a rock surface that represents a gap in the geologic record. It is like the place where pages are missing from a book. An unconformity can be a buried surface where there was a pause in sedimentation, a time between two lava flows, or a surface that was eroded before more sediment was deposited on top of it.

There are three kinds of unconformities (Figure 8.1). A **disconformity** is an unconformity between *parallel* strata or lava flows. Most disconformities are very irregular surfaces, and pieces of the underlying rock are often included in the strata above them. An **angular unconformity** is an unconformity between two sets of strata that are not parallel to one another. It forms when new horizontal layers cover up older layers folded by mountain-building processes and eroded down to a nearly level surface. A **nonconformity** is an

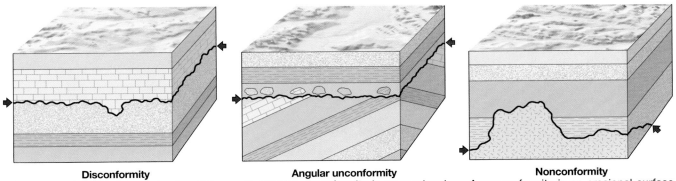

Disconformity

In a succession of rock layers (sedimentary strata or lava flows) parallel to one another, the disconformity surface is a gap in the layering. The gap may be a non-depositional surface where some layers never formed for a while, or the gap may be an erosional surface where some layers were removed before younger layers covered up the surface.

Angular unconformity

An angular unconformity is an erosional surface between two bodies of layered sedimentary strata or lava flows that are not parallel. The gap is because the older body of layered rock was tilted and partly eroded (rock was removed) before a younger body of horizontal rock layers covered the eroded surface.

Nonconformity

A nonconformity is an erosional surface between older igneous and/or metamorphic rocks and younger rock layers (sedimentary strata or lava flows). The gap is because some of the older igneous and/or metamorphic rocks were partly eroded (rock was removed) before the younger rock layers covered the eroded surface.

FIGURE 8.1 Three kinds of unconformities—surfaces that represent gaps (missing layers) in the geologic record; analogous to a gap (place where pages are missing) in a book. Red arrows point to the unconformity surface (bold black line) in each block diagram.

unconformity between younger sedimentary rocks and subjacent metamorphic or igneous rocks. It forms when stratified sedimentary rocks or lava flows are deposited on eroded igneous or metamorphic rocks.

Analyze and evaluate Figures 8.2–8.8 to learn how the above laws of relative age dating are applied in cross sections of Earth's crust. These are the kinds of two-dimensional cross sections of Earth's crust that are exposed in road cuts, quarry walls, and mountain sides. *Ignore the symbols for fossils until Part 8B. Be sure that you consider all of these examples before proceeding*.

Questions

1. Refer to the geologic cross sections in Figures 8.9 and 8.10. The colors and symbols for rock types, contacts, faults, unconformities, and zones of contact metamorphism are the same as the symbols used in Figure 8.4. For each figure, determine the relative ages of rock units and other features labeled with letters. Indicate the sequence in which the labeled features developed by writing the letters from oldest (first) to youngest (latest) in the blanks provided. Refer back to Figures 8.1–8.8 and the laws of relative age dating based on physical relationships, as needed.

2. Refer to Figure 8.10, which is a cross section of the inner gorge of Grand Canyon, Arizona. Notice the names and relative ages of the formations

(named rock units). Based on your determination of the relative age relationships of these formations (which you listed from oldest to youngest, by letter, in Question 1) and associated other features (unconformities, zones of contact metamorphism), write three paragraphs about the Grand Canyon.

a. In the first paragraph, use names and rock types of the formations to describe events that occurred in the region during Precambrian time.

b. In the second paragraph, use names and rock types of the formations to describe events that occurred in the region during the Cambrian Period of time.

c. In the third paragraph, describe what geologic events are occurring at the present time and infer what and where different kinds of unconformities (disconformity, angular unconformity, nonconformity) are forming.

3. Refer to the geologic cross sections in Figures 8.11 and 8.12. For each figure, determine the relative ages of rock units and other features labeled with letters. Indicate the sequence in which the labeled features developed by writing the letters from oldest (first) to youngest (latest) in the blanks provided. Refer back to Figures 8.1–8.8 and the laws of relative age dating based on physical relationships, as needed. *More than one solution is possible for both of these figures so be prepared to justify your reasoning if asked to do so.*

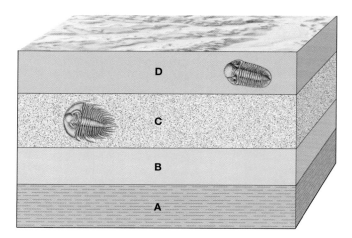

FIGURE 8.2 This is a sequence of strata that has maintained its original horizontality and does not seem to be disturbed. Therefore, Formation **A** is the oldest, because it is on the bottom of a sedimentary sequence of rocks. **D** is the youngest, because it is at the top of the sedimentary sequence. The sequence of events was deposition of **A**, **B**, **C**, and **D**, in that order and stacked one atop the other.

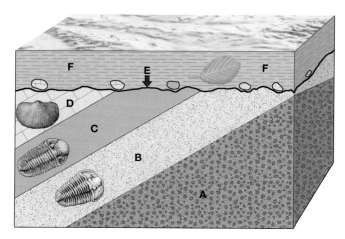

FIGURE 8.3 This is another sequence of strata, some of which do not have their original horizontality. Formation **A** is the oldest, because it is at the bottom of the sedimentary sequence. Formation **F** is youngest, because it forms the top of that sequence. Tilting and erosion of the sequence occurred after **D** but before deposition of Formation **F**. **E** is an angular unconformity.

The sequence of events began with deposition of **A**, **B**, **C**, and **D** in that order and stacked one atop the other. The sequence was then tilted, and its top was eroded. Siltstone **F** was deposited horizontally on top of the erosional surface, which is now an angular unconformity.

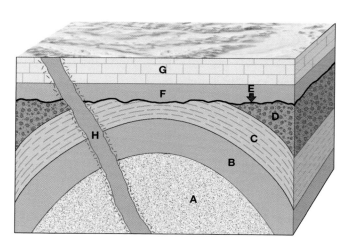

FIGURE 8.4 The body of igneous rock **H** is the youngest rock unit, because it cuts across all of the others. (When a narrow body of igneous rock cuts across strata in this way, it is called a **dike**.) **A** is the oldest formation, because it is at the bottom of the sedimentary rock sequence that is cut by **H**. Folding and erosion occurred after **D** was deposited, but before **F** was deposited. **E** is an angular unconformity.

The sequence of events began with deposition of formations **A** through **D** in alphabetical order and one atop the other. The sequence was folded, and the top of the fold was eroded. Formation **F** was deposited horizontally atop the folded sequence and the erosional surface, which became angular unconformity **E**. **G** was deposited atop **F**. Lastly, a magma intruded across all of the strata and cooled to form basalt dike **H**.

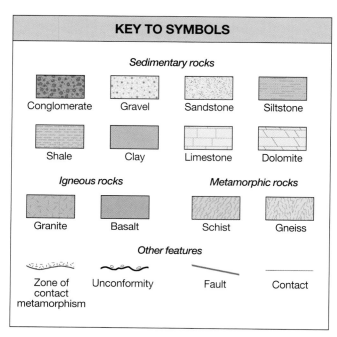

KEY TO SYMBOLS

Sedimentary rocks

Conglomerate Gravel Sandstone Siltstone

Shale Clay Limestone Dolomite

Igneous rocks   *Metamorphic rocks*

Granite Basalt Schist Gneiss

Other features

Zone of contact metamorphism Unconformity Fault Contact

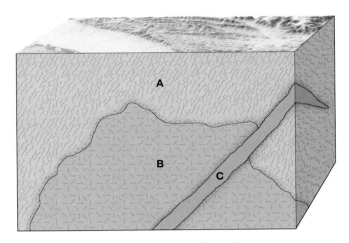

FIGURE 8.5 The body of granite **B** must have formed from the cooling of a body of magma that intruded the pre-existing rock **A**, called **country rock.** The country rock is schist **A** containing a zone of contact metamorphism adjacent to the granite. Therefore, the sequence of events began with a body of country rock **A**. The country rock was intruded by a body of magma, which caused development of a zone of contact metamorphism and cooled to form granite **B**. Lastly, another body of magma intruded across both **A** and **B**. It caused development of a second zone of contact metamorphism and cooled to form basalt dike **C**.

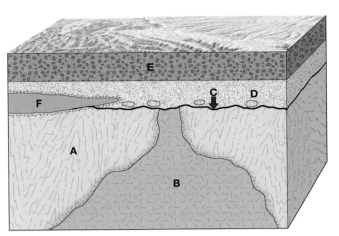

FIGURE 8.6 At the base of this rock sequence there is gneiss **A**, which is separated from granite **B** by a zone of contact metamorphism. This suggests that a body of magma intruded **A**, then cooled to form the contact zone and granite **B**. There must have been erosion of both **A** and **B** *after* this intrusion (to form surface **C**), because there is no contact metamorphism between **B** and **D**. Formation **D** was deposited horizontally atop the eroded igneous and metamorphic rocks, forming nonconformity **C**. Sometime after **D** was deposited (before or after deposition of **E**), a second body of magma **F** intruded across **A**, **C**, and **D**. Such an intrusive igneous body that is intruded along (parallel to) the strata is called a **sill** (**F**).

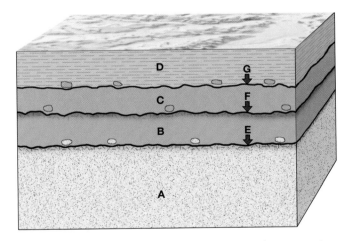

FIGURE 8.7 Notice that this is a sequence of strata and basalt lava flows (that have cooled to form the basalt). There are zones of contact metamorphism beneath both of the basalt lava flows. The sequence of events must have begun with deposition of sandstone **A**, because it is on the bottom. A lava flow was deposited atop **A** and cooled to form basalt **B**. This first lava flow caused development of the zone of contact metamorphism in **A** and the development of disconformity **E**. A second lava flow was deposited atop **B** and cooled to form basalt **C**. This lava flow caused the development of a zone of contact metamorphism and a disconformity at the top of **B**. An erosional surface developed atop **C**, and the surface became a disconformity **G** when shale **D** was deposited on top of it.

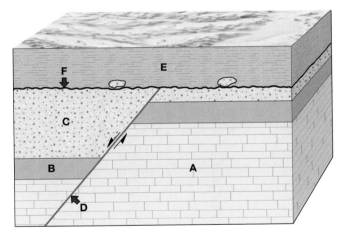

FIGURE 8.8 This is a sequence of relatively horizontal strata: **A**, **B**, **C**, and **E**. **A** must be the oldest of these formations, because it is on the bottom. **E** is the youngest of these formations, because it is on top. Formations **A**, **B**, and **C** are cut by a fault, which does not cut **E**. This means that the fault **D** must be younger than **C** and older than **E**. **F** is a disconformity. The sequence of events began with deposition of formations **A**, **B**, and **C**, in that order and one atop the other. This sequence was then cut by fault **D**. After faulting, the land surface was eroded. When siltstone **E** was deposited on the erosional surface, it became disconformity **F**.

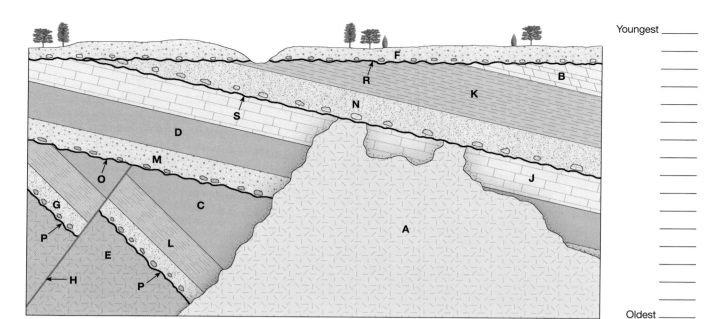

Youngest _____
_____
_____
_____
_____
_____
_____
_____
_____
_____
_____
_____
_____
_____
_____
_____
_____
_____
_____
_____
Oldest _____

FIGURE 8.9 Geologic cross section for relative age analysis. Place letters on the lines along the right side of the cross section to indicate the relative ages of the rock units and other features (unconformities, fault), from oldest (first) to youngest (last).

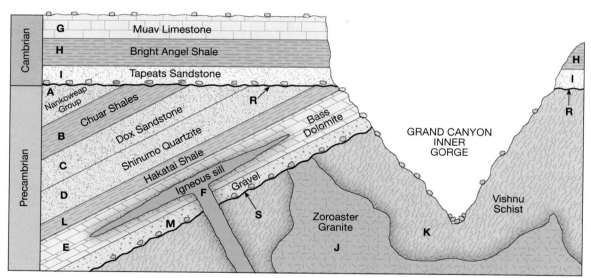

Youngest _____
_____
_____
_____
_____
_____
_____
_____
_____
_____
_____
_____
_____
_____
_____
Oldest _____

FIGURE 8.10 Geologic cross section of the Grand Canyon for relative age analysis. Place letters on the lines along the right side of the cross section to indicate the relative ages of the rock units and unconformities, from oldest (first) to youngest (last).

KEY TO SYMBOLS

Sedimentary rocks

Conglomerate Gravel Sandstone Siltstone

Shale Clay Limestone Dolomite

Igneous rocks *Metamorphic rocks*

Granite Basalt Schist Gneiss

Other features

Zone of contact metamorphism Unconformity Fault Contact

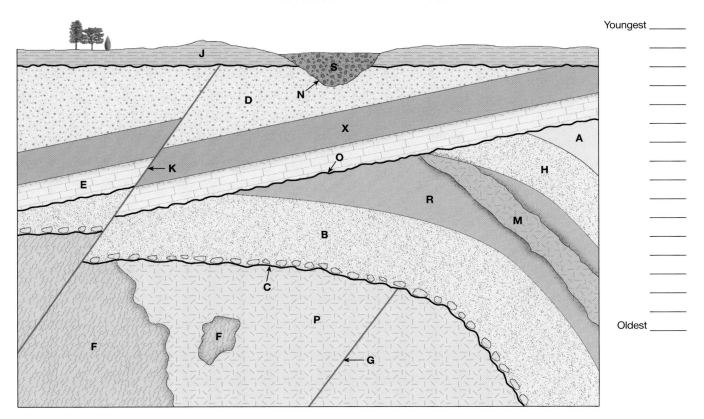

Youngest _____
_____
_____
_____
_____
_____
_____
_____
_____
_____
_____
_____
_____
_____
_____
_____
_____
_____
Oldest _____

FIGURE 8.11 Geologic cross section for relative age analysis. Place letters on the lines along the right side of the cross section to indicate the relative ages of the rock units and other features (unconformities, faults), from oldest (first) to youngest (last).

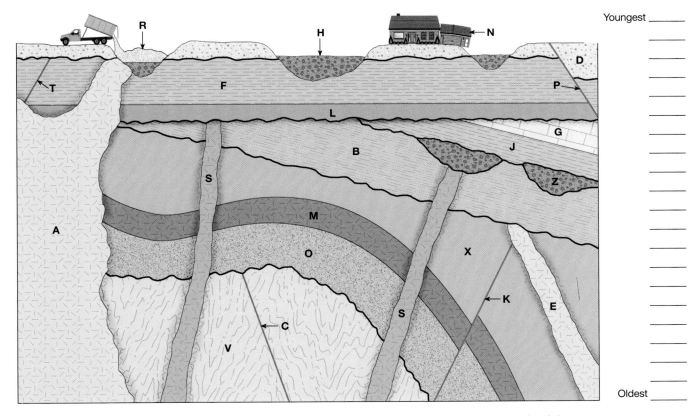

Youngest _____
_____
_____
_____
_____
_____
_____
_____
_____
_____
_____
_____
_____
_____
_____
_____
_____
_____
_____
_____
_____
_____
_____
Oldest _____

FIGURE 8.12 Geologic cross section for relative age analysis. Place letters on the lines along the right side of the cross section to indicate the relative ages of the rock units and other features (unconformities, faults), from oldest (first) to youngest (last).

157

PART 8B: USING FOSSILS TO DETERMINE AGE RELATIONSHIPS

The sequence of strata that makes up the geologic record is a graveyard filled with the fossils of millions of kinds of organisms that are now extinct. Geologists know that they existed only on the basis of their fossilized remains or the traces of their activities (like tracks and trails). Geologists have also determined that fossil organisms originate, co-exist, or disappear from the geologic record in a definite sequential order recognized throughout the world, so *any rock layer containing a group of fossils can be identified and dated in relation to other layers on the basis of its fossils.* This is known as the **Principle of Fossil Succession.**

The sequence of strata in which fossils of a particular organism are found is called a **range zone,** which represents a chron of time. Organisms whose range zones have been used to represent named divisions of the geologic time scale are called **index fossils.**

The range zones of some well-known Phanerozoic index fossils are presented on the right side of Figure 8.13. Relative ages of the rocks containing these fossils are presented as *periods* and *eras* on the left side of Figure 8.13. By noting the range zone of a fossil (vertical black line), you can determine the corresponding era(s) or period(s) of time in which it lived. For example, the fossil record indicates that sharks have lived from (and are an index fossil for) late in the Devonian Period of the Paleozoic Era to the present time. All of the different species of dinosaurs lived and died during the Mesozoic Era of time, long before (many layers below) the time when humans first existed and left a record of fossils. Notice that Figure 8.13 also includes the following groups:

- **Brachiopods** (pink on chart): marine invertebrate animals with two symmetrical seashells of unequal size. They range throughout the Paleozoic, Mesozoic, and Cenozoic Eras, but they were most abundant in the Paleozoic Era. Only a few species exist today, so they are nearly extinct.

- **Trilobites** (orange on chart): an extinct group of marine invertebrate animals related to lobsters. They are only found in Paleozoic rocks, so they are a good index fossil for the Paleozoic Era and its named subdivisions.

- **Plants** (dark green on chart)

- **Reptiles** (pale green on chart): the group of vertebrate animals that includes lizards, snakes, turtles, and dinosaurs. **Dinosaurs** are only found in Mesozoic rocks, so they are an index fossil for the Mesozoic and its subdivisions.

- **Mammals** (gray on chart): the group of vertebrate animals (including humans) that are warm blooded, nurse their young, and have hair.

- **Amphibians** (brown on chart): the group of vertebrate animals that includes frogs and salamanders.

- **Sharks** (blue on chart).

Notice that absolute ages in millions of years are also presented on Figure 8.13. Determining absolute ages will be addressed in Part C, but you will need to use the absolute ages in this figure to answer some of the questions below.

Questions

4. Analyze the fossiliferous rock in Figure 8.14.

    a. Based on Figure 8.13, what is the *relative age* of the rock in Figure 8.14? Explain your reasoning.

    b. Based on Figure 8.13, what is the *absolute age* of the rock in Figure 8.14? Explain your reasoning.

5. Analyze the fossiliferous rock in Figure 8.15.

    a. Based on Figure 8.13, what is the *relative age* of the rock in Figure 8.15? Explain your reasoning.

    b. Based on Figure 8.13, what is the *absolute age* of the rock in Figure 8.15? Explain your reasoning.

6. Re-examine the geologic cross section in Figure 8.2 on the basis of its fossils.

    a. Which one of the contacts between lettered layers is a disconformity?

    b. What is missing at the disconformity?

    c. If the present landscape in this cross section were covered today with a layer of sediment, then how much time would the resulting disconformity represent? Explain your reasoning.

7. What geologic events occurred during the Mesozoic Era in the region where Figure 8.3 is located? Explain your reasoning.

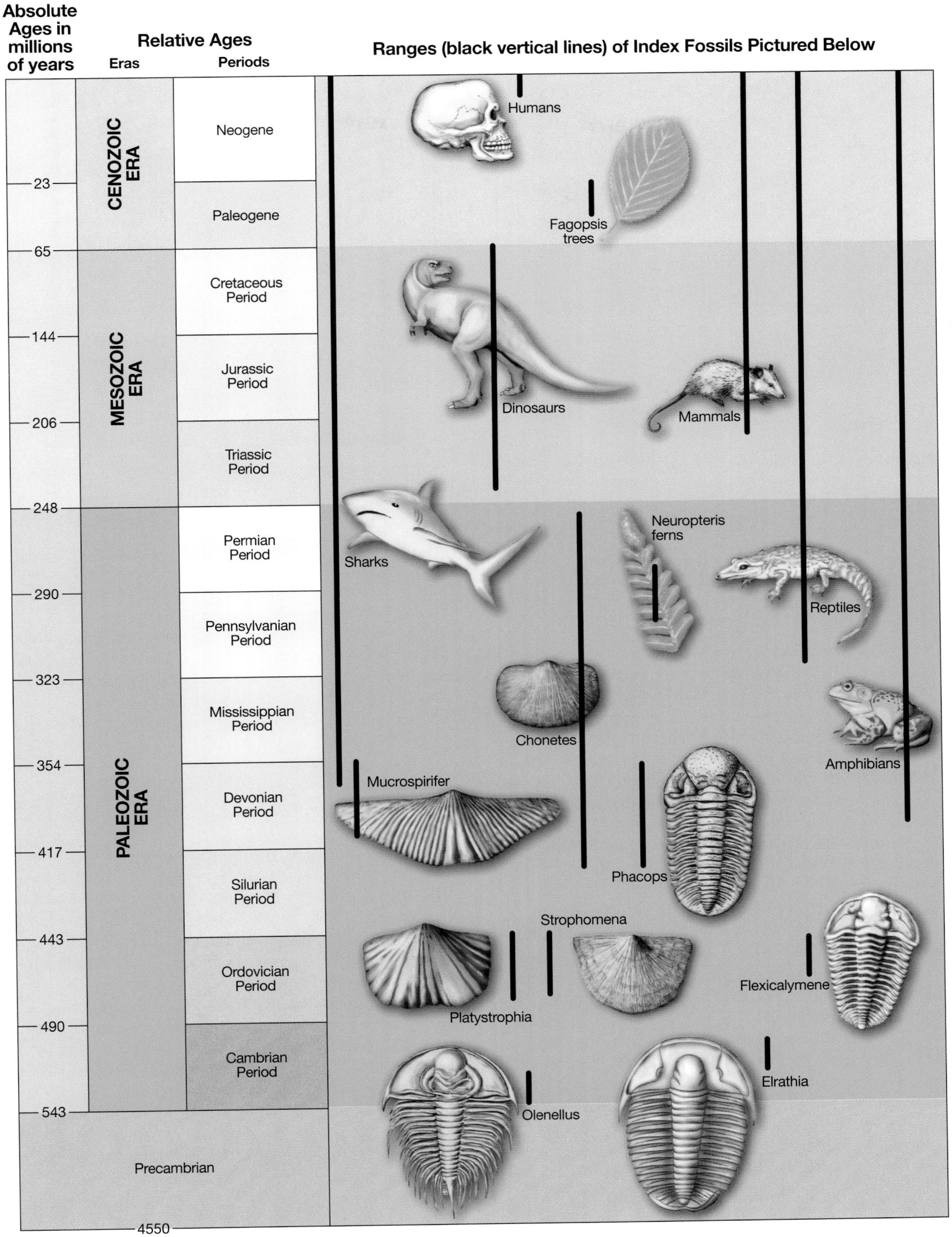

FIGURE 8.13 Range zones (black lines) of some well-known index fossils relative to the geologic time scale.

FIGURE 8.14 Fossiliferous rock sample for age analysis.

FIGURE 8.15 Fossiliferous rock sample for age analysis.

PART 8C: DETERMINING ABSOLUTE AGES BY RADIOMETRIC DATING

You measure the passage of time on the basis of the rates and rhythms at which regular changes occur around you. For example, you are aware of the rate at which hands move on a clock, the rhythm of day and night, and the regular sequence of the four seasons. These regular changes allow you to measure the passage of minutes, hours, days, and years.

Another way to measure the passage of time is by the regular rate of decay of radioactive isotopes. This technique is called **radiometric dating** and is one way that geologists determine absolute ages of some geologic materials.

You may recall that **isotopes** of an element are atoms that have the same number of protons and electrons but different numbers of neutrons. This means that the different isotopes of an element vary in atomic weight (mass number) but not in atomic number (number of protons).

There are about 350 different isotopes that occur naturally. Some of these are *stable isotopes*, meaning that they are not radioactive and do not decay through time. The others are *radioactive isotopes* that decay spontaneously, at regular rates through time. When a mass of atoms of a radioactive isotope is incorporated into the structure of a newly formed crystal or seashell, it is referred to as a **parent isotope.** When atoms of the parent isotope decay to a stable form, they have become a **daughter isotope.** A parent isotope and its corresponding daughter are called a **decay pair.**

Atoms of a parent isotope always decay to atoms of their stable daughter isotope at an exponential rate that does not change. The rate of decay can be expressed in terms of **half-life**—the time it takes for half of the parent atoms in a sample to decay to stable daughter atoms.

Radiometric Dating of Geologic Materials

The decay parameters for all radioactive isotopes can be represented graphically as in Figure 8.16. Notice that the decay rate is exponential (not linear)—during the second half-life interval, only half of the remaining half of parent atoms will decay. All radioactive isotopes decay in this way, but each decay pair has its own value for half-life.

Half-lives for some isotopes used for radiometric dating have been experimentally determined by physicists and chemists, as noted in the top chart of Figure 8.16. For example, Uranium-238 is a radioactive isotope (parent) found in crystals of the mineral zircon. It decays to Lead-206 (daughter) and has a half-life of about 4,500 million years (4.5 billion years).

To determine the age of an object, it must contain atoms of a radioactive decay pair that originated when the object formed. You must then measure the percent of those atoms that is parent atoms (**P**) and

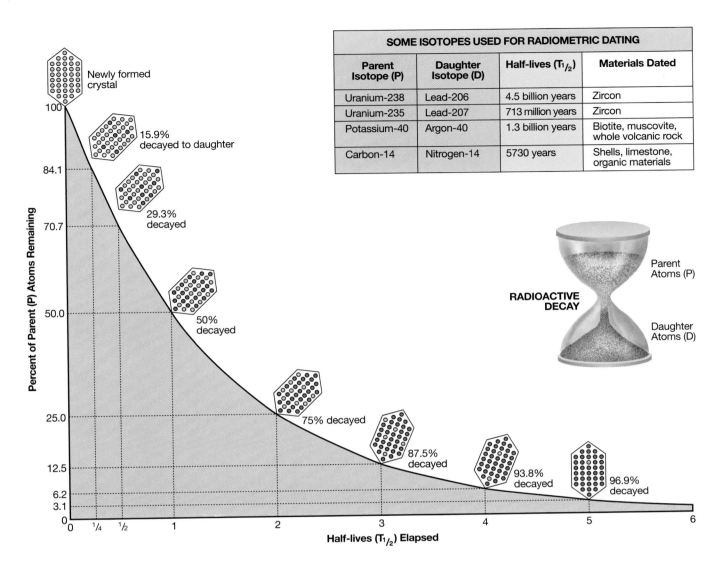

SOME ISOTOPES USED FOR RADIOMETRIC DATING			
Parent Isotope (P)	Daughter Isotope (D)	Half-lives ($T_{1/2}$)	Materials Dated
Uranium-238	Lead-206	4.5 billion years	Zircon
Uranium-235	Lead-207	713 million years	Zircon
Potassium-40	Argon-40	1.3 billion years	Biotite, muscovite, whole volcanic rock
Carbon-14	Nitrogen-14	5730 years	Shells, limestone, organic materials

DECAY PARAMETERS FOR ALL RADIOACTIVE DECAY PAIRS			
Percent of Parent Atoms (P)	Percent of Daughter Atoms (D)	Half-lives Elapsed	Age
100.0	0.0	0	$0.000 \times T_{1/2}$
98.9	1.1	1/64	$0.015 \times T_{1/2}$
97.9	2.1	1/32	$0.031 \times T_{1/2}$
95.8	4.2	1/16	$0.062 \times T_{1/2}$
91.7	8.3	1/8	$0.125 \times T_{1/2}$
84.1	15.9	1/4	$0.250 \times T_{1/2}$
70.7	29.3	1/2	$0.500 \times T_{1/2}$
50.0	50.0	1	$1.000 \times T_{1/2}$
35.4	64.6	1½	$1.500 \times T_{1/2}$
25.0	75.0	2	$2.000 \times T_{1/2}$
12.5	87.5	3	$3.000 \times T_{1/2}$
6.2	93.8	4	$4.000 \times T_{1/2}$
3.1	96.9	5	$5.000 \times T_{1/2}$

FIGURE 8.16 Some isotopes useful for radiometric dating and their decay parameters. The half-life of each decay pair is different (top chart), but the graph and decay parameters (bottom charts) are the same for all decay pairs.

161

the percent that is daughter atoms (**D**). This is generally done in a chemistry laboratory with an instrument called a *mass spectrometer*. Based on **P** and **D** and the chart at the bottom of Figure 8.16, find the number of half-lives that have elapsed and the object's corresponding age in number of half-lives. Finally, multiply that number of half-lives by the known half-life for that decay pair (noted in the top chart of Figure 8.16).

For example, a sample of Precambrian granite contains biotite mineral crystals, so it can be dated using the Potassium-40 to Argon-40 decay pair. If there are 3 Argon-40 atoms in the sample for every 1 Potassium-40 atom, then the sample is 25.0% Potassium-40 parent atoms (**P**) and 75.0% Argon-40 daughter atoms (**D**). This means that 2 half-lives have elapsed, so the age of the biotite (and the granite) is 2.0 times 1.3 billion years, which equals 2.6 billion years.

Questions

8. A solidified lava flow containing zircon mineral crystals is present in a sequence of rock layers that are exposed in a hillside. A mass spectrometer analysis was used to count the atoms of Uranium-235 and Lead-207 isotopes in zircon samples from the lava flow. The analysis revealed that 71% of the atoms were Uranium-235, and 29% of the atoms were Lead-207.

    a. About how many half-lives of the Uranium-235 to Lead-207 decay pair have elapsed in the zircon crystals?

    b. What is the absolute age of the lava flow based on its zircon crystals? Show your calculations.

    c. What is the age of the rock layers beneath the lava flow?

    d. What is the age of the rock layers above the lava flow?

9. Astronomers think that the Earth probably formed at the same time as all of the other rocky materials in our solar system, including the oldest meteorites. The oldest meteorites ever found on Earth contain nearly equal amounts of both Uranium-238 and Lead-206. Based on Figure 8.16, what is Earth's age? Explain your reasoning.

10. If you assume that the global amount of radiocarbon (formed by cosmic-ray bombardment of atoms in the upper atmosphere and then dissolved in rain and seawater) is constant, then decaying Carbon-14 is continuously replaced in organisms while they are alive. However, when an organism dies, the amount of its Carbon-14 decreases as it decays to Nitrogen-14.

    a. The carbon in a buried peat bed has about 6% of the Carbon-14 of modern shells. What is the age of the peat bed? Explain.

    b. In sampling the peat bed you must be careful to avoid any young plant roots or old limestone. Why?

11. Layers of sand on a New Jersey beach contain common zircon crystals.

    a. Could the zircon crystals be used to date exactly when the layers of sand were deposited? Explain.

    b. Suggest a rule that geologists should follow when they date rocks according to radiometric ages of crystals inside the rocks.

PART 8D: INFER THE GEOLOGIC HISTORY OF TWO FIELD SITES

Questions

12. Refer to Figure 8.17.

    a. What is the relative age of the sedimentary rocks in this rock exposure? Explain.

    b. What is the absolute age of the sill? Explain.

    c. Locate the fault. How much displacement has occurred along this fault?

    d. Explain the geologic history of this region, starting with deposition of the sandstone and ending with the time this picture was taken. Use names of relative ages of geologic time and absolute ages in your explanation. *Assume that the fault occurred after emplacement of the sill.*

13. Carefully examine Figure 8.18, a surface mine (strip mine) in northeastern Pennsylvania's anthracite coal mining district. Describe the age and all of the events that have happened to the fossil plants from the time when they were alive to the time when they were exposed by bulldozers. *Your reasoning may differ from that of other students, because more than one inference is possible about the geologic history of the site. Be prepared to discuss your reasoning with other members of your class.*

FIGURE 8.17 Surface mine (strip mine) in northern New Mexico, from which bituminous coal is being extracted. Note the sill, sedimentary rocks, fossils, whole-rock data, and fault.

FIGURE 8.18 Surface mine (strip mine) in northeastern Pennsylvania, from which anthracite coal was extracted. Close-ups show plant fossils that were found at the site.

164

PART 8E: CONSTRUCT AND INTERPRET A SUBSURFACE GEOLOGIC CROSS SECTION

The top of Figure 8.19 is a cross section of five wells drilled along a west–east line. At the bottom of Figure 8.19 are well logs for these wells. These logs are a record of the faults and rock units (layers) intersected by each well. The dip (inclination) of faults or any rock units that are no longer horizontal is also noted. You also need to know the lithologic descriptions of the rock units:

Unit 1: Cross-bedded eolian (wind-blown) sandstone

Unit 2: Brown-to-gray siltstone with shale zones and some coal seams

Unit 3: Parallel-bedded, poorly sorted sandstone

Unit 4: Conglomerate

Unit 5: Poorly sorted sandstone with some clay, silt, pebbles

Unit 6: Black, clayey shale

Unit 7: Parallel-bedded, well-sorted, coarse-grained sandstone

Unit 8: Black shale

Unit 9: Gray limestone

Complete the cross section in Figure 8.19. On each well (vertical lines), mark with ticks the elevations of the contacts between units (lightly in pencil). For example, in well A, unit 1 extends from the surface (2400 feet) to 2100 feet, so make tick marks at these points; unit 2 extends from 2100 to 2050 feet, so make ticks at these points; and so on. Label each unit number lightly beside each column, between the ticks.

Pay careful attention to the *dip* (inclination) *angles* indicated for faults and some rock units that are not horizontal. When you make tick marks, it is very helpful to angle them approximately to indicate dip (use a protractor). This is especially true if you encounter any *faults* in the cross section.

When you have all units plotted in the five wells, connect corresponding points between wells. (You are *correlating* well logs when you do this. You are also preparing a subsurface cross section of the type actually constructed by petroleum-exploration geologists.)

From the lithologic descriptions given, you can fill in some of the rock units with patterns—for example, sandstone (dots), conglomerate (tiny circles), and coal (solid black). Use the symbols given in Figure 8.4. Then complete the items below.

Questions

14. What is the nature and geologic origin of the bottom contact of Unit 2?

15. Why is coal not found in wells A and B, whereas two coal seams are found in well E?

16. Wells A and E are **dry holes,** so-called because they produced no petroleum. But the others produce petroleum. An oil pool is penetrated in well B from 750 feet to 650 feet, in well C from 550 down to 500 feet, and in well D from 750 down to 500 feet. Sketch and label the outline of the oil pools on the cross section and explain why the oil was trapped there.

17. Why is there no oil in either well A or well E?

18. Using the laws of original horizontality, superposition, and cross-cutting (refer back to part 8A of this laboratory if necessary), describe the sequence of events that developed this geologic situation.

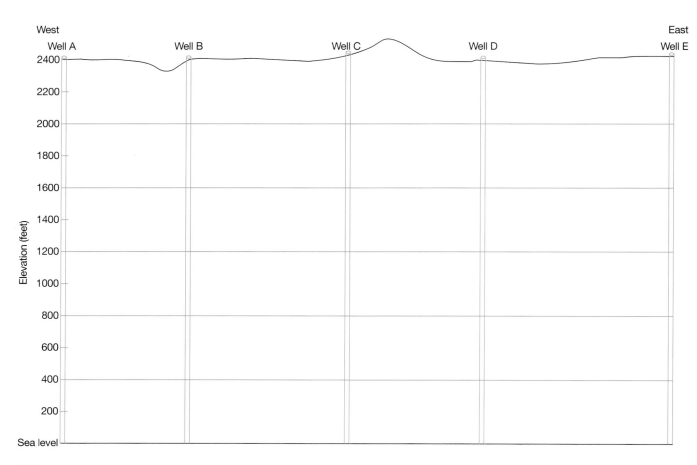

West   East

Well A

2400–2100	Unit 1, horizontal
2100–2050	Unit 2, horizontal
2050–1700	Unit 3, dips westward 15°
1700–1150	Unit 4, dips westward 15°
1150– 800	Unit 5, dips westward 15°
800– 550	Unit 6, dips westward 15°
550– 200	Unit 7, dips westward 15°
Bottom of well	

Well B

2300–2100	Unit 1, horizontal
2100–1980	Unit 2, horizontal
1980–1650	Unit 3, dips westward 15°
1650	Fault, dips eastward 60°
1650–1350	Unit 4, dips westward 15°
1350–1000	Unit 5, dips westward 15°
1000– 750	Unit 6, dips westward 15°
750– 200	Unit 7, dips westward 15°
200–sea level	Unit 8, dips westward 15°
Bottom of well	

Well C

2350–2100	Unit 1, horizontal
2100–1900	Unit 2, horizontal; coal seam at 1950
1900–1700	Unit 3, dips westward 15°
1700–1150	Unit 4, dips westward 15°

1150– 800	Unit 5, dips westward 15°
800– 550	Unit 6, dips westward 15°
550– 200	Unit 7, dips westward 15°
200	Fault; dips eastward 60°
200– 100	Unit 9, dips westward 15°
Bottom of well	

Well D

2300–2100	Unit 1, horizontal
2100–1800	Unit 2, horizontal; coal seam at 1950
1800–1350	Unit 4, horizontal
1350–1000	Unit 5, horizontal
1000– 750	Unit 6, horizontal
750– 200	Unit 7, horizontal
200– 100	Unit 8, horizontal
Bottom of well	

Well E

2400–2100	Unit 1, horizontal
2100–1650	Unit 2, horizontal; coal seams at 1950 and 1850
1650–1450	Unit 3, dips eastward 21°
1450– 900	Unit 4, dips eastward 21°
900– 550	Unit 5, dips eastward 21°
550– 300	Unit 6, dips eastward 21°
300– 200	Unit 7, dips eastward 21°
Bottom of well	

FIGURE 8.19 Cross section of wells and their well logs. Refer to text for descriptions of the rock units.

Topographic Maps, Aerial Photographs, and Satellite Images

•CONTRIBUTING AUTHORS•

Charles G. Higgins • *University of California*

Evelyn M. Vandendolder • *Arizona Geological Survey*

John R. Wagner • *Clemson University*

James R. Wilson • *Weber State College*

OBJECTIVES

A. Be able to locate features on topographic maps using map symbols and colors, latitude and longitude, the U.S. Public Land Survey System (PLS), the Universal Transverse Mercator System (UTM), and compass bearings.

B. Be able to interpret four kinds of map scales (ratio scales, fractional scales, verbal scales, graphic bar scales) and convert one scale to another.

C. Be able to construct topographic maps by drawing contour lines based on points of known elevations for areas of Earth's surface.

D. Be able to interpret contour lines to measure gradients and relief and identify hills, saddles, ridges, spurs, valleys, closed depressions, steep slopes, gentle slopes, vertical cliffs, and overhanging cliffs.

E. Be able to construct topographic profiles and calculate their vertical exaggeration.

F. Understand how stereo pairs (stereograms) of aerial photographs are obtained and used in geological studies.

MATERIALS

Pencil, eraser, laboratory notebook, topographic quadrangle map (obtained by you or provided by your instructor), calculator, and pocket stereoscope (optional); millimeter ruler, protractor, and UTM templates from GeoTools Sheets 2–4 at the back of the manual.

INTRODUCTION

In 1937, American aviator Amelia Earhart and her navigator Fred Noonan attempted to make the first round-the-world flight. But two-thirds of the way around the globe, they disappeared in the South Pacific Ocean. Earhart and Noonan were trying to reach tiny Howland Island, a mere speck of land just north of the Equator, when they vanished. It appears that their flight plan gave the wrong coordinates for the island.

Earhart's flight plan listed the island's coordinates as 0°49' north latitude, 176°43' west longitude. But the actual coordinates are 0°48' north latitude, 176°38' west longitude (Barker, V., *New Haven Register*, Dec. 21, 1986:A48). In the open ocean, with nothing else to guide them and limited fuel, such a miss was fatal.

Investigators who researched their disappearance thought that Earhart and Noonan were on course and would certainly have reached Howland Island—had they been given the correct coordinates. Thus, their demise probably was due to a mapmaker's mistake or to the flight planner's inability to correctly read a map.

Earhart's story illustrates that map errors, or errors in map use, can have drastic effects. Your ability to construct, read, and interpret maps is essential for conducting many geologic studies. Geologists generally use aerial photographs in combination with maps to provide additional visual information not given on maps. When used in pairs taken from slightly different angles, aerial photographs allow the geologist to see the ground in stereo.

PART 9A: INTRODUCTION TO TOPOGRAPHIC MAPS

A **topographic map** is a two-dimensional (flat) representation (model) of a three-dimensional land surface (landscape). It shows landforms (hills, valleys, slopes, coastlines, gullies) and their **relief** (difference in elevation) by using **contour lines** to represent elevations of hills and valleys. The contour lines are the distinguishing features of a topographic map. They are what makes a topographic map different from the more familiar *planimetric* map, such as a highway map, which has no contour lines and does not show relief of the land. The three-dimensional aspect of topographic maps makes them a valuable tool in geological and engineering studies. They also are used by hikers, hunters, campers, developers, planners, and anyone else who needs to know about the topography of a region.

Topographic Quadrangles and Declination

Most United States topographic maps are published by the U.S. Geological Survey (**http://www.usgs.gov**). Most Canadian topographic maps are produced by the Centre for Topographic Information of Natural Resources Canada (**http://maps.nrcan.gc.ca**). Although some topographic maps cover areas defined by political boundaries (such as a state, county, or city), most topographic maps depict rectangular sections of Earth's surface, called quadrangles. A **quadrangle** is a section of Earth's surface that is bounded by lines of *latitude* at the top (north) and bottom (south) and by lines of *longitude* on the left (west) and right (east)—see Figure 9.1.

Latitude and longitude are both measured in *degrees* (°). Latitude is measured from 0° at the Equator to 90°N (at the North Pole) or 90°S (at the South Pole). Longitude is measured in degrees east or west of the *prime meridian*, a line that runs from the North Pole to the South Pole through Greenwich, England. Locations in Earth's Eastern Hemisphere are east of the prime meridian, and locations in the Western Hemisphere are west of the prime meridian. For finer measurements each degree can be subdivided into 60 equal subdivisions called *minutes* ('), and the minutes can be divided into 60 equal subdivisions called *seconds* (").

Quadrangle maps are published in several sizes, but two are most common: 15-*minute* quadrangle maps and $7\frac{1}{2}$-*minute* quadrangle maps. The numbers refer to the amount of area that the maps depict, in degrees of latitude and longitude. A *15-minute topographic map* represents an area that measures 15 minutes of latitude by 15 minutes of longitude. A $7\frac{1}{2}$-minute topographic map represents an area that measures $7\frac{1}{2}$ minutes of latitude by $7\frac{1}{2}$ minutes of longitude. Each 15-minute map can be divided into four $7\frac{1}{2}$-minute maps (Figure 9.1).

Because longitude lines form the left and right boundaries of a topographic map, north is always at the top of the quadrangle. This is called grid north (GN) and is usually the same direction as *true north* on the actual Earth. Unfortunately, magnetic compasses are not attracted to true north (the geographic North Pole). Instead, they are attracted to the *magnetic north pole* (MN), currently located northwest of Hudson Bay in Northern Canada, about 700 km (450 mi) from the true North Pole. A compass-like symbol on the bottom margin of topographic maps shows the **declination** (difference in degrees) between compass north (MN) and true north (usually a *star* symbol). Also shown is the declination between compass north (*star* symbol) and grid north (GN). The magnetic pole migrates very slowly, so the declination is exact only for the year listed on the map.

Map Symbols and Revisions

Topographic maps have colors, patterns, and symbols (Figure 9.2) that are used to depict water bodies, vegetation, roads, buildings, political boundaries, place-names, and other natural and cultural features of the landscape represented by the map (Figure 9.3).

Additional information is presented in the margins of these maps, including the revision date. Because people constantly change the cultural features on Earth's surface, and because Earth's surface itself occasionally changes rapidly from events such as

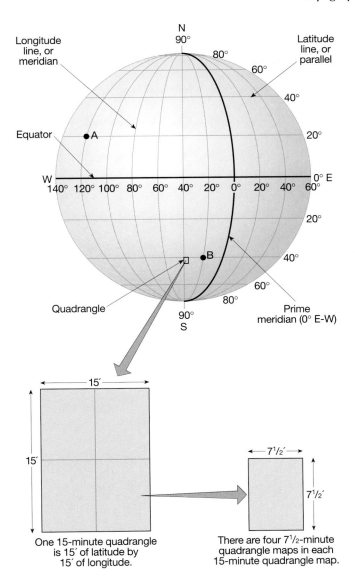

N
90°

Longitude
line, or
meridian

80°

60°

40°

Latitude
line, or
parallel

20°

Equator

●A

0° E

W
140° 120° 100° 80° 60° 40° 20° 0° 20° 40° 60°

20°

●B

40°

60°

Quadrangle

80°

90°
S

Prime
meridian (0° E-W)

15′

15′

One 15-minute quadrangle
is 15′ of latitude by
15′ of longitude.

7¹⁄₂′

7¹⁄₂′

There are four 7¹⁄₂-minute
quadrangle maps in each
15-minute quadrangle map.

FIGURE 9.1 Latitude and longitude geographic grid and coordinate system. Earth's spherical surface is divided into lines of latitude (*parallels*) that go around the world parallel to the Equator, and lines of longitude (*meridians*) that go around the world from pole to pole. There are 360 degrees (360°) around the entire Earth, so the distance from the Equator to a pole (one-fourth of the way around Earth) is 90° of latitude. The Equator is assigned a value of zero degrees (0°) latitude, the North Pole is 90 degrees North latitude (90° N), and the South Pole is 90 degrees south latitude (90° S). The *prime meridian* is zero degrees of longitude and runs from pole to pole through Greenwich, England. Locations in Earth's Eastern Hemisphere are located in degrees east of the prime meridian, and points in the Western Hemisphere are in located in degrees west of the prime meridian. Therefore, any point on Earth (or a map) can be located by its latitude-longitude coordinates. The latitude coordinate of the point is its position in degrees north or south of the Equator. The longitude coordinate of the point is its position in degrees east or west of the Prime Meridian. For example, point **A** is located at coordinates of: 20° North latitude, 120° West longitude.

For greater detail, each degree of latitude and longitude can also be subdivided into 60 minutes (60′), and each minute can be divided into 60 seconds (60″). Note that a 15-minute (15′) quadrangle map represents an area of Earth's surface that is 15 minutes of longitude wide (E-W) and 15 minutes of latitude long (N-S). A 7.5-minute quadrangle map is one-fourth of a 15-minute quadrangle map.

landslides and floods, the maps must be updated. This is done by *photorevision*. Aerial photographs are used to discover changes on the landscape, and the changes are overprinted on the maps in a standout color like purple, red, or gray.

Contour Lines

Examine the image of one of the Galapagos Islands in Figure 9.4, a perspective view of the landscape that has been false colored to show relief. It was made by transmitting imaging radar from an airplane (flying at a constant altitude). Timed pulses of the radar measured the distance between the airplane (flying at a constant elevation) and the ground. Overlapping

pulses of the radar produced the three-dimensional perspective similar to the way that overlapping lines of sight from your eyes enable you to see in stereo. Notice that the island has a distinct coastline, which has the same elevation all of the way around the island (zero feet above sea level). Similarly, all points at the very top of the green (including yellow-green) regions form a line at about 300 ft above sea level. These lines of equal elevation (i.e., the coastline and 300-ft line) are called **contour lines.** Unfortunately, the 1200-ft contour line (located at the boundary between yellow and pink) is not visible behind Darwin and Wolf Volcanoes in this perspective view. The only way that you could see all of the 0-ft, 300-ft, and 1200-ft contour lines at the same time would be if you viewed

FIGURE 9.2 Symbols used on topographic quadrangle maps produced by the U.S. Geological Survey.

Primary highway, hard surface . . . .

Secondary highway, hard surface . . . .

Light-duty road, hard or improved surface . . . .

Unimproved road . . . .

Road under construction, alinement known . . . .

Proposed road . . . .

Dual highway, dividing strip 25 feet or less . . . .

Dual highway, dividing strip exceeding 25 feet . . . .

Trail . . . .

Railroad: single track and multiple track . . . .

Railroads in juxtaposition . . . .

Narrow gage: single track and multiple track . . . .

Railroad in street and carline . . . .

Bridge: road and railroad . . . .

Drawbridge: road and railroad . . . .

Footbridge . . . .

Tunnel: road and railroad . . . .

Overpass and underpass . . . .

Small masonry or concrete dam . . . .

Dam with lock . . . .

Dam with road . . . .

Canal with lock . . . .

Buildings (dwelling, place of employment, etc.) . . . .

School, church, and cemetery . . . . Cem

Buildings (barn, warehouse, etc.) . . . .

Power transmission line with located metal tower . . . .

Telephone line, pipeline, etc. (labeled as to type) . . . .

Wells other than water (labeled as to type) . . . . oOil . . . . oGas

Tanks: oil, water, etc. (labeled only if water) . . . . Water

Located or landmark object; windmill . . . .

Open pit, mine, or quarry; prospect x

Shaft and tunnel entrance . . . . Y

Horizontal and vertical control station:

  Tablet, spirit level elevation . . . . BM △ 5653

  Other recoverable mark, spirit level elevation . . . . △ 5455

Horizontal control station: tablet, vertical angle elevation VABM △ 95I9

  Any recoverable mark, vertical angle or checked elevation   △ 3775

Vertical control station: tablet, spirit level elevation . . . . . BM X 957

  Other recoverable mark, spirit level elevation . . . . X 954

Spot elevation . . . . x 7369

Water elevation 670   670

Boundaries: National . . . .

  State . . . .

  County, parish, municipio . . . .

  Civil township, precinct, town, barrio . . . .

  Incorporated city, village, town, hamlet . . . .

  Reservation, National or State . . . .

  Small park, cemetery, airport, etc. . . . .

  Land grant . . . .

Township or range line, United States land survey . . . . . .

Township or range line, approximate location . . . .

Section line, United States land survey . . . .

Section line, approximate location . . . .

Township line, not United States land survey . . . .

Section line, not United States land survey . . . .

Found corner: section and closing . . . .

Boundary monument: land grant and other . . . .

Fence or field line . . . .

Index contour . . . .   Intermediate contour . .

Supplementary contour . . . . .   Depression contours . .

Fill . . . .   Cut . . . .

Levee . . . .   Levee with road . . . . .

Mine dump . . . .   Wash . . . .

Tailings . . . .   Tailings pond . . . .

Shifting sand or dunes . . . .   Intricate surface . . . . .

Sand area . . . .   Gravel beach . . . .

Perennial streams . . . .   Intermittent streams . .

Elevated aqueduct . . . .   Aqueduct tunnel . . . . .

Water well and spring . . . . . .   Glacier . . . .

Small rapids . . . .   Small falls . . . .

Large rapids . . . .   Large falls . . . .

Intermittent lake . . . . .   Dry lake bed . . . .

Foreshore flat . . . .   Rock or coral reef . . . .

Sounding, depth curve .   Piling or dolphin . . . . .

Exposed wreck . . . .   Sunken wreck . . . .

Rock, bare or awash; dangerous to navigation . . . .

Marsh (swamp) . . . .   Submerged marsh . . . .

Wooded marsh . . . .   Mangrove . . . .

Woods or brushwood . .   Orchard . . . .

Vineyard . . . .   Scrub . . . .

Land subject to
controlled inundation   Urban area . . . .

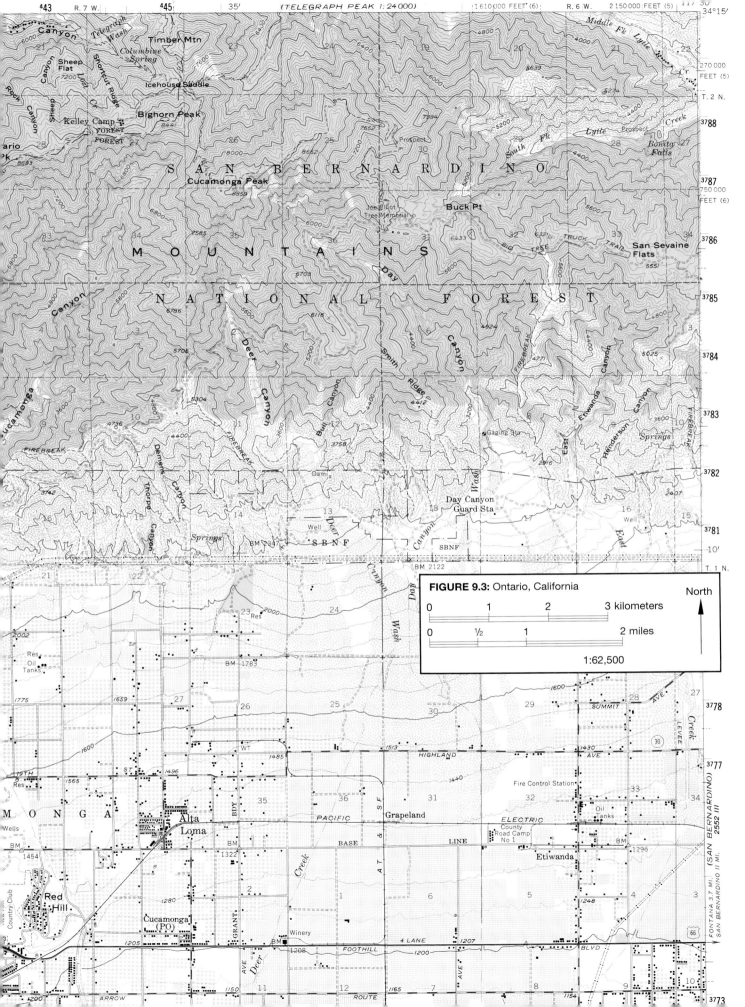

FIGURE 9.3: Ontario, California

| North |
| 0   1   2   3 kilometers |
| 0   ½   1   2 miles |
| 1:62,500 |

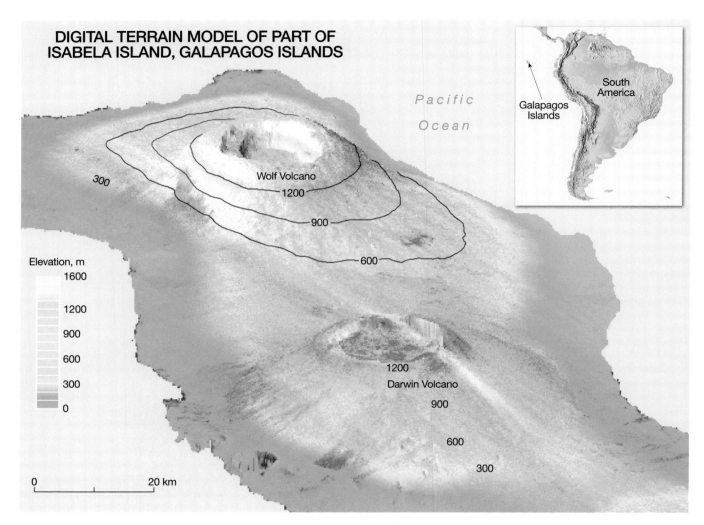

FIGURE 9.4 Digital model (AIRSAR/TOPSAR 3-dimensional perspective) of part of Isabela Island, Galapagos Islands. The image has been false-colored to show relief. It was made with imaging radar transmitted from an airplane (flying at a constant altitude of 33,000 ft). Timed pulses of the radar measured the distance between the airplane (flying at a constant elevation) and the ground. Overlapping pulses of the radar produced the three-dimensional perspective. See if you can draw the remaining 300-, 600-, 900-, and 1200-foot contour lines. (Image courtesy of NASA/JPL–Caltech)

the island from directly above. This is how topographic maps are constructed (Figure 9.5).

Topographic maps are made from overlapping pairs of photographs, called *stereo pairs*. Each stereo pair is taken from an airplane making two closely-spaced passes over a region at the same elevation. The passes are flown far enough apart to provide the stereo effect, yet close enough to be almost directly above the land that is to be mapped. After the stereo pairs are used to define contour lines and construct a first draft of the topographic map, angular distortion is removed and the exact elevations of the contour lines on the map are "ground truthed" (checked on the ground) using very precise altimeters and GPS.

Therefore, topographic maps are miniature models of Earth's three-dimensional surface, printed on two-dimensional pieces of paper. Two of the dimensions are the lengths and widths of objects and landscape features. But the third dimension, elevation (height), is shown using contour lines. Each **contour line** connects all points on the map that have the same elevation above sea level (Figure 9.6, rule 1). Look at the topographic map in Figure 9.3 and notice the light brown and heavy brown contour lines. The heavy brown contour lines are called **index contours,** because they have elevations printed on them (whereas the lighter contour lines do not; Figure 9.6, rule 6). Index contours are your starting point when reading

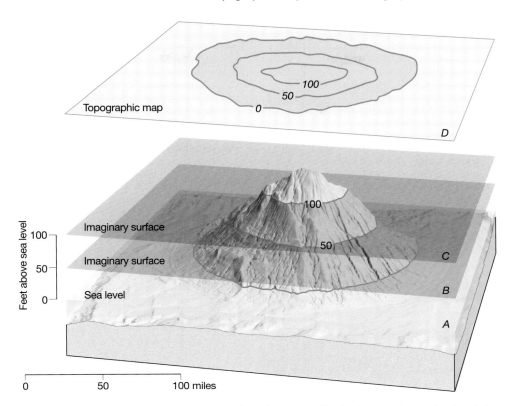

FIGURE 9.5 Topographic map construction. A *contour line* is drawn where a horizontal plane (such as **A**, **B**, or **C**) intersects the land surface. Where sea level (plane **A**) intersects the land, it forms the 0-ft contour line. Plane **B** is 50 ft above sea level, so its intersection with the land is the 50-ft contour line. Plane **C** is 100 ft above sea level, so its intersection with the land is the 100-foot contour line. **D** is the resulting topographic map of the island. It was constructed by looking down onto the island from above and tracing the 0, 50, and 100-ft contour lines. The elevation change between any two contour lines is 50 ft, so the map is said to have a 50-ft *contour interval*. The topographic datum (reference level) is sea level, so all contour lines on this map represent elevations in feet above sea level and are *topographic contour lines*. (Contours below sea level are called *bathymetric contour lines* and are generally shown in blue).

elevations on a topographic map. For example, notice that every fifth contour line on Figure 9.3 is an index contour. Also notice that the index contours are labeled with elevations in increments of 400 ft. This means that the map has five contours for every 400 ft of elevation, or a **contour interval** of 80 ft. The contour interval is specified on most topographic maps in feet or meters. All contour lines are multiples of the contour interval above a specific surface (almost always sea level). For example, if a map uses a 10-ft contour interval, then the contour lines represent elevations of 0 ft (sea level), 10 ft, 20 ft, 30 ft, 40 ft, and so on. Most maps use the smallest contour interval that will allow easy readability and provide as much detail as possible.

Additional rules for contour lines are provided in Figure 9.6. For example, contour lines never cross, except in the rare case where an overhanging cliff is present. If contour lines merge into one line, then that line indicates a cliff. The spacing of contour lines can be used to interpret the steepness of a slope and whether it is uniform or variable in steepness. The apex (tip) of a V-shaped notch in a contour line always points up hill.

Be sure to review all of the rules for contour lines in Figure 9.6 and the common kinds of landforms represented by contour lines on topographic maps (Figure 9.7). Your ability to use a topographic map is based on your ability to interpret what the contour lines mean.

RULES FOR CONTOUR LINES

1. Every point on a contour line is of the exact same elevation; that is, contour lines connect points of equal elevation. The contour lines are constructed by surveying the elevation of points, then connecting points of equal elevation.

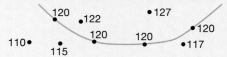

2. Interpolation is used to estimate the elevation of a point B located in line between points A and C of known elevation. To estimate the elevation of point B:

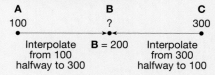

A	B	C
100	?	300

Interpolate from 100 halfway to 300   **B = 200**   Interpolate from 300 halfway to 100

3. Extrapolation is used to estimate the elevations of a point C located in line beyond points A and B of known elevation. To estimate the elevation of point C, use the distance between A and B as a ruler or graphic bar scale to estimate in line to elevation C.

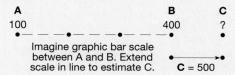

A	B	C
100	400	?

Imagine graphic bar scale between A and B. Extend scale in line to estimate C.   **C = 500**

4. Contour lines always separate points of higher elevation (uphill) from points of lower elevation (downhill). You must determine which direction on the map is higher and which is lower, relative to the contour line in question, by checking adjacent elevations.

5. Contour lines always close to form an irregular circle. But sometimes part of a contour line extends beyond the mapped area so that you cannot see the entire circle formed.

6. The elevation between any two adjacent contour lines of different elevation on a topographic map is the *contour interval*. Often every fifth contour line is heavier so that you can count by five times the contour interval. These heavier contour lines are known as *index contours*, because they generally have elevations printed on them.

7. Contour lines never cross each other except for one rare case: where an overhanging cliff is present. In such a case, the hidden contours are dashed.

Dashed contour — Overhanging cliff

8. Contour lines can merge to form a single contour line only where there is a vertical cliff or wall.

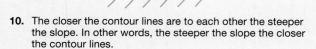
Vertical cliff

9. Evenly spaced contour lines of different elevation represent a uniform slope.

10. The closer the contour lines are to each other the steeper the slope. In other words, the steeper the slope the closer the contour lines.

Steep   Less steep

11. A concentric series of closed contours represents a hill:

12. *Depression contours* have hachure marks on the downhill side and represent a closed depression:

See Figure 9.8

13. Contour lines form a V pattern when crossing streams. The apex of the V always points upstream (uphill):

Uphill   downstream (downhill)
Apex (tip) of the V

14. Contour lines that occur on opposite sides of a valley or ridge always occur in pairs. See Figure 9.9.

FIGURE 9.6   Rules for constructing and interpreting contour lines on topographic maps.

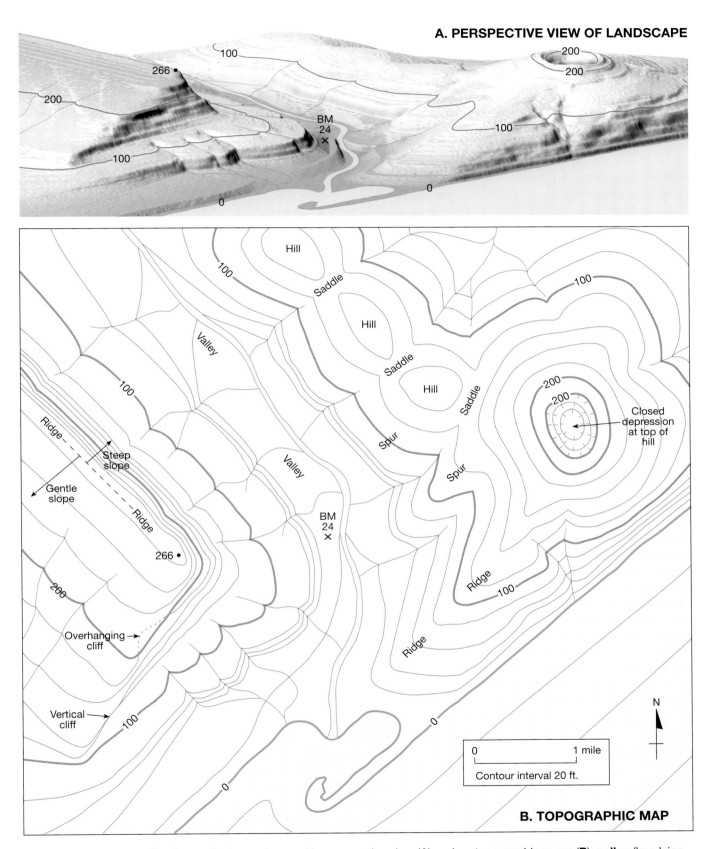

FIGURE 9.7 Names of landscape features observed in perspective view (**A**) and on topographic maps (**B**): **valley** (low-lying land bordered by higher ground), **hill** (rounded elevation of land; mound), **ridge** (linear or elongate elevation or crest of land), **spur** (short ridge or branch of a main ridge), **saddle** (low point in a ridge or line of hills; it resembles a horse saddle), **closed depression** (low point/area in a landscape from which surface water cannot drain; contour lines with hachure marks), **steep slope** (closely-spaced contour lines), **gentle slope** (widely-spaced contour lines), **vertical cliff** (merged contour lines), **overhanging cliff** (dashed contour line that crosses a solid one; the dashed line indicates what is under the overhanging cliff).

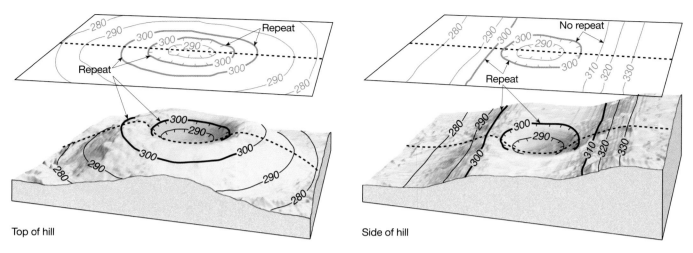

Top of hill

Side of hill

FIGURE 9.8 Contour lines repeat on opposite sides of a depression (left illustration), except when the depression occurs on a slope (right illustration).

Reading Elevations

If a point lies on an index contour, you simply read its elevation from that line. If the point lies on an un-numbered contour line, then its elevation can be determined by counting up or down from the nearest index contour. For example, if the nearest index contour is 300 ft, and your point of interest is on the fourth contour line *above* it, and the contour interval is 20 ft, then you simply count up by 20s from the index

contour: 320, 340, 360, 380. The point is 380 ft above sea level. (Or, if the point is three contour lines *below* the index contour, you count down: 280, 260, 240; the point is 240 ft above sea level.)

If a point lies between two contour lines, then you must estimate its elevation by interpolation (Figure 9.6, rule 2). For example, on a map with a 20-ft contour interval, a point might lie between the 340 and 360-ft contours, so you know it is between 340 and 360 ft above sea level. If a point lies between a contour line

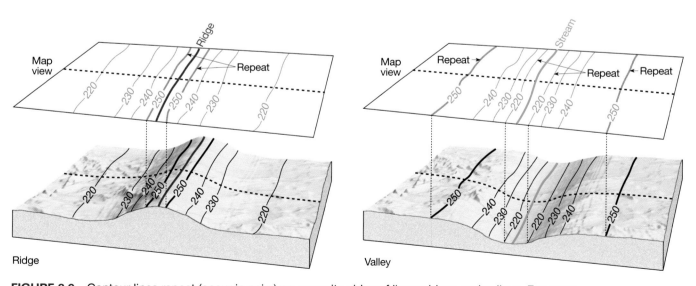

Ridge

Valley

FIGURE 9.9 Contour lines repeat (occur in pairs) on opposite sides of linear ridges and valleys. For example, in the left illustration, if you walked the dashed line from left to right, you would cross the 220, 230, 240 and 250-ft contour lines, go over the crest of the ridge, and cross the 250, 240, 230, and 220-ft contour lines again as you walk down the other side. Note that the 250-ft contour lines on these maps are heavier than the other lines because they are *index contours*. On most maps every fifth contour line above sea level is an index contour, so you can count by five times the contour interval. The *contour interval* (elevation between any two contour lines) of these maps is 10 ft, so the index contours are every 50 feet of elevation.

and the margin of the map, then you must estimate its elevation by extrapolation (Figure 9.6, rule 3).

Figure 9.8 shows how to read topographic contour lines in and adjacent to a depression. *Hachure marks* (short line segments pointing downhill) on some of the contour lines in these maps indicate the presence of a closed depression (a depression from which water cannot drain) (Figure 9.6, rule 12). At the top of a hill, contour lines repeat on opposite sides of the rim of the depression. On the side of a hill, the contour lines repeat only on the downhill side of the depression.

Figure 9.9 shows how topographic contour lines represent linear ridge crests and valley bottoms. Ridges and valleys are roughly symmetrical, so individual contour lines repeat on each side (Figure 9.6, rule 14). To visualize this, picture yourself walking along an imaginary trail across the ridge or valley (dashed lines in Figure 9.9). Every time you walk up the side of a hill or valley, you cross contour lines. Then, when you walk down the other side of the hill or valley, you recross contour lines of the same elevations as those crossed walking uphill.

Elevations of specific points on topographic maps (tops of peaks, bridges, survey points, etc.) sometimes are indicated directly on the maps beside the symbols indicated for that purpose. The notation "BM" denotes a **benchmark,** a permanent marker (usually a metal plate) placed by the U.S. Geological Survey or Bureau of Land Management at the point indicated on the map (Figure 9.2). Elevations usually are given. For example, look at the middle map in Figure 9.10. At the top center part of the map is an "x" symbol labeled "BM 463," indicating that "x" marks the location of the benchmark that was exactly 463 ft above sea level at the time of its placement. Two other benchmarks also appear on this map: BM 360 and BM 261.

The elevations of prominent hilltops, peaks, or other features are sometimes identified specifically, even if there is no benchmark on the ground. For example, the highest point on the ridge in the west central part of Figure 9.7B has an elevation of 266 ft above sea level.

Relief and Gradient

Recall that **relief** is the difference in elevation between landforms, specific points, or other features on a landscape or map. *Regional relief* is the difference in elevation between the highest and lowest points on a topographic map. The highest point is the top of the highest hill or mountain; the lowest point is generally where the major stream of the area leaves the map, or a coastline. **Gradient** is a measure of the steepness of

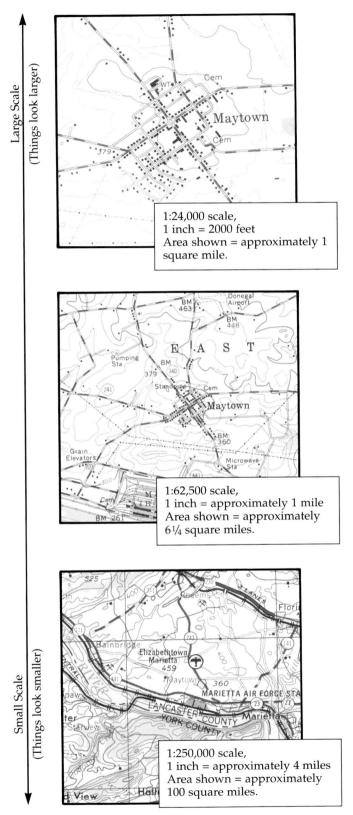

Large Scale (Things look larger)

1:24,000 scale,
1 inch = 2000 feet
Area shown = approximately 1 square mile.

1:62,500 scale,
1 inch = approximately 1 mile
Area shown = approximately 6¼ square miles.

Small Scale (Things look smaller)

1:250,000 scale,
1 inch = approximately 4 miles
Area shown = approximately 100 square miles.

FIGURE 9.10 Three scales of maps, centered on Maytown. Notice that everything looks small on the small-scale map. Everything looks larger on the large-scale map.

a slope. One way to determine and express the gradient of a slope is by measuring its steepness as an angle of ascent or descent (expressed in degrees). On a topographic map, gradient is usually determined by dividing the relief (rise or fall) between two points on the map by the distance (run) between them (expressed as a fraction in feet per mile or meters per kilometer). For example, if points **A** and **B** on a map have elevations of 200 ft and 300 ft, and the points are located two miles apart, then:

$$gradient = \frac{\text{relief (amount of rise or fall between } \mathbf{A} \text{ and } \mathbf{B})}{\text{distance between } \mathbf{A} \text{ and } \mathbf{B}}$$

$$= \frac{100 \text{ ft}}{2 \text{ mi}} = 50 \text{ ft/mi}$$

Scales of Maps and Models

Maps are scale models, like toy cars or boats. To make a model of anything, you must first establish a model scale. This is the proportion by which you will reduce the real object to the model size. For example, if you make a $\frac{1}{4}$-scale model of a 16-ft car, your model would be 4 ft long. The ratio scale of model-to-object is 4:16, which reduces to 1:4. A house floorplan, which really is a map of a house, commonly is drawn so that one foot on the plan equals 30 or 40 ft of real house, or a **ratio scale** of 1:30 or 1:40.

Topographic maps often model large portions of Earth's surface, so the ratio scale must be much greater—like 1:24,000. This ratio scale can also be expressed as a **fractional scale** (1/24,000), indicating that the portion of Earth represented has been reduced to the fraction of 1/24,000th of its actual size.

Therefore, a *ratio scale* of 1:24,000 equals a *fractional scale* of 1/24,000. They both are ways of indicating that any unit (inch, centimeter, foot, etc.) on the map represents 24,000 of the same units (inches, centimeters, feet) on Earth's surface. For example, 1 cm on the map represents 24,000 cm on the ground, or your thumb width on the map represents 24,000 thumb widths on the ground. Other common map scales are 1:25,000, 1:50,000, 1:62,500, 1:63,360, 1:100,000, 1:125,000, and 1:250,000. Figure 9.10 shows how maps at different scales show the same region but present different amounts of detail and area.

Drawing a map at 1:24,000 scale provides a very useful amount of detail. But knowing that 1 inch on the map = 24,000 inches on the ground is not very convenient, because no one measures big distances in inches! However, if you divide the 24,000 inches by 12 to get 2000 ft, the scale suddenly becomes useful: "1 in. on the map = 2000 ft on the ground." An

American football field is 100 yards (300 ft) long, so: "1 in. on the map = $6\frac{2}{3}$ football fields." Such scales expressed with words are called **verbal scales.**

On a map with a scale of 1:63,360, 1 in. = 63,360 in., again not meaningful in daily use. But there are 63,360 in. in a mile. So, the verbal scale, "1 in. = 1 mi," is very meaningful. A standard 1:62,500 map (15-minute quadrangle map) is very close to this scale, so it is common practice to say that "one inch equals approximately one mile" on such a map. Note that verbal scales are often approximate because their sole purpose is to increase the convenience of using a map.

Finally, all topographic maps have one or more **graphic bar scales** printed in their lower margin. They are essentially rulers for measuring distances on the map. U.S. Geological Survey topographic maps generally have four different bar scales: miles, feet, kilometers, and meters.

PLS—Public Land Survey System

The **U.S. Public Land Survey System (PLS)** was initiated in the late 1700s. All but the original thirteen states, and a few states derived from them, are covered by this system. Other exceptions occur in the southwestern United States, where land surveys may be based upon Spanish land grants, and in areas of rugged terrain that were never surveyed.

The PLS scheme was established in each state by surveying **principal meridians,** which are north–south lines, and **base lines,** which are east–west lines (Figure 9.11A). Once the initial principal meridian and base lines were established, additional lines were surveyed parallel to them and six miles apart. This created a grid of 6 mi by 6 mi squares of land. The north–south squares of the grid are called **townships** and are numbered relative to the base line (Township 1 North, Township 2 North, etc.). The east–west squares of the grid are **ranges** and are numbered relative to the principal meridian (Range 1 West, Range 2 West, etc.). Each 6 mi by 6 mi square is, therefore, identified by its township and range position in the PLS grid. For example, the square in Figure 9.11B is located at T1S (Township 1 South) and R2W (Range 2 West). Although each square like this is identified as both a township and a range within the PLS grid, it is common practice to refer to the squares as townships rather than township-and-ranges.

Townships are used as political subdivisions in some states and are often given placenames. Each township square is also divided into 36 small squares,

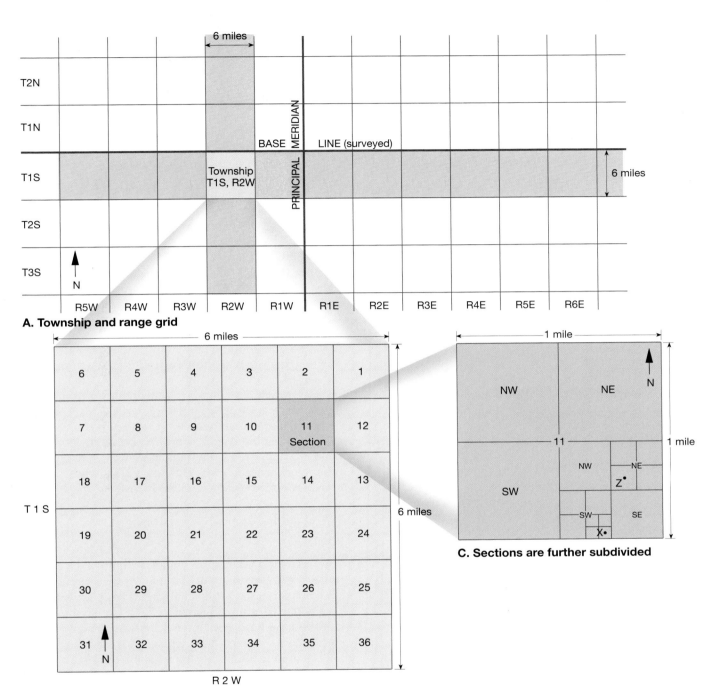

A. Township and range grid

B. A township contains 36 sections

C. Sections are further subdivided

FIGURE 9.11 U.S. Public Land Survey System (PLS) is based on a grid of townships and ranges unique to specific states or regions. **A**. Townships and ranges are located relative to a state's *principal meridian* (N-S line) of longitude and its *base line* (E-W line, surveyed perpendicular to the principal meridian). *Township* strips (columns) of land are 6 miles long and parallel to the base line. North of the base line, townships are numbered T1N, T2N, and so on. South of the base line, townships are numbered T1S, T2S, and so on. *Range* strips (rows) of land are 6 miles wide and parallel to the principal meridian. East of the principal meridian, ranges are numbered R1E, R2E, and so on. West of the principal meridian, ranges are numbered R1W, R2W, and so on. Each intersection of a township strip of land with a range strip of land forms a square, called a *township*. Note the location of Township T1S, R2W. **B**. Each township is 6 miles wide and 6 miles long, so it contains 36 square miles. Each square mile (640 acres) is called a *section,* and each section is numbered exactly as shown above. **C**. Sections are subdivided according to a hierarchy of square *quarters* listed in order of increasing size and direction. For example, point **x** is located in the southeast quarter, of the southeast quarter, of the southwest quarter, of the southeast quarter, of section 11. This is written: $SE\frac{1}{4}$, $SE\frac{1}{4}$, $SW\frac{1}{4}$, $SE\frac{1}{4}$, sec. 11, T1S, R2W.

each having an area of 1 square mile (640 acres). These squares are called **sections.**

Sections are numbered from 1 to 36, beginning in the upper right corner (Figure 9.11B). Sometimes these are shown on topographic quadrangle maps (Figure 9.3, red grid). Any point can be located precisely within a section by dividing the section into quarters (labeled NW, NE, SW, SE). Each of these quarters can itself be subdivided into quarters and labeled (Figure 9.11C).

GPS—Global Positioning System

The **Global Positioning System (GPS)** is a constellation of 28 navigational communication satellites in 12-hour orbits approximately 12,000 miles above Earth (about 24 of these are operational at any given time). The GPS constellation is maintained by the United States (NOAA and NASA) for operations of the U.S. Department of Defense, but it is free for anyone to use. Since GPS receivers can be purchased for as little as $100, they are widely used by airplane navigators, automated vehicle navigation systems, ship captains, hikers, and scientists to map locations on Earth. More expensive and accurate receivers with millimeter accuracy are used for space-based geodesy measurements that reveal plate motions over time (Laboratory 2).

Each GPS satellite communicates simultaneously with fixed ground-based Earth stations and other GPS satellites, so it knows exactly where it is located relative to the center of Earth and Universal Time Coordinated (UTC, also called Greenwich Mean Time). Each GPS satellite also transmits its own radio signal on a different channel, which can be detected by a fixed or handheld GPS receiver. If you turn on a handheld GPS receiver in an unobstructed outdoor location, then the receiver immediately *acquires* (picks up) the radio channel of the strongest signal it can detect from a GPS satellite. It downloads the navigational information from that satellite channel, followed by a second, third, and so on. A receiver must acquire and process radio transmissions from at least four GPS satellites to triangulate a determination of its exact position and elevation—this is known as a *fix*.

Most newer models of GPS receivers are *12 channel parallel receivers,* which means they can receive and process radio signals from as many as twelve satellites at the same time (the maximum possible number for any point on Earth). Older models cycle through the channels one at a time, or have fewer parallel channels, so they take longer to process data and usually give less accurate results. An unobstructed view is also best (GPS receivers cannot operate indoors). If the path from satellite to receiver is obstructed by

trees, canyon walls, or buildings, then the receiver has difficulty acquiring that radio signal. It is also possible that more or fewer satellites will be nearly overhead at one time than another, because they are in constant motion within the constellation. Therefore, if you cannot obtain a fix at one time (because four satellite channels cannot be acquired), you may be able to obtain a fix in another half hour or so. Acquiring more than four satellite channels will provide more navigational data and more accurate results. Most handheld, 12-channel parallel receivers have an accuracy of about 10–15 meters.

When using a GPS receiver for the first time in a new region, it generally takes about one to three minutes for it to triangulate a fix. This information is stored in the receiver, so readings taken over the next few hours at nearby locations normally take only seconds. Consult the operational manual for your receiver so you know the time it normally takes for a *cold* fix (first time) or *warm* fix (within a few hours of the last fix).

GPS navigation does not rely on the latitude-longitude or the public land survey system. It relies on an Earth-centered geographic grid and coordinate system called the *World Geodetic System 1984* or *WGS 84. WGS 84* is a **datum** (survey or navigational framework) based on the Universal Transverse Mercator (UTM) grid described below.

UTM—Universal Transverse Mercator System

The U.S. National Imagery and Mapping Agency (NIMA) developed a global military navigation grid and coordinate system in 1947 called the **Universal Transverse Mercator System (UTM).** Unlike the latitude-longitude grid that is spherical and measured in degrees, minutes, seconds, and nautical miles (1 nautical mile = 1 minute of latitude), the UTM grid is rectangular and measured in decimal-based metric units (meters).

The UTM grid (top of Figure 9.12) is based on sixty north–south **zones,** which are strips of longitude having a width of 6°. The zones are consecutively numbered from Zone 01 (between 180° and 174° west longitude) at the left margin of the grid, to Zone 60 (between 174° and 180° east longitude) at the east margin of the grid. The location of a point within a zone is defined by its **easting** coordinate—its distance within the zone measured in meters from west to east, and a **northing** coordinate—its distance from the Equator measured in meters. In the Northern Hemisphere, northings are given in meters north of the Equator. To avoid negative numbers for northings in

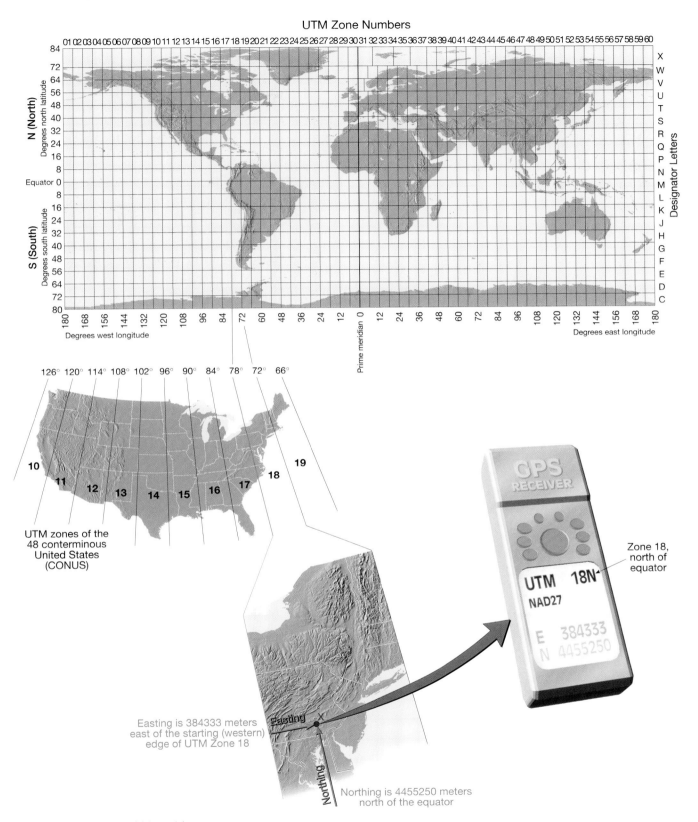

UTM Zone Numbers

01 02 03 04 05 06 07 08 09 10 11 12 13 14 15 16 17 18 19 20 21 22 23 24 25 26 27 28 29 30 31 32 33 34 35 36 37 38 39 40 41 42 43 44 45 46 47 48 49 50 51 52 53 54 55 56 57 58 59 60

N (North) Degrees north latitude

84 72 64 56 48 40 32 24 16 8

Equator 0

S (South) Degrees south latitude

8 16 24 32 40 48 56 64 72 80

Degrees west longitude

180 168 156 144 132 120 108 96 84 72 60 48 36 24 12

Prime meridian 0

12 24 36 48 60 72 84 96 108 120 132 144 156 168 180

Degrees east longitude

Designator Letters

X W V U T S R Q P N M L K J H G F E D C

126° 120° 114° 108° 102° 96° 90° 84° 78° 72° 66°

10 11 12 13 14 15 16 17 18 19

UTM zones of the 48 conterminous United States (CONUS)

GPS RECEIVER

Zone 18, north of equator

UTM 18N

NAD27

E 384333
N 4455250

Easting is 384333 meters east of the starting (western) edge of UTM Zone 18

Easting X

Northing

Northing is 4455250 meters north of the equator

FIGURE 9.12 UTM and GPS. A handheld Global Positioning System (GPS) receiver operated at point **X** indicates its location according to the Universal Transverse Mercator (UTM) grid and coordinate system, *North American Datum 1927 (NAD27)*. Refer to text for explanation.

the Southern Hemisphere, NIMA assigned the Equator a reference northing of 10,000,000 meters.

Since satellites did not exist until 1957, and GPS navigational satellites did not exist until decades later, the UTM grid was applied for many years using regional ground-based surveys to determine locations of the grid boundaries. Each of these regional or continental surveys is called a **datum** and is identified on the basis of its location and the year it was surveyed. Examples include *North American Datum 1927 (NAD27)* and *North American Datum 1983 (NAD83)*, which appear on many Canadian and U.S. Geological Survey topographic quadrangle maps. The Global Positioning System relies on an Earth-centered UTM datum called the *World Geodetic System 1984* or *WGS 84*, but GPS receivers can be set up to display regional datums like *NAD27*. When using GPS with a topographic map, be sure to set the GPS receiver to display the UTM datum of that map.

Study the illustration of a GPS receiver in Figure 9.12. Notice that the receiver is displaying UTM coordinates (based on *NAD27*) for a point **X** in Zone 18 (north of the Equator). Point **X** has an easting coordinate of E384333, which means that it is located 384333 meters east of the starting (west) edge of Zone 18. Point **X** also has a northing coordinate of N4455250, which means that it is located 4455250 meters north of the Equator. Therefore, Point **X** is located in southeast Pennsylvania. To plot Point **X** on a 1:24,000 scale, $7\frac{1}{2}$ minute topographic quadrangle map, see Figure 9.13.

Point **X** is located within the Lititz, PA $7\frac{1}{2}$ Minute Series (USGS, 1:24,000 scale) topographic quadrangle map (Figure 9.13). Information printed on the map margin indicates that the map has blue ticks spaced 1000 m apart along its edges that conform to *NAD27*, Zone 18. Notice how the ticks for northings (blue) and eastings (green) are represented on the northwest corner of the Lititz map—Figure 9.13B. One northing label is written out in full ($^{44}56^{000m}$ N) and one easting label is written out in full ($^{3}84^{000m}$ E), but the other values are given in UTM shorthand for thousands of meters (i.e., do not end in^{000m}). Since Point **X** has an easting of E384333 within Zone 18, it must be located 333 m east of the tick mark labeled $^{3}84^{000m}$ E along the top margin of the map. Since Point **X** has a northing of N4455250, it must be located 250 m north of the tick mark labeled as $^{44}55$ in UTM shorthand. Distances east and north can be measured using a ruler and the map's graphic bar scale as a reference (333 m = 0.333 km, 250 m = 0.250 km). However, you can also use the graphic bar scale to construct a UTM grid like the one in Figure 9.13C. If you construct such a grid and print it onto a transparency, then you can use it as a UTM *grid overlay*. To plot a

point or determine its coordinates, place the grid overlay on top of the square kilometer in which the point is located. Then use the grid as a two-dimensional ruler for the northing and easting. Grid overlays for many different scales of UTM grids are provided in GeoTools Sheets 2–4 at the back of the manual for you to cut out and use.

The UTM system described above is known as the *civilian UTM grid and coordinate system*, and it is the system you should use in your work in this manual. The U.S. Department of Defense has modified this civilian UTM grid to form a Military Grid Reference System (MGRS) that divides the zones into horizontal sections identified by *designator letters* (Figure 9.12). These sections are 8° wide and lettered consecutively from **C** (between 80° and 72° south latitude) through **X** (between 72° and 84° north latitude). Letters I and O are not used.

Compass Bearings

A **bearing** is the *compass direction* along a line from one point to another. If expressed in degrees east or west of true north or south, it is called a *quadrant bearing*. Or it may be expressed in degrees between 0 and 360, called an *azimuth bearing*, where north is 0° (or 360°), east is 90°, south is 180°, and west is 270°. Linear geologic features (faults, fractures, dikes), lines of sight and travel, and linear property boundaries are all defined on the basis of their bearings.

Remember that a compass points to Earth's *magnetic north* (MN) pole rather than the *grid north* (GN) pole that was used to construct the UTM and latitude-longitude grids of a map. Therefore, a diagram at the margin of every topographic map shows the *declination* (degrees of difference) between MN and GN. If the MN arrow is to the right of GN, then subtract the degrees of declination from your compass reading. If the MN arrow is to the left of GN, then add the degrees of declination to your compass reading. These adjustments will mean that your compass readings are synchronized with the map. However, the magnetic pole migrates very slowly, so the declination is exact only for the year listed on the map.

To determine a compass bearing on a map, draw a straight line from the starting point to the destination point and also through any one of the map's borders. Align a protractor (left drawing, Figure 9.14) or the N–S or E–W directional axis of a compass (right drawing, Figure 9.14) with the map's border, and read the bearing in degrees toward the direction of the destination. Imagine that you are buying a property for your dream home. The boundary of the property is marked by four metal rods driven into the ground,

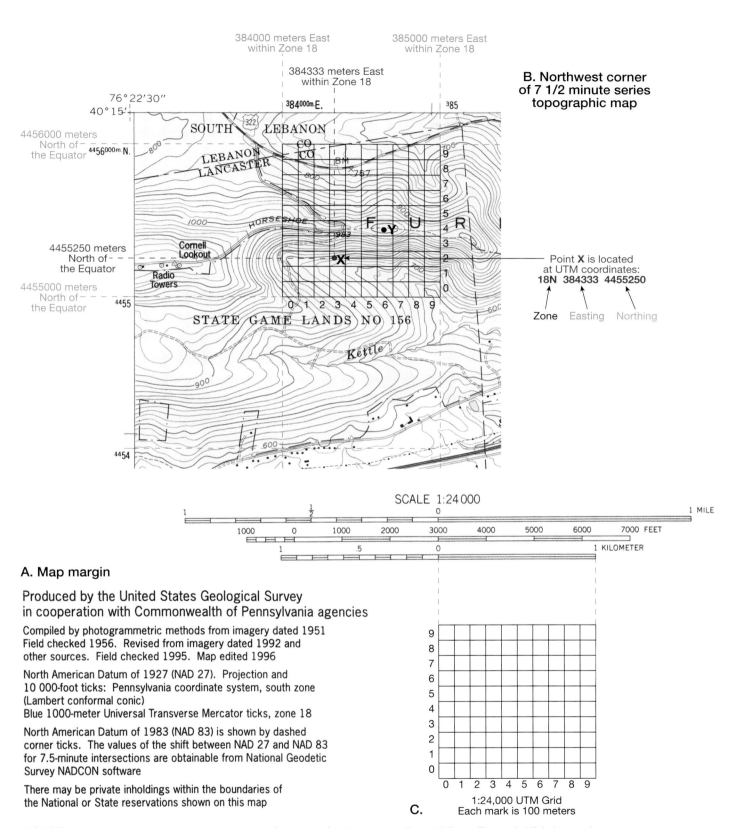

384000 meters East
within Zone 18

385000 meters East
within Zone 18

384333 meters East
within Zone 18

4456000 meters
North of
the Equator

4455250 meters
North of
the Equator

4455000 meters
North of
the Equator

Point **X** is located
at UTM coordinates:
18N 384333 4455250

Zone   Easting   Northing

SCALE 1:24 000

A. Map margin

Produced by the United States Geological Survey
in cooperation with Commonwealth of Pennsylvania agencies

Compiled by photogrammetric methods from imagery dated 1951
Field checked 1956. Revised from imagery dated 1992 and
other sources. Field checked 1995. Map edited 1996

North American Datum of 1927 (NAD 27). Projection and
10 000-foot ticks: Pennsylvania coordinate system, south zone
(Lambert conformal conic)
Blue 1000-meter Universal Transverse Mercator ticks, zone 18

North American Datum of 1983 (NAD 83) is shown by dashed
corner ticks. The values of the shift between NAD 27 and NAD 83
for 7.5-minute intersections are obtainable from National Geodetic
Survey NADCON software

There may be private inholdings within the boundaries of
the National or State reservations shown on this map

C.

1:24,000 UTM Grid
Each mark is 100 meters

FIGURE 9.13 UTM and topographic maps—refer to text for discussion. Point **X** (from Figure 9.12) is located
within the Lititz, PA $7\frac{1}{2}$ Minute Series (USGS, 1:24,000 scale) topographic quadrangle map. **A.** Map margin
indicates that the map includes UTM grid data based on *North American Datum 1927* (*NAD27*, Zone 18)
and represented by blue ticks spaced 1000 meters (1 km) apart along the map edges. **B.** Connect the blue
1000 m ticks to form a grid square, each representing 1 square kilometer. Northings (blue) are read along
the N-S map edge, and eastings (green) are located along the E-W map edge. **C.** You can construct a 1 km
grid (1:24,000 scale) from the map's bar scale, then make a transparency of it to form a grid overlay (see
GeoTools Sheet 2 and 4 at back of manual). Place the grid overlay atop the 1-kilometer square on the map
that includes point **X**, and determine the *NAD27* coordinates of **X** as shown (red).

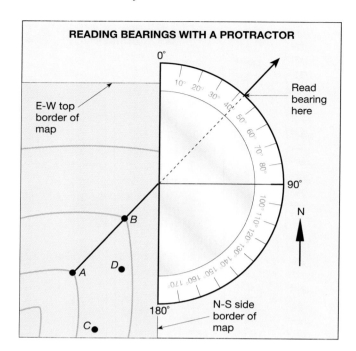

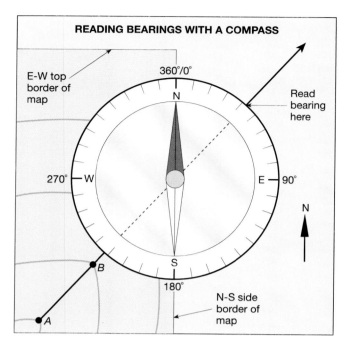

FIGURE 9.14 Examples of how to read the *bearing* (compass direction) from one point to another using a map and protractor (left) or compass (right). Remember that a compass points to Earth's *magnetic north* pole (MN) rather than the North Pole upon which the UTM and latitude-longitude grid of map is based, called *grid north* (GN). A diagram at the margin of each map shows the *declination* (difference) between MN and GN. If the MN arrow is to the right of GN then subtract the degrees of declination from your compass reading. If the MN arrow is to the left of GN then add the degrees of declination to your compass reading.

To determine a compass bearing on a map, draw a straight line from the starting point to the destination point and also through any one of the map's borders. For example, to find the bearing from *A* to *B*, a line was drawn through both points and the east edge of the map. Align a protractor (left drawing) or the N-S or E-W directional axis of a compass (right drawing) with the map's border and read the bearing in degrees toward the direction of the destination. For example above, notice that the *quadrant bearing* from point *A* to *B* is North 43° East (left map, using protractor) or an *azimuth bearing* of 43°. If you walked in the exact opposite direction, from *B* to *A*, then you would walk along a quadrant bearing of South 43° West or an azimuth bearing of 223° (i.e., 43° + 180° = 223°).

one at each corner of the property. The location of these rods is shown on the map in Figure 9.14 (left side) as points *A*, *B*, *C*, and *D*. The property deed notes the distances between the points *and* bearings between the points. This defines the shape of the property. Notice that the northwest edge of your property lies between two metal rods located at points *A* and *B*. You can measure the distance between the points using a tape measure. How can you measure the bearing?

First, draw a line (very lightly in pencil so that it can be erased) through the two points, *A* and *B*. Make sure the line also intersects an edge of the map. In both parts of Figure 9.14, a line was drawn through points *A* and *B* so that it also intersects the east edge of the map. Next, orient a protractor so that its 0° and 180° marks are on the edge of the map, with the 0°

end toward geographic north. Place the origin of the protractor at the point where your line *A–B* intersects the edge of the map. You can now read a bearing of 43° east of north. We express this as a quadrant bearing of "North 43° East" (written N43°E) or as an azimuth bearing of 43°. If you were to determine the opposite bearing, from *B* to *A*, then the bearing would be pointing southwest and would be read as "South 43° West," or as an azimuth of 223°.

You also can use a compass to read bearings, as shown in Figure 9.14 (right). Ignore the compass needle and use the compass as if it were a circular protractor. Some compasses are graduated in degrees, from 0–360, in which case you read an azimuth bearing from 0–360°. Square azimuth protractors for this purpose are provided in GeoTools Sheets 3 and 4 at the back of this manual.

Questions

1. Draw the 300-foot contour line on Figure 9.4. Draw the 1200-foot contour line on Darwin Volcano. Then use interpolation (Figure 9.6, rule 2) to draw the 600- and 900-foot contour lines on Darwin Volcano.

2. What are the latitude-longitude coordinates of point **B** in Figure 9.1?

3. What are the three forms of vegetation shown by the green patterns in sec. 29, T1N, R6W, in Figure 9.3?

4. The map in Figure 9.3 is contoured in feet above sea level. What is its contour interval?

5. Refer to Figure 9.3 and fill in the blanks below.

    a. The ratio scale of this map is: _____.

    b. One inch on this map equals exactly _____ inches in real life.

    c. One inch on this map equals exactly _____ mile(s) in real life, so we can say that 1 inch on the map equals approximately 1 mi on the ground.

6. Review how the location of point **x** in Figure 9.11C was determined using PLS shorthand (see the caption for Figure 9.11). Then, determine the location of point **z** in Figure 9.11C using PLS shorthand.

7. How many acres are present in the township in Figure 9.11B? (*Hint:* There are 640 acres in 1 mi^2.)

8. Imagine that you wanted to purchase the $NE\frac{1}{4}$ of the $SE\frac{1}{4}$ of section 11 in Figure 9.11C. If the property costs $500 per acre, then how much must you pay for the entire property? Explain.

9. Examine Figure 9.3.

    a. In what UTM Zone is this map located? Explain your reasoning.

    b. What are the exact UTM coordinates of the northeast corner of this map?

10. What is the bearing from point C to point D in Figure 9.14?

11. What is the bearing from point D to point C in Figure 9.14? (Refer to Figure 9.6 as needed.)

12. Most handheld 12-channel parallel GPS receivers have an error of about 15 m when they fix on their position, and most geologists plot their data on $7\frac{1}{2}$-minute topographic quadrangle maps that have a ratio scale of 1:24,000. If an object is 15 meters long in real life, then exactly how long (in millimeters) would it be on the 1:24,000 scale map?

13. Contour Figure 9.15 using a contour interval of 100 ft. (Refer to Figure 9.6 as needed.)

14. Contour Figure 9.16 using a contour interval of 10 ft. (Refer to Figure 9.6 as needed.)

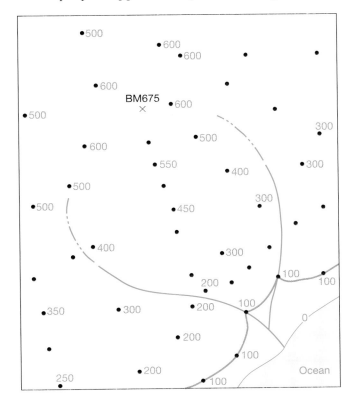

FIGURE 9.15 Use interpolation and extrapolation to estimate elevations of points that are not labeled (see Figure 9.6). Then add contour lines with a 100-foot contour interval. Note how the 0-foot and 100-foot contour lines have already been drawn.

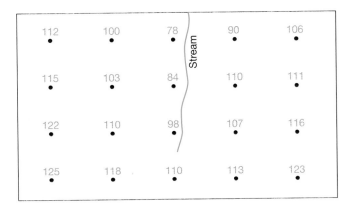

FIGURE 9.16 Construct a topographic map by contouring these elevations. Use a contour interval of 10 feet. (Refer to Figure 9.5 as needed.)

Contour Interval = 20 feet

Contour Interval = 20 feet

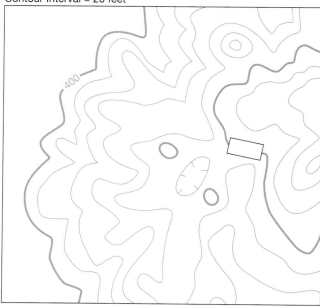

FIGURE 9.17 Topographic map interpretation. Use your pencil to lightly shade in the portion of this map that represents the highest elevation of land. Label a hill, "H." Label a ridge, "R." Label a saddle, "S." Use an arrow to label the lowest contour line in the map and label the arrow with the elevation of the contour. (Refer to Figures 9.5–9.8 as needed.)

FIGURE 9.18 Topographic map interpretation. Use your pencil to lightly shade in the portion of this map that represents the lowest elevation. Label a closed depression, "CD." In the small box, write the elevation of the index contour on which it lies. (Refer to Figures 9.5–9.8 as needed.)

15. Using your pencil, color in the area of Figure 9.17 that represents the top of the highest hill.

16. In Figure 9.18, fill in the box on the index contour with its correct elevation in feet above sea level. (*Note:* Contour interval is 20 feet.) Using your pencil, color in the area of Figure 9.18 that represents the lowest elevations. Label a closed depression with the letters "CD."

17. Using a contour interval of 10 ft, label every contour line on the topographic map in Figure 9.19 with its exact elevation above sea level.

18. Analyze the topographic map in Figure 9.20.

   a. The contour lines on this map are labeled in meters. What is the contour interval of this map?

   b. What is the regional relief of the land represented in this map?

   c. What is the gradient from **Y** to **X**?

   d. How could you find the areas of this map that have a gradient of 20 meters per kilometer or greater? (*Hint:* Think of the contour interval and how many contour lines of map elevation must occur along one kilometer of map distance.)

   e. Imagine that you need to drive a truck from point **A** to point **B** in this mapped area and that your truck cannot travel up any slopes having a gradient over 20 m/km. Trace a route that you could drive to get from point **A** to point **B**. (More than one solution is possible.)

PART 9B: TOPOGRAPHIC PROFILES AND VERTICAL EXAGGERATION

A topographic map provides an overhead (aerial) view of an area, depicting features and relief by means of its symbols and contour lines. Occasionally a cross section of the topography is useful. A **topographic profile** is a cross section that shows the elevations and slopes along a given line (Figure 9.21).

Follow these steps and Figure 9.21 to construct a topographic profile:

Step 1: On the map, draw a **line of section** along which the profile is to be constructed. Label the section line **A–A′**. Be sure that the line intersects all of

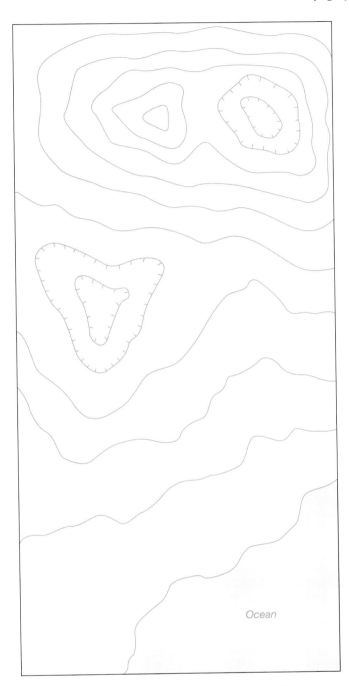

Ocean

FIGURE 9.19 Complete this topographic map. Use a contour interval of 10 ft and label the elevation of every contour on the map. (*Hint*: Start at sea level and refer to Figures 9.8 and 9.9.)

the features (ridges, valleys, streams, etc.) that you wish the profile to show.

Step 2: On a strip of paper placed along section line **A—A′,** make tick marks at each place where a contour line intersects the section line and note the elevation

at the tick marks. Also note the location and elevation of points **A, A′,** and any streams crossed.

Step 3: Draw the profile. On a separate sheet of paper, draw a series of equally spaced parallel lines that are the same length as the line of section (graph paper can be used). Each horizontal line on this sheet represents a *constant elevation* and therefore corresponds to a contour line. The total number of horizontal lines that you need, and their elevations, depends on the total relief along the line of section and on whether you make the space between the lines equal to the contour interval, or to multiples of it (vertical exaggeration, which will be discussed shortly). Label your lines so that the highest and lowest elevations along the line of section will be within the grid.

Then, take the strip of paper you marked in Step 2 and place it along the base of your profile. Mark a dot on the grid above it for each elevation. Smoothly connect these dots to complete the topographic profile. (This line should not make angular bends. Make it a smoothly curving line that reflects the relief of the land surface along the line of section.)

Step 4: The vertical scale of your profile will vary greatly depending on how you draw the grid. It almost certainly will be larger than the horizontal scale of the map. This difference causes an exaggeration in the vertical dimension. Such exaggeration almost always is necessary to construct a readable profile, for without vertical exaggeration, the profile might be so shallow that only the highest peaks would be visible. Calculate the **vertical exaggeration** by one of two methods. *You can divide the horizontal ratio scale* (1:24,000) *by the vertical ratio scale* (1:1,440), which reduces to 24,000/1,440, which reduces to 16.7 (Method 1, Step 4, Figure 9.21). Or *you can divide the vertical fractional scale* (1/1,440) *by the horizontal fractional scale* (1/24,000), which reduces to 24,000/1,440, which reduces to 16.7 (Method 2, Step 4, Figure 9.21). The number 16.7 (usually written 16.7 ×) indicates that the relief shown on the profile is 16.7 times greater than the true relief. This makes the slopes on the profile 16.7 times steeper than the corresponding real slopes on the ground.

Question

19. In Figure 9.22, construct a topographic profile (using the graph paper provided beneath it) and calculate its vertical exaggeration.

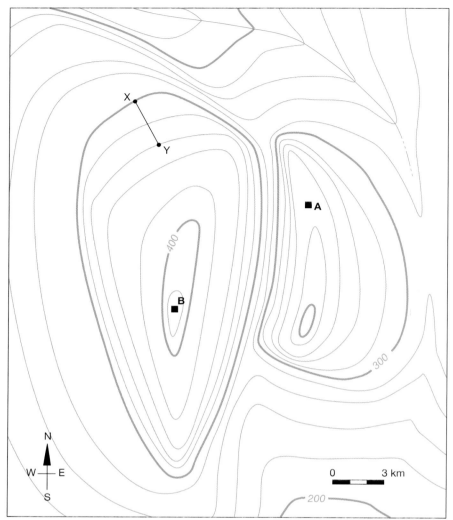

Elevations are in meters.

FIGURE 9.20 *Gradient* is a measure of the steepness of a slope, expressed in feet per mile or meters per kilometer. To determine the gradient of a slope, divide the *relief* (difference in elevation between two points on a map) by the distance measured between the two points. This is sometimes called *rise over run*. For example, this topographic map is contoured in meters. Can you determine the contour interval? Can you determine the gradient from point **X** to point **Y**? Can you plot a path from point **A** to point **B** that does not cross any slopes with a gradient above 20 meters per kilometer? Explain your reasoning.

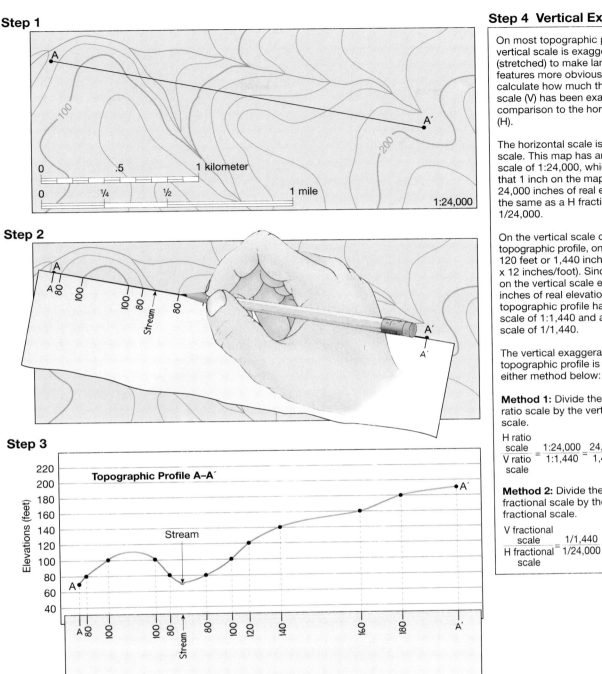

Step 1

A

A'

100

200

| 0 | .5 | 1 kilometer |

| 0 | 1/4 | 1/2 | 1 mile |

1:24,000

Step 2

A
A'
80
100
100
80
Stream
80
A'
A'

Step 3

Topographic Profile A–A´

Elevations (feet)

220
200
180
160
140
120
100
80
60
40

A

Stream

A´

A 80 100   100 80 Stream 80 100 120 140   160   180   A´

Paper strip with elevations noted beside tick marks.

Step 4 Vertical Exaggeration

On most topographic profiles, the vertical scale is exaggerated (stretched) to make landscape features more obvious. One must calculate how much the vertical scale (V) has been exaggerated in comparison to the horizontal scale (H).

The horizontal scale is the map's scale. This map has an H ratio scale of 1:24,000, which means that 1 inch on the map equals 24,000 inches of real elevation. It is the same as a H fractional scale of 1/24,000.

On the vertical scale of this topographic profile, one inch equals 120 feet or 1,440 inches (120 feet x 12 inches/foot). Since one inch on the vertical scale equals 1,440 inches of real elevation, the topographic profile has a V ratio scale of 1:1,440 and a V fractional scale of 1/1,440.

The vertical exaggeration of this topographic profile is calculated by either method below:

Method 1: Divide the horizontal ratio scale by the vertical ratio scale.

$$\frac{\text{H ratio scale}}{\text{V ratio scale}} = \frac{1{:}24{,}000}{1{:}1{,}440} = \frac{24{,}000}{1{,}440} = 16.7\times$$

Method 2: Divide the vertical fractional scale by the horizontal fractional scale.

$$\frac{\text{V fractional scale}}{\text{H fractional scale}} = \frac{1/1{,}440}{1/24{,}000} = \frac{24{,}000}{1{,}440} = 16.7\times$$

FIGURE 9.21 Topographic profile construction and vertical exaggeration. Shown are a topographic map (Step 1), topographic profile constructed along line **A–A'** (Steps 2 and 3), and calculation of vertical exaggeration (Step 4). **Step 1**—Select two points (**A**, **A'**), and the line between them (line **A–A'**), along which you want to construct a topographic profile. **Step 2**—To construct the profile, the edge of a strip of paper was placed along line **A–A'** on the topographic map. A tick mark was then placed on the edge of the paper at each point where a contour line and stream intersected the edge of the paper. The elevation represented by each contour line was noted on its corresponding tick mark. **Step 3**—The edge of the strip of paper (with tick marks and elevations) was placed along the bottom line of a piece of lined paper, and the lined paper was graduated for elevations (along its right margin). A black dot was placed on the profile above each tick mark at the elevation noted on the tick mark. The black dots were then connected with a smooth line to complete the topographic profile. **Step 4**—*Vertical exaggeration* of the profile was calculated using either of two methods. Thus, the vertical dimension of this profile is exaggerated (stretched) to 16.7 times greater than it actually appears in nature compared to the horizontal/map dimension.

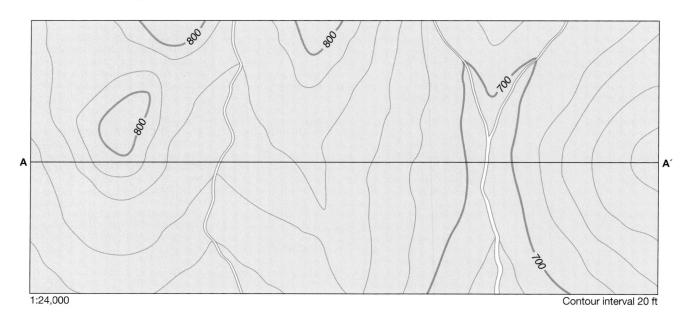

1:24,000   Contour interval 20 ft

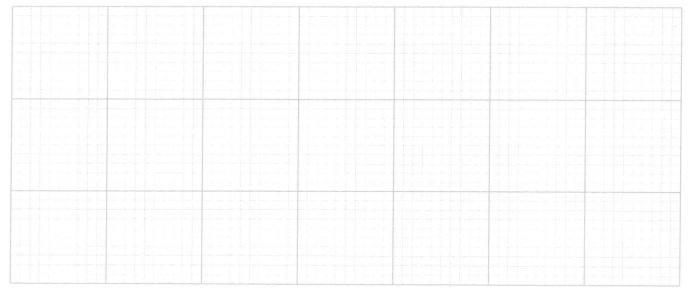

FIGURE 9.22 Topographic profile construction and vertical exaggeration. Can you construct a topographic profile for line **A–A'** and calculate its vertical exaggeration?

PART 9C: ANALYSIS OF THE ONTARIO, CALIFORNIA, TOPOGRAPHIC MAP

Questions

Refer to Figure 9.3 to answer the following questions.

20. What natural feature is located in the $SW\frac{1}{4}$, $NE\frac{1}{4}$, sec. 13, T1N, R7W? (*Hint:* It is a small body of water.)

21. Locate Bighorn Peak in the northwest corner of the map.

    a. Use the PLS numbering system to describe the location of Bighorn Peak.

    b. Use UTM coordinates to describe the location of Bighorn Peak.

    c. What is the exact elevation of Bighorn Peak?

22. What is the distance in kilometers from the railroad intersection in Grapeland (sec. 31, T1N,

R6W) to the Day Canyon Guard Station (sec. 17, T1N, R6W)?

23. What is the difference in elevation between Grapeland and the Day Canyon Guard Station?

24. What is the average gradient of the land between Grapeland and the Day Canyon Guard Station?

25. In what general direction does the stream in Day Canyon flow (sec. 31, T2N, R6W and sec. 6, T1N, R6W)? How can you tell?

26. What is the total relief on this map area? Show your calculation.

PART 9D: ANALYSIS OF YOUR TOPOGRAPHIC QUADRANGLE MAP

Questions

Refer to the quadrangle map provided by your instructor (or otherwise obtained by you) and answer the following questions. (*Do not mark the map with any dark lines or pen marks!*)

27. What latitude line marks the northern boundary of the quadrangle?

28. What latitude line marks the southern boundary of the quadrangle?

29. What longitude line marks the eastern boundary of the quadrangle?

30. What longitude line marks the western boundary of the quadrangle?

31. What is the distance (in degrees, minutes, and seconds) from the southern to the northern boundary of the quadrangle?

32. What is the distance (in degrees, minutes, and seconds) from the western to the eastern boundary of the quadrangle?

33. Is this a $7\frac{1}{2}$-minute quadrangle or a 15-minute quadrangle?

34. What UTM datum(s) is/are represented on the map?

35. In what UTM Zone is this map located?

36. Give the location of the northwest corner of the map in UTM coordinates. Be sure to specify the datum used if there is more than one datum represented on the map.

37. What is the magnetic declination (between true north and magnetic north) of this quadrangle?

38. In what year was the map originally published?

39. What is the fractional scale of this quadrangle?

40. How can this scale be expressed as a verbal scale in miles? In kilometers?

41. Two inches on this quadrangle map represent how many feet on the ground? How many miles? How many kilometers?

42. What is the name of this quadrangle?

43. What is the name of the quadrangle map directly adjacent to the south?

44. What color was used to indicate these features?

   a. water

   b. vegetation (mainly woods or forests)

   c. contour lines

   d. buildings

45. Was this map ever photorevised? If so, when?

46. What is the elevation of the highest point on the quadrangle?

47. What is the elevation of the lowest point on the quadrangle?

48. What is the total relief within the quadrangle?

PART 9E: AERIAL PHOTOGRAPHS

Aerial photographs are pictures of Earth taken from airplanes, with large cameras that generally make 9-by-9-inch negatives. Most of these photographs are black and white, but color pictures sometimes are available. The photographs may be large scale or small scale, depending on the elevation at which they were taken, on the focal length of the camera lens, and on whether the pictures have been enlarged or reduced from the negatives.

Air photographs can be taken nearly straight down from the plane, termed *vertical*, or they may be taken at an angle to the vertical, termed *oblique*. Oblique views help reveal geological features and landforms; however, vertical air photographs are even more useful in geological studies. The photographs used in this exercise are verticals.

Vertical air photos are taken in a series during a flight so that the images form a continuous view of the area below. They are taken so that approximately 60% image overlap occurs between any two adjacent photos. The view is straight down at the very center of each picture (called the *center point* or *principal point*), but all other portions of the landscape are viewed at an angle that becomes increasingly oblique away from the center of the picture.

The scale of any photographic image cannot be uniform, because it differs with the distance of the camera lens from the ground. Thus, in photos of flat terrain, the scale is largest at the center of the photo, where the ground is closest to the camera lens, and decreases away from the center. Also, hilltops and other high points that are closer to the camera lens are shown at larger scales than are valley bottoms and other low places.

Air photos commonly are overlapped to form a **stereogram,** or **stereo pair,** to be viewed with a *stereoscope* (Figures 9.23, 9.24). When the photos are viewed through the stereoscope, the image appears three-dimensional (stereo). This view is startling, dramatic, and reveals surprises about the terrain, as you shall see shortly. Stereoscopes can be of many types, but the most commonly used variety is a *pocket stereoscope* such as the one shown in Figure 9.23.

Figure 9.24 shows parts of three overlapping vertical air photos. They have been cropped (trimmed) and mounted in sequence. The view is of Garibaldi Provincial Park in British Columbia, Canada. All three pictures show a dark volcanic cinder cone bulging out into Garibaldi Lake from the edge of Mount Price. Each photo shows it from a different overhead viewpoint. These landscape features are depicted by contours on the topographic map in Figure 9.25.

The center point of each photo is marked with a circled **X.** In the right-hand photo (BC 866:50), the center point is in the lake. In the middle photo (BC 866:49), it is near the cinder cone. In the left-hand photo (BC 866:48), it is near the margin of the page. By locating these center points on the map and connecting them with straight lines, you can see the **flight line,** or route flown by the photographing aircraft.

Locating the centers of these photos on the topographic map (Figure 9.25) is not easy, because the map and the photos show different types of features. However, you can plot the centers fairly accurately by referring to recognizable topographic features such as stream valleys and angles in the lakeshore.

Questions

49. Notice how the image of the cinder cone is distorted when you compare the three successive pictures. Not only is the base of the cone different in each picture, but the round patch of snow in the central crater appears to shift position relative to the base. In which photograph (left, middle, or right) does the image of the cone appear to be least

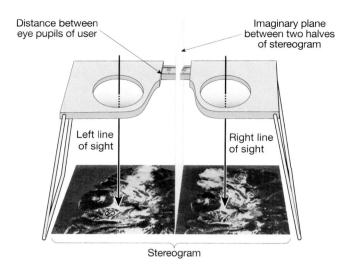

FIGURE 9.23 How to use a pocket stereoscope. First, have a partner measure the distance between the pupils of your eyes, in millimeters. Set the distance on the stereoscope. Then, position the stereoscope so that your lines of sight are aimed at a common point on each half of the stereogram (stereopair). As you look through the stereoscope, move it around slightly until the image "pops" into three dimensions. Be patient during this first attempt so that your eyes can focus correctly.

Most people do not need a stereoscope to see the stereograms in stereo (three-dimensions). Try holding the stereogram at a comfortable distance (one foot or so) from your eyes. Cross your eyes until you see four photographs (two stereograms), then relax your eyes to let the two center photographs merge into one stereo image.

distorted? Why? The same varying-perspective view that distorts features in air photos also makes it possible to view them **stereoscopically.** Thus, when any two overlapping photos in a sequence are placed side-by-side and viewed with the stereoscope, you see the overlap area as a vertically exaggerated three-dimensional image of the landscape. (With practice, you can train your eyes to do this without a stereoscope.)

50. The scale of the original negatives of the Garibaldi Lake photos is approximately 1:31,680, or 1 inch $= \frac{1}{2}$ mile. However, the pictures reproduced here have been reduced in size. Calculate their scale by measuring corresponding distances on the photos and on the topographic map, setting these distances as a ratio, and then multiplying the map scale by this ratio. What is the nominal scale thus derived? (You must write "nominal scale" because the actual scale differs with elevation and distance from the camera lens, as mentioned above.)

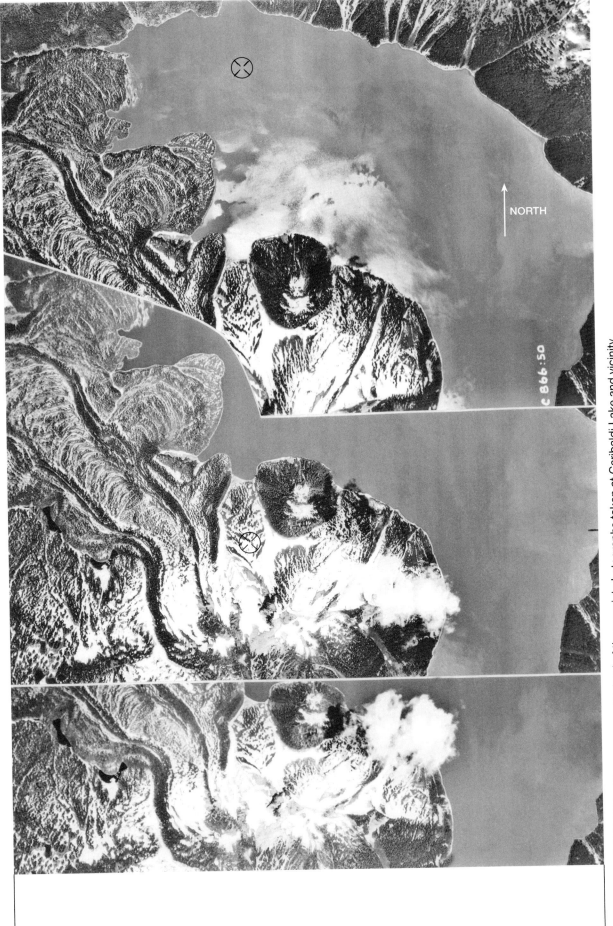

FIGURE 9.24 Stereogram comprised of three aerial photographs taken at Garibaldi Lake and vicinity, British Columbia, on 13 July 1949. (Photos BC 866:48–50, reproduced courtesy of Surveys & Resource Mapping Branch, Ministry of Environment, Government of British Columbia, Canada.)

NORTH

C 866:50

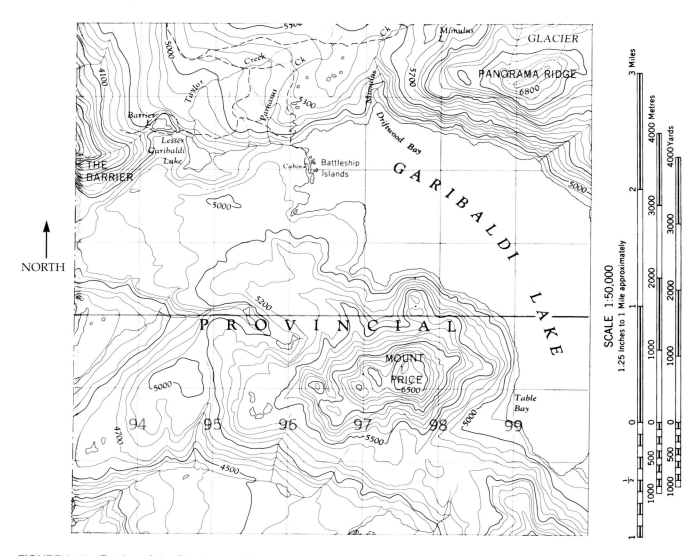

FIGURE 9.25 Portion of the Cheakamus River, East, topographic quadrangle map, British Columbia, on 13 July 1949; the same time at which the stereogram in Figure 9.24 was made. (Reproduced courtesy of Surveys & Resource Mapping Branch, Ministry of Environment, Government of British Columbia, Canada.)

51. Photograph BC 866:48 shows two broad pathways that sweep down the slopes of Mount Price from a point near the summit. One trends down to the south and west and the other to the north and west. Both are bordered by narrow ridges. What are these features? Of what rock type are they probably formed?

52. On the map and stereogram, examine both sides of the outlet channel where Garibaldi Lake over-

flows into Lesser Garibaldi Lake, near the northwest corner of BC 866:50. Is the rock that forms the slopes on the north side of the outlet channel the same or different from that on the south side? Name two features in the photos that lead you to this interpretation.

53. Lakes, and the basins they occupy, can be formed in many ways. On the basis of these photos, how do you think Garibaldi Lake formed?

Geologic Structures, Maps, and Block Diagrams

•CONTRIBUTING AUTHORS•

Michael J. Hozik • *Stockton State College*

William R. Parrott, Jr. • *Stockton State College*

Raymond W. Talkington • *Stockton State College*

OBJECTIVES

A. Be able to identify common kinds of geologic structures in three dimensional block diagrams and know the symbols used to represent them on geologic maps.

B. Be able to construct geologic cross sections from geologic maps and interpret them.

C. Be able to read and interpret geologic maps.

MATERIALS

Pencil, eraser, laboratory notebook, ruler, set of colored pencils, scissors, Cardboard Models 1–6 (located at the back of this laboratory manual), and a geologic map (provided by your instructor, or obtained as noted by your instructor).

INTRODUCTION

Structural geology is the study of how *geologic units* (bodies of rock or sediment) are arranged when first formed and how they are deformed afterward. When a body of rock or sediment is subjected to severe *stress* (directed pressure), then it may eventually *strain* (undergo deformation, such as a change in shape). Therefore, deformed formations are geologic units that have adjusted to a severe stress. Much of the

study of structural geology involves deciphering stress and strain relationships.

Generally, geologists can see how bodies of rock or sediment are positioned where they *crop out* (stick out of the ground as an outcrop) at Earth's surface (Figure 10.1A). Geologists record this outcrop data on flat (two-dimensional) **geologic maps** using different colors and symbols to represent the different units of rock or sediment and their positions (Figures 10.1B, 10.2, 10.3). They apply information from geologic maps to infer the three-dimensional arrangement of the units. The structural geology of an area can be described and interpreted from this three-dimensional arrangement, viewed as a conceptual model in your mind, or as a physical model. You will interpret as many as six different (physical) cardboard models of structural geology in this laboratory.

PART 10A: STRUCTURAL GEOLOGY

Three representations of Earth are commonly used by structural geologists. These are the geologic map, cross section, and block diagram:

• **Geologic map**—shows the distribution of rocks at Earth's surface. The rocks commonly are divided into mappable rock units that can be recognized and traced across the map area. This division is made on the basis of color, texture, or composition.

195

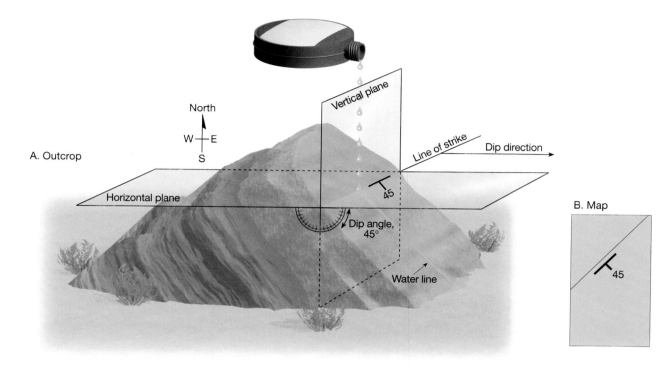

A. Outcrop

B. Map

FIGURE 10.1 Strike and dip of a rock layer as directly observed in nature **(A)** and as represented on a geologic map **(B)**. *Strike* is the direction of a line formed by the intersection of the surface of an inclined (tilted) rock layer and a horizontal plane. *Dip* is the maximum angle of inclination (tilting) of the rock layer, always measured perpendicular to the line of strike (looking straight down on it, in map view) and in the direction that the rock layer tilts down into the ground. Water poured onto a dipping rock layer drains along the angle of dip. The **"T"** and **45** together form the standard strike-and-dip symbol. The long top of the **"T"** is the line of strike, the short upright of the **"T"** shows the dip direction, and **"45"** is the dip angle in degrees.

Such mappable rock units are called **formations.** They may be subdivided into **members** comprised of **beds** (individual layers of rock or sediment). The boundaries between geologic units are **contacts,** which form lines on geologic maps. A geologic map also shows the topography of the land surface with contour lines, so it is both a geologic *and* topographic map.

• **Geologic cross section**—a drawing of a vertical slice through Earth, with the material in front of it removed: a cutaway view. It shows the arrangement of formations and their contacts. A good cross section also shows the topography of the land surface, like a topographic profile.

• **Block diagram**—a combination of the geologic map and cross section. It looks like a solid block, with a geologic map on top and a geologic cross section on each of its visible sides (e.g., Figure 10.4). Each block diagram is a small three-dimensional model of a portion of Earth's crust.

FIGURE 10.2 Geologic maps with strike and dip symbols indicating the attitude of rock layers. Note that strike and dip can be expressed in quadrant or azimuth form. When expressing strike and dip directions as azimuth bearings, they should be expressed as three digits in order to distinguish them from two-digit dip angles. Note also that a line of strike can be expressed as a bearing in either direction. For example **(A)**, a line of strike with a quadrant bearing of North 45° West also has a bearing of South 45° East. A line of strike with an azimuth bearing of 335° also has a bearing of 155° (i.e., 180° less than 335°).

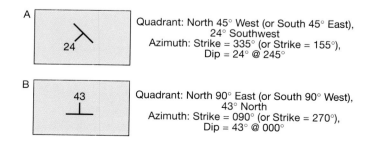

A

Quadrant: North 45° West (or South 45° East), 24° Southwest
Azimuth: Strike = 335° (or Strike = 155°), Dip = 24° @ 245°

B

Quadrant: North 90° East (or South 90° West), 43° North
Azimuth: Strike = 090° (or Strike = 270°), Dip = 43° @ 000°

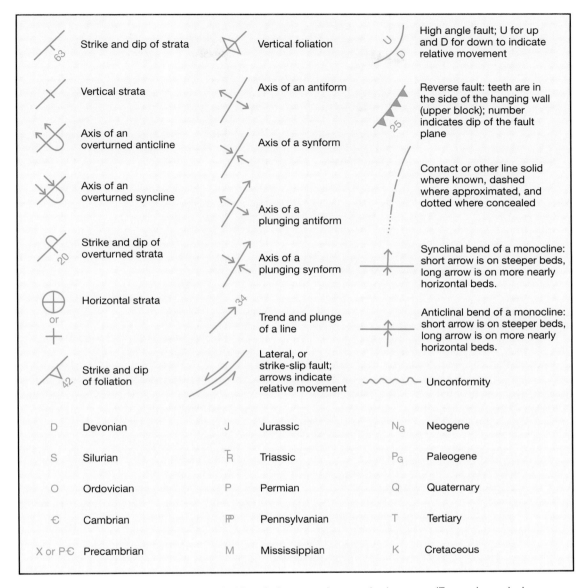

FIGURE 10.3 Structural symbols and abbreviations used on geologic maps. (For rock symbols, see Figure 8.4.)

Measuring the Attitude of Rock Units

Attitude is the orientation of a rock unit or surface. Geologists have devised a system for measuring and describing attitude to understand three-dimensional relationships of formations and geologic structures. Strike and dip serve this purpose (see Figure 10.1):

- **Strike**—the *compass bearing* (direction) of a line formed by the intersection of a horizontal plane (such as the surface of a lake) and an inclined layer (bed, stratum) of rock, fault, fracture, or other surface (Figure 10.1). If strike is expressed in

degrees east or west of true north or true south, it is called a *quadrant bearing*. Strike can also be expressed as a three-digit *azimuth bearing* in degrees between 000 and 360. In azimuth form, north is 000° (or 360°), east is 090°, south is 180°, and west is 270°.

- **Dip**—the *angle* between a horizontal plane and the inclined (tilted) stratum, fault, or fracture. As you can see in Figure 10.1, a thin stream of water poured onto an inclined surface always runs downhill along the **dip direction,** which is always perpendicular to the line (bearing) of strike. The

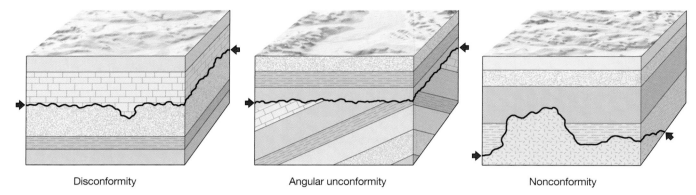

Disconformity   Angular unconformity   Nonconformity

FIGURE 10.4 Unconformities. Arrows point to the unconformity surface (black line). A *disconformity* is an unconformity between relatively *parallel* strata. An *angular unconformity* is an unconformity between *nonparallel* strata. A *nonconformity* is an unconformity between sedimentary rock/sediment and igneous or metamorphic rock.

inclination of the water line, down from the horizontal plane, is the **dip angle.**

Dip is always expressed in terms of its dip angle and dip direction. The dip angle is always expressed in two digits (e.g., 45° in Figure 10.1). The dip direction can be expressed as a three-digit azimuth direction or as a quadrant direction (e.g., North, Northeast, East).

Strike and dip are shown on maps by use of "T"-shaped symbols (see Figures 10.1, 10.2, and 10.3). The long line (top of the "T") shows strike direction, and the short line (upright of the "T") shows dip direction. Note that dip is always perpendicular to the line of strike. The short line of the "T" points *downdip.* The accompanying numerals indicate the dip angle in degrees. Refer to Figure 10.2 for examples of how to read and express strike and dip in quadrant or azimuth form. Also note that special symbols are used for horizontal strata (rock layers) and vertical strata (Figure 10.3).

Unconformities

Structural geologists must locate, observe, and interpret many different structures. Fundamentally, these include unconformities, faults, and folds. There are three common types of *unconformities* (see Figure 10.4), which you may recall from Laboratory 8:

- **Disconformity**—an unconformity between relatively *parallel* strata.

- **Angular unconformity**—an unconformity between *nonparallel* strata.

- **Nonconformity**—an unconformity between sedimentary rock/sediment and *non-sedimentary* (igneous or metamorphic) rock.

Any unconformity may be a very irregular surface, because it is usually a surface where erosion has occurred (before it was buried to form the unconformity). For example, bedrock surfaces exposed on the slopes of hills and mountains in your region are part of a regional surface of erosion that could become an unconformity. If sea level were to rise and cover your region with a fresh layer of mud or sand, then the uneven regional surface of erosion would become a regional unconformity.

Faults

Faults in rock units are breaks along which movement has occurred. Faults form when brittle rocks experience three kinds of severe stress: *tension* (pulling apart or lengthening), *compression* (pushing together, compacting, and shortening), and *shear* (smearing or tearing). The three kinds of stress force the rocks to fault in distinctive ways (Figure 10.5).

Normal faults, reverse faults, and *thrust faults* all involve vertical motions of rocks. These faults are named by noting the *sense of motion* of the top surface of the fault (top block) relative to the bottom surface (bottom block), regardless of which one actually has moved. The top surface of the fault is called the **hanging wall** and is the base of the **hanging wall** (top) **block** of rock. The bottom surface of the fault is called the **footwall** and forms the top of the **footwall block.** The headwall block sits on top of the footwall block.

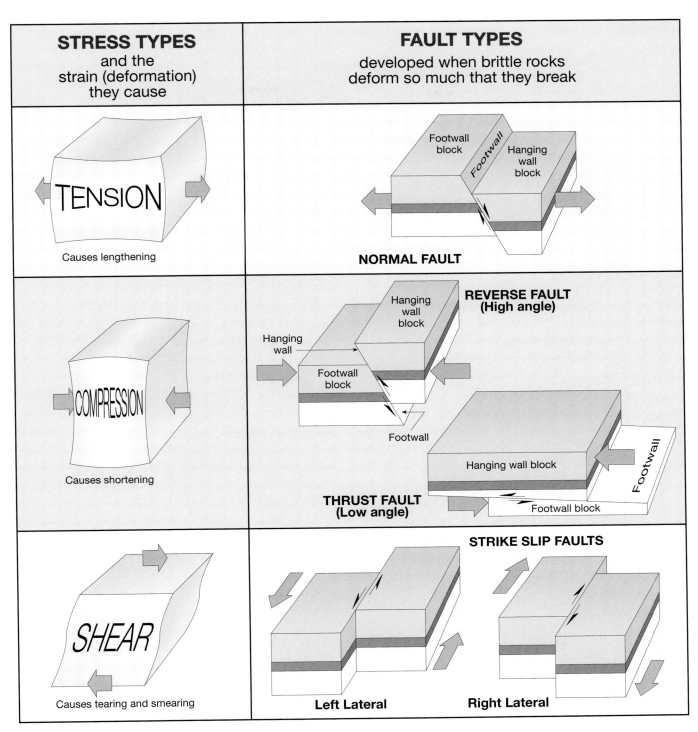

STRESS TYPES and the strain (deformation) they cause	FAULT TYPES developed when brittle rocks deform so much that they break
TENSION Causes lengthening	**NORMAL FAULT**
COMPRESSION Causes shortening	**REVERSE FAULT** (High angle) / **THRUST FAULT** (Low angle)
SHEAR Causes tearing and smearing	**STRIKE SLIP FAULTS** Left Lateral / Right Lateral

FIGURE 10.5 Three types of stress and strain and the fault types they produce.

Normal faults are caused by tension (rock lengthening). As tensional stress pulls the rocks apart, gravity pulls down the hanging wall block. Therefore, normal faulting gets its name because it is a normal response to gravity. You can recognize normal faults by recognizing the motion of the hanging wall block relative to the footwall block. If the hanging wall block has moved downward in relation to the footwall block, then the fault is a normal fault.

Reverse faults are caused by compression (rock shortening). As compressional stress pushes the rocks together, one block of rock gets pushed atop another. You can recognize reverse faults by recognizing the motion of the hanging wall block relative to the footwall block. If the hanging wall block has moved upward in relation to the footwall block, then the fault is a reverse fault. **Thrust faults** are reverse faults that develop at a very low angle and may be very difficult to recognize (Figure 10.5). Reverse faults and thrust faults generally place older strata on top of younger strata.

Strike slip faults (**lateral faults**) are caused by shear and involve horizontal motions of rocks (Figure 10.5). If you stand on one side of a strike slip fault and look across it, then the rocks on the opposite side of the fault will appear to have slipped to the right or left. Along a *right-lateral (strike slip) fault*, the rocks on the opposite side of the fault appear to have moved to the right. Along a *left-lateral (strike slip) fault*,

the rocks on the opposite side of the fault appear to have moved to the left.

Folded Structures

Folds are upward or downward bends of rock layers (Figures 10.6, 10.7, 10.8, and 10.9). **Antiforms** are "upfolds" or "convex folds." If the *oldest* rocks are in the middle, then they are called **anticlines. Synforms** are "downfolds" or "concave folds." If the *youngest* rocks are in the middle, then they are called **synclines.**

In a fold, each stratum is bent around an imaginary axis, like the crease in a piece of folded paper. This is the **fold axis** (or **hinge line**). For all strata in a fold, the fold axes lie within the **axial plane** of the fold (Figures 10.6 and 10.7).

The fold axis may not be horizontal, but rather it may plunge into the ground. This is called a **plunging fold** (Figure 10.7B). **Plunge** is the angle between the fold axis and horizontal. The **trend** of the plunge is the bearing (compass direction), measured in the direction that the axis is inclined downward. You can also think of the trend of a plunging fold as the direction a marble would roll if it were rolled down the plunging axis of the fold.

Folds normally have two sides, or **limbs,** one on each side of the axial plane (see Figure 10.7). If a fold is tilted so that one limb is upside down, then the entire fold is called an **overturned fold** (Figure 10.8).

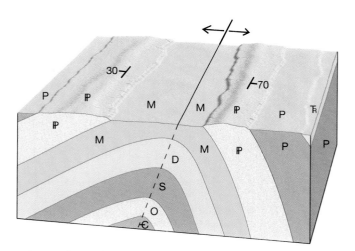

A. ANTICLINE (asymmetrical): oldest rocks (Є) occur in the center of the fold

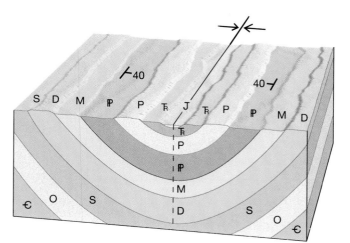

B. SYNCLINE (symmetrical): youngest rocks (J) occur in the center of the fold

FIGURE 10.6 Folds—the two common types. Letters on rock layers indicate their relative ages on the geologic time scale (Figures 10.3,1.3). Note that solid lines (dashed where underground) are used to show the position of the axial planes of the folds. Note the symbols for axis of an anticline and axis of a syncline in Figure 10.3. Also note the orientation of symbols for strike and dip in relation to the attitude (orientation) of rock layers (strata).

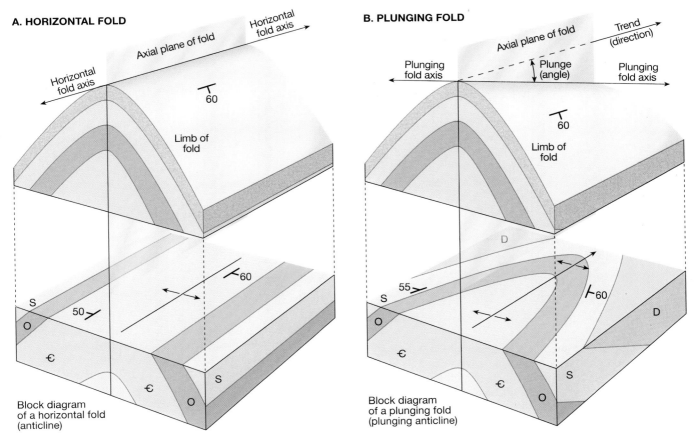

A. HORIZONTAL FOLD

B. PLUNGING FOLD

FIGURE 10.7 Fold terminology and block diagrams. **A.** Simple horizontal fold (anticline). **B.** More complex plunging fold (plunging anticline). Note that the fold axis plunges into Earth, but the *trend* is the compass direction (bearing) on the surface. Also note the orientation of rock layers, symbols for strike and dip, and symbols for the fold axes in the block diagrams.

Monoclines have two axial planes that separate two nearly horizontal limbs from a single, more steeply inclined limb (Figure 10.9).

Domes and **basins** (Figure 10.10) are large, somewhat circular structures formed when strata are warped upward, like an upside-down bowl (dome) or downward, like a bowl (basin). Strata are oldest at the center of a dome, and youngest at the center of a basin.

Cardboard Block Diagrams

Six cardboard block diagrams (Cardboard Models 1–6) are provided at the back of this laboratory manual. Carefully remove them from the book and cut off the torn edges, so you can fold them into blocks. To fold them, follow the procedure in Figure 10.11. *Be sure that you have cut out and folded your cardboard models before you proceed to the items below.* You will also need to understand and apply the symbols for geologic structures from Figure 10.3 and follow the set of simple rules for interpreting geologic maps on the tops of the block diagrams. Symbols for rock types are the same as in Figure 8.4, although colors may vary.

Cardboard Model 1

This model shows Devonian (green), Mississippian (brown), Pennsylvanian (yellow), and Permian (salmon) formations striking due north and dipping 25° to the west. Cambrian (tan) and Ordovician (gray) formations strike due north and are vertical (dip angle = 90°). Provided are a complete geologic map (the top of the diagram) and three of the four vertical cross sections (the south, east, and west sides of the block diagram).

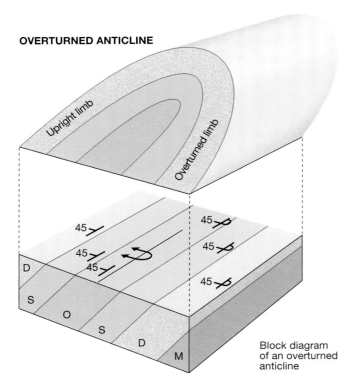

OVERTURNED ANTICLINE

Upright limb

Overturned limb

45

45

45

45

45

45

D

S

O

S

D

M

Block diagram
of an overturned
anticline

FIGURE 10.8 Overturned fold. Note that one limb of the
fold has been turned under the other, so it is overturned
(upside down). Also note the symbols used for strike and
dip of strata (rock layers), strike and dip of overturned
strata, and axis of an overturned anticline in the block
diagram and Figure 10.3.

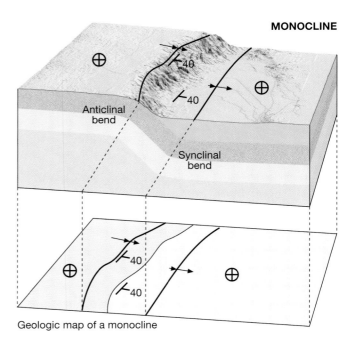

MONOCLINE

40

40

Anticlinal
bend

Synclinal
bend

40

40

Geologic map of a monocline

FIGURE 10.9 Monocline. Not all folds have two limbs. The
monocline is a fold inclined in only one direction. A monocline has two axial planes (shown as dashed lines in bends)
that separate two nearly horizontal limbs from a more
steeply inclined limb. Note the symbols used to indicate
horizontal strata (rock layers) and the axes of a monocline
in the block diagram and Figure 10.3.

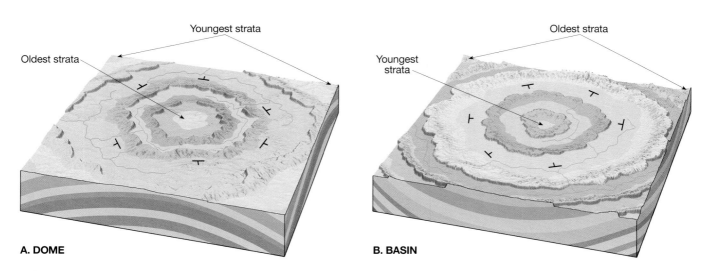

Youngest strata

Oldest strata

Oldest strata

Youngest
strata

A. DOME

B. BASIN

FIGURE 10.10 Dome and basin. Both of these structures are bowl-shaped in three dimensions and
appear as relatively circular "bull's eye" patterns on maps. **A.** A *dome* is convex (bowed upward, like an
upside-down bowl) and has the oldest strata in its center. Rocks dip away from the center of the dome
(note strike-dip symbols). **B.** A *basin* is concave (bowed downward; bowl shaped) and has the youngest
strata in its center. Rocks dip toward the center of the basin (note strike-dip symbols).

FORMING THE STRUCTURE MODELS

1. Lay the paper with the model on it face down in front of you. Orient the long dimension of the paper up and down, as if you were going to read a normal typewritten page.

2. Carefully curl back one side until you can see the solid black line that runs all the way from the top to the bottom of the page. Crease the paper exactly along that line.

3. Now repeat this process for the other side of the paper.

4. Unfold the two sides, and curl back the top until you see the solid black line that runs across the page. Crease the paper exactly along that line.

5. Now repeat this process for the bottom of the paper.

6. To make a block, you still need to do something about the extra material where the corners are. In each corner there is a dashed line. Start at one corner and push that line toward the inside of the block. Fold the sides down so that they match, and crease the flap you folded in. Your crease should be approximately along the dashed line. Do the same thing with the other three corners.

7. If the block is folded correctly, the top will be flat and the strata will match on the map (top) and on the cross sections (sides).

8. The block will not really stay together without tape, but do not tape it. You will find that it is easier to draw on the block if you can unfold it and lay it out flat.

9. Write your name on the blank inside of the block so that your instructor can identify your work when you hand it in.

FIGURE 10.11 Directions for forming the structure models.

Questions

1. Finalize Cardboard Model 1 as follows. First construct the vertical cross section on the north side of the block so it shows the formations and their attitudes (dips). On the map, draw a strike and dip symbol on the Mississippian sandstone that dips 25° to the west and on the Ordovician gray shale that is vertical (see Figure 10.3 for the strike and dip symbol for a vertical bed). Also draw in the symbol for an unconformity (Figure 10.3) at the contact between the Ordovician gray shale and the Devonian and Mississippian formations plus everywhere else that the unconformity occurs in the north and south cross sections of the diagram.

2. Complete the following questions:

   a. Note that both the yellow Pennsylvanian sandstone and Ordovician gray shale formations have the same thickness, but the yellow sandstone makes a much wider band on the geologic map (top of block). Why?

   b. What kind of unconformity is present in this block diagram, and how can you tell?

   c. Explain the sequence of events that led to the relationships that now exist among the formations in this block diagram.

Cardboard Model 2

This model is slightly more complicated than the previous one. The geologic map is complete, but only two of the cross sections are available.

Questions

3. Finalize Cardboard Model 2 as follows. First, complete the north and east sides of the block. Notice that the rock units define a fold. This fold is an antiform, because the strata are convex upward. It is nonplunging, because its axis is horizontal. (Refer back to Figure 10.7 for the differences between plunging and nonplunging folds if you are uncertain about this.) On the geologic map, draw strike and dip symbols to indicate the attitudes of formation **E** (gray formation) at points **I**, **II**, **III**, and **IV**. Also draw the proper symbol on the map (top of model) along the axis of the fold (refer to Figure 10.3).

4. How do the strikes at all four locations compare with each other?

5. How does the dip direction at points **I** and **II** compare with the dip direction at points **III** and **IV**? *In your answer, include the dip direction at all four points.*

204 • *Laboratory Ten*

Cardboard Model 3

This cardboard model has a complete geologic map. However, only one side and part of another are complete.

Questions

6. Finalize Cardboard Model 3 as follows. Complete the remaining two-and-a-half sides of this model, using as guides the geologic map on top of the block and the one-and-a-half completed sides. On the map, draw strike and dip symbols showing the orientation of formation **C** at points **I, II, III,** and **IV.** Also draw the proper symbol along the axis of the fold (refer to Figure 10.3).

7. How do the strikes of all four locations compare with each other?

8. How does the dip direction (of formation **C**) at points **I** and **II** compare with the dip direction at points **III** and **IV**? *Include the dip direction at all four points in your answer.*

9. Is this fold plunging or nonplunging? Is it an antiform or a synform?

10. On the basis of this example, how much variation is there in the strike at all points in a nonplunging fold?

Cardboard Model 4

This model shows a plunging antiform and an unconformity. The antiform plunges to the north, following the general rule that *anticlines plunge in the direction in which the fold closes* (refer to rules, Figure 10.12).

Questions

11. Finalize Cardboard Model 4 as follows. Complete the north and east sides of the block. Draw strike and dip symbols on the map at points **I, II, III, IV,** and **V.** Draw the proper symbol on the map along the axis of the fold, including its direction of plunge. Also draw the proper symbol on the geologic map to indicate the orientation of beds in formation **J.**

12. How do the directions of strike and dip differ from those in Model 3?

13. What type of unconformity is at the base of formation **J?**

A SET OF SIMPLE RULES FOR INTERPRETING GEOLOGIC MAPS

1. Anticlines have their oldest beds in the center.

2. Synclines have their youngest beds in the center.

3. Anticlines plunge toward the nose (closed end) of the structure.

4. Synclines plunge toward the open end of the structure.

5. Contacts between horizontal beds "V" upstream and are parallel to topographic contour lines.

6. Contacts of horizontal beds, or of beds that have a dip lower than stream gradient, "V" upstream.

7. Contacts of beds that have a dip greater than stream gradient "V" downstream if they are dipping downstream.

8. Contacts of beds that have a dip greater than stream gradient will "V" upstream if they are dipping upstream.

9. Vertical beds do not "V" or migrate with erosion.

10. The upthrown blocks of faults tend to be eroded more than downthrown blocks.

11. Contacts migrate downdip upon erosion.

12. True dip angles can only be seen in cross section if the cross section is perpendicular to the fault or to the strike of the beds.

FIGURE 10.12 Simple rules used by geologists to interpret geologic maps.

Cardboard Model 5

This model shows a plunging synform. Two of the sides are complete and two remain incomplete.

Questions

14. Finalize Cardboard Model 5 as follows. Complete the north and east sides of the diagram. Draw strike and dip symbols on the map at points **I, II, III, IV,** and **V** to show the orientation of layer **G.** *Synforms plunge in the direction in which the fold opens* (refer to rules, Figure 10.11). Draw the proper symbol along the axis of the fold to indicate its location and direction of plunge.

15. In which direction does this synform plunge?

Cardboard Model 6

This model shows a fault that strikes due west and dips 45° to the north. Three sides of the diagram are complete, but the east side is incomplete.

Questions

16. Finalize Cardboard Model 6 as follows. At point **I,** draw a strike and dip symbol showing the *orientation of the fault.* On the west edge of the block, draw arrows parallel to the fault, indicating relative motion. Label the hanging wall and the footwall. Complete the east side of the block. Draw arrows parallel to the fault, indicating relative motion. Now look at the geologic map and at points **II** and **III.** Write **U** on the side that went up and **D** on the side that went down. At points **IV** and **V,** draw strike and dip symbols for formation **B.**

17. Is the fault in this model a normal fault or a reverse fault? Why?

18. On the geologic map, what happens to the contact between units **A** and **B** where it crosses the fault?

19. There is a general rule that, as erosion of the land proceeds, *contacts migrate downdip.* Is this true in this example? Explain.

20. Could the same offset along this fault have been produced by strike-slip motion?

PART 10B: BLOCK DIAGRAMS, GEOLOGIC CROSS SECTIONS, AND GEOLOGIC MAPS

Illustrated block diagrams and geologic maps are provided for you to develop your skills of identifying, describing, and interpreting geologic structures. You will need to understand and apply the symbols from Figure 10.3 and follow the set of simple rules for interpreting geologic maps (Figure 10.12). Refer back to Figures 10.4–10.10 as needed.

Questions

21. Complete each block diagram in Figure 10.13 as directed. On the line provided, indicate what kind of geologic structure is represented in the diagram.

22. Complete the geologic cross section in Figure 10.14 (you will need pencil, scratch paper, ruler, and protractor; colored pencils are optional). Place the edge of a piece of scratch paper along line X – Y and mark it to record the exact width of each colored formation. Transfer this information to the topographic surface line in the geologic cross section. Use a protractor to extend the colored formations with known dips into the subsurface of the geologic cross section (lightly in pencil). Draw in the remaining colored formations parallel to the ones with known dips, and smooth the contacts to form the geologic structure(s) beneath the surface. Finally, project the geologic structure up above the topographic surface line to show how the geologic structure(s) existed there before being eroded.

    a. Label your geologic cross section to indicate the kind(s) of geologic structure(s) revealed by your work. Then add the appropriate symbols from Figure 10.3 to the geologic map in order to show the axes of the folds revealed in your geologic cross section.

    b. Add half-arrows to the fault near the center of the geologic map to show the relative motions of its two sides. Exactly what kind of fault is it (Refer to Figure 10.5)?

23. Complete the geologic cross section in Figure 10.15 (you will need pencil, scratch paper, ruler, and protractor; colored pencils are optional). Place the edge of a piece of scratch paper along line X – Y and mark it to record the exact width of each colored formation. Transfer this information to the topographic surface line in the geologic cross section. Use a protractor to extend the colored formations with known dips into the subsurface of the geologic cross section (lightly in pencil). Draw in the remaining colored formations parallel to the ones with known dips, and smooth the contacts to form the geologic structure(s) beneath the surface. Finally, project the geologic structure up above the topographic surface line to show how the geologic structure(s) existed there before being eroded.

    a. What kind of geologic structure is present, and how do you think it formed?

    b. Modify the geologic map by adding the appropriate symbols from Figure 10.3 to show the position of the structure on the map the best you can.

24. Refer to the geologic maps in Figure 10.16. Do the following for each of these maps.

    a. Make a list of the ages of rocks present in the map, from oldest at the bottom of the list to youngest at the top of the list. Indicate in your list, and on the map, where there are unconformities (gaps or missing intervals of rock in the sequence), if any.

    b. Make a list of other geologic structures that you can identify on the map, then describe where each structure is located.

    c. Write a paragraph or outline of the general geologic history of the region (i.e., describe when the rock layers formed and how they were deformed or eroded).

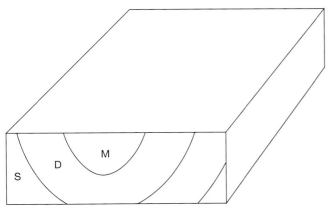

A. Complete top and side. Add appropriate symbols from Figure 10.3. What geologic structure is present?

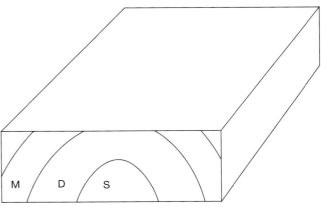

B. Complete top and side. Add appropriate symbols from Figure 10.3. What geologic structure is present?

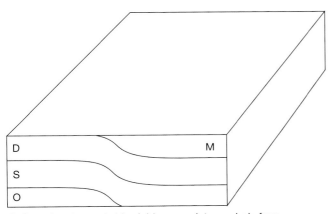

C. Complete top and side. Add appropriate symbols from Figure 10.3. What geologic structure is present?

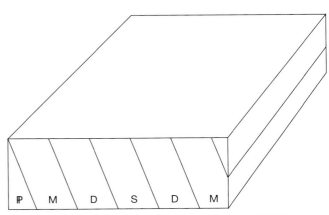

D. Complete top of diagram. Add appropriate symbols from Figure 10.3. What geologic structure is present?

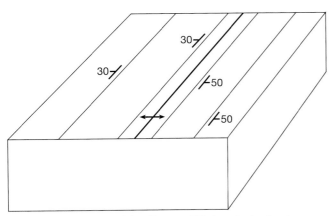

E. Complete the sides of the diagram. What geologic structure is present?

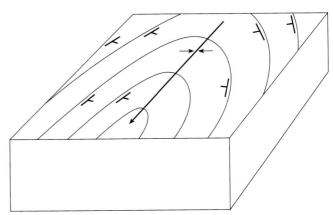

F. Complete the sides of the diagram. What geologic structure is present?

FIGURE 10.13 Block diagrams to complete (Question 21).

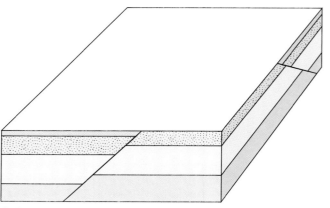

G. Complete top of diagram. Add appropriate symbols from Figure 10.3. What geologic structure is present?

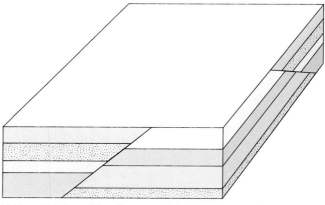

H. Complete top of the diagram. Add appropriate symbols from Figure 10.3. What geologic structure is present?

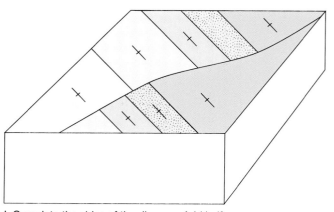

I. Complete the sides of the diagram. Add half-arrows. What geologic structure is present?

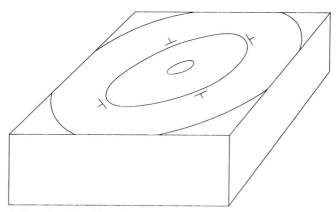

J. Complete the sides of the diagram. What geologic structure is present?

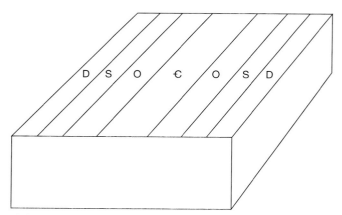

K. Complete sides of the diagram. Add appropriate symbols from Figure 10.3. What geologic structure is present?

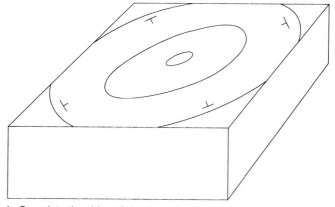

L. Complete the sides of the diagram. What geologic structure is present?

FIGURE 10.13 (CONTINUED) Block diagrams to complete (Question 21).

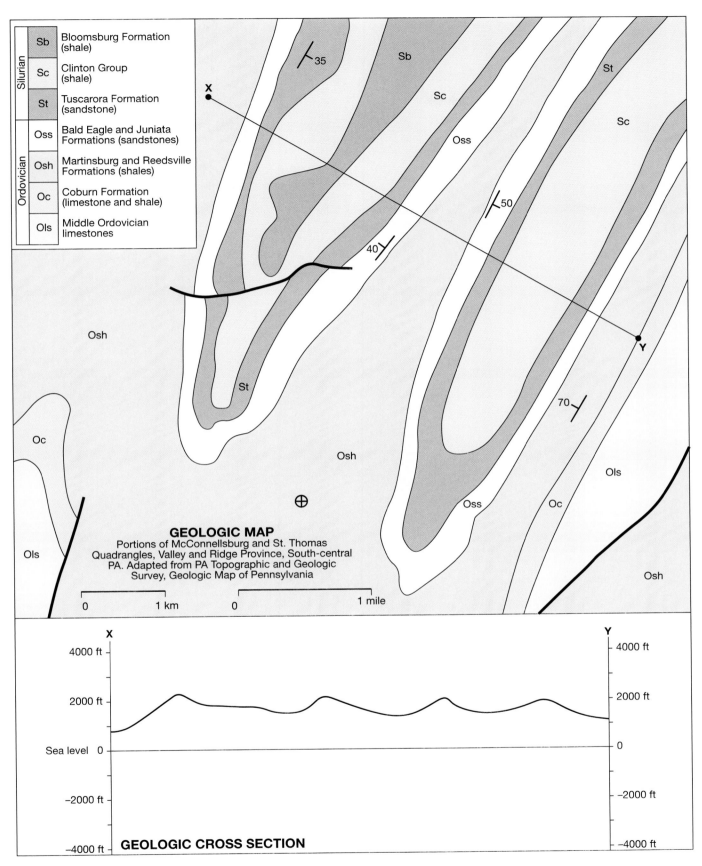

Legend

Silurian	Sb	Bloomsburg Formation (shale)
	Sc	Clinton Group (shale)
	St	Tuscarora Formation (sandstone)
Ordovician	Oss	Bald Eagle and Juniata Formations (sandstones)
	Osh	Martinsburg and Reedsville Formations (shales)
	Oc	Coburn Formation (limestone and shale)
	Ols	Middle Ordovician limestones

GEOLOGIC MAP

Portions of McConnellsburg and St. Thomas Quadrangles, Valley and Ridge Province, South-central PA. Adapted from PA Topographic and Geologic Survey, Geologic Map of Pennsylvania

0   1 km

0   1 mile

GEOLOGIC CROSS SECTION

X — 4000 ft — 2000 ft — Sea level 0 — −2000 ft — −4000 ft

Y — 4000 ft — 2000 ft — 0 — −2000 ft — −4000 ft

FIGURE 10.14 Geologic map and cross section to complete (Question 22).

209

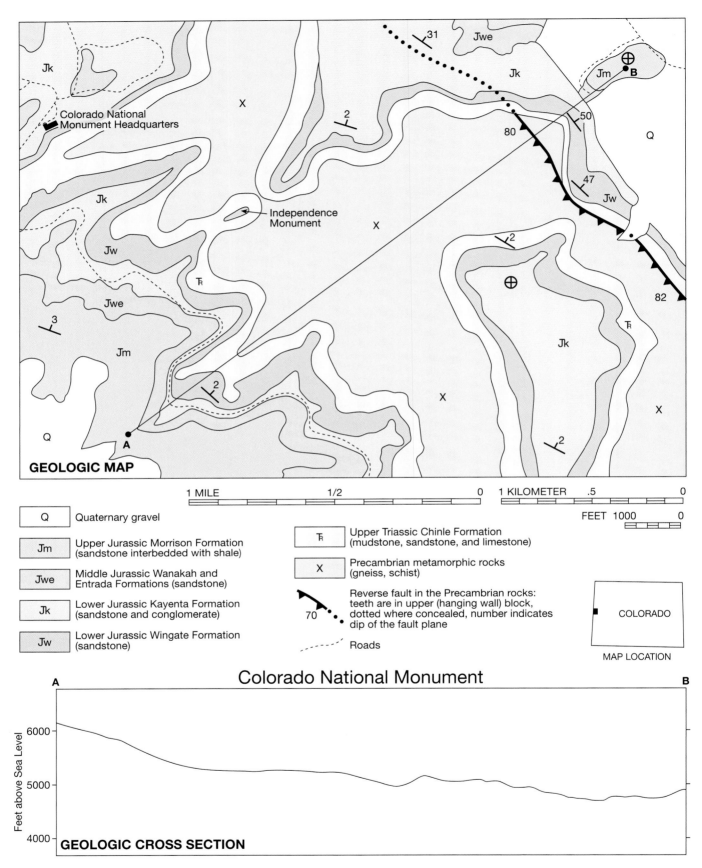

FIGURE 10.15 Geologic map and cross section to complete (Question 23). Map data adapted from USGS Geologic Investigations Series I-2740 by Robert B. Scott *et al.*

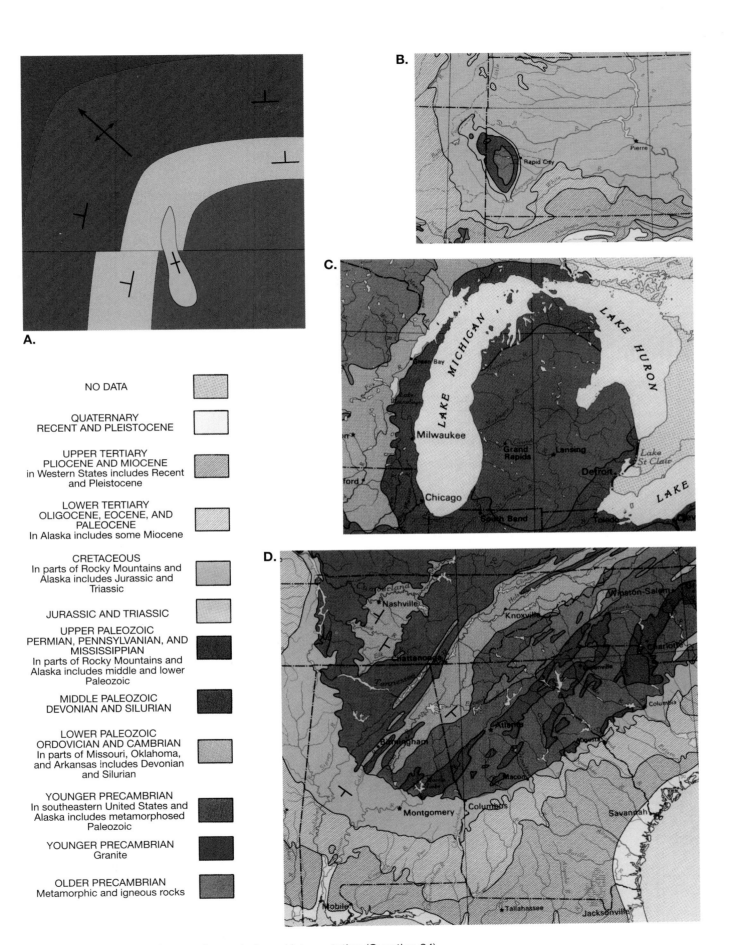

NO DATA

QUATERNARY
RECENT AND PLEISTOCENE

UPPER TERTIARY
PLIOCENE AND MIOCENE
in Western States includes Recent
and Pleistocene

LOWER TERTIARY
OLIGOCENE, EOCENE, AND
PALEOCENE
In Alaska includes some Miocene

CRETACEOUS
In parts of Rocky Mountains and
Alaska includes Jurassic and
Triassic

JURASSIC AND TRIASSIC

UPPER PALEOZOIC
PERMIAN, PENNSYLVANIAN, AND
MISSISSIPPIAN
In parts of Rocky Mountains and
Alaska includes middle and lower
Paleozoic

MIDDLE PALEOZOIC
DEVONIAN AND SILURIAN

LOWER PALEOZOIC
ORDOVICIAN AND CAMBRIAN
In parts of Missouri, Oklahoma,
and Arkansas includes Devonian
and Silurian

YOUNGER PRECAMBRIAN
In southeastern United States and
Alaska includes metamorphosed
Paleozoic

YOUNGER PRECAMBRIAN
Granite

OLDER PRECAMBRIAN
Metamorphic and igneous rocks

FIGURE 10.16 Geologic maps for analysis and interpretation (Question 24).

PART 10C: ANALYSIS OF A GEOLOGIC MAP

Refer to the geologic map provided by your instructor. Otherwise, obtain a geologic map as directed to do so by your instructor. (*Do not mark the map with any dark lines or pen marks!*)

Questions

25. What geographic area (quadrangle or other geographic division) does this geologic map represent?

26. What is the general topography of the area? *Use elevations, if available, and describe orientations of hills and valleys.*

27. What is the name and age of the oldest formation represented on the map?

28. What is the name and age of the youngest formation represented on the map?

29. Make three list headings: igneous formations, sedimentary formations, and metamorphic formations. Under each heading, list the names of the formations that are the indicated type of rock. If there are no formations of one rock type, then write *none* under the heading.

30. What unconformities (see Figure 10.4) have developed, or are currently developing, in the sequence of formations of this area? Describe where they occur and how much time/rock is missing at each unconformity.

31. What formations form the hilltops? Why?

32. What formations form the valleys? Why?

33. Where are modern sediments accumulating, and what symbol(s) is/are used to map them?

34. How does the geology of the map area influence:
    a. the topography?
    b. the location of streams or other bodies of water?
    c. the location of natural vegetation, orchards, farms, and ranches?
    d. the location of communities?
    e. the location of quarries or mines, if any?

35. List and describe any folds, domes, basins, or significant faults in the map area.

36. List, describe, and give the ages of any igneous intrusions in the map area.

37. Write a one- or two-page summary of the geologic history of this region. Start with the oldest formation and events and end with the present time. Mention all of the formations by name and how they developed (sedimentary, igneous, or metamorphic processes). Mention when any formations were eroded or deformed into geologic structures (unconformities, folds, domes, basins, faults). Also describe the kinds of stresses that may have caused the structures to develop.

LABORATORY ELEVEN

Stream Processes, Landscapes, Mass Wastage, and Flood Hazards

---

•CONTRIBUTING AUTHORS•

Pamela J.W. Gore • *Georgia Perimeter College*

Richard W. Macomber • *Long Island University–Brooklyn*

Cherukupalli E. Nehru • *Brooklyn College (CUNY)*

OBJECTIVES

A. Be able to use topographic maps and stereograms to describe and interpret streams: their valley shapes, channel configurations, drainage patterns, and the erosional landscapes and depositional features they create.

B. Understand erosional and mass wastage processes that occur at Niagara Falls, and be able to evaluate rates at which the falls is retreating upstream.

C. Explore and infer meander evolution on the Rio Grande.

D. Be able to construct a flood magnitude/frequency graph, map flooded and flood hazard zones, and assess the extent of flood hazards along the Flint River, Georgia.

E. Be able to determine the flood hazard level where you live and what to do if you are caught in a flood (using online resources of the Federal Emergency Management Agency).

MATERIALS

Pencil, eraser, laboratory notebook, ruler, calculator, 12-inch length of string, pocket stereoscope (optional), and highlighter pen (optional for Question 6).

INTRODUCTION

It all starts with a single raindrop, then another and another. As water drenches the landscape, some soaks into the ground and becomes *groundwater* (Laboratory 12). Some flows over the ground and into streams and ponds of *surface water.* The streams will continue to flow for as long as they receive a water supply from additional rain, melting snow, or *base flow* (groundwater that seeps into a stream via porous rocks, fractures, and springs).

Perennial streams flow continuously throughout the year and are represented on topographic maps as blue lines. *Intermittent streams* flow only at certain times of the year, such as rainy seasons or when snow melts in the Spring. They are represented on topographic maps as blue line segments separated by blue dots (three blue dots between each line segment). All streams, perennial and intermittent, have the potential to **flood** (overflow their banks). Floods damage more human property in the United States than any other natural hazard.

Streams are also the single most important natural agent of *land erosion* (wearing away of the land). They erode more sediment from the land than wind, glaciers, or ocean waves. The sediment is transported and eventually deposited, whereupon it is called **alluvium.** Alluvium consists of gravel, sand, silt, and clay deposited in floodplains, point bars, channel bars, deltas, and alluvial fans (Figure 11.1).

Therefore, stream processes (or *fluvial processes*) are among the most important agents that shape Earth's surface and cause damage to humans and their property.

PART 11A: STREAM PROCESSES AND LANDSCAPES

Recall the last time you experienced a drenching rainstorm. Where did all the water go? During drenching rainstorms, some of the water seeps slowly into the ground. But most of the water flows over the ground before it can seep in. It flows over fields, streets, and sidewalks as sheets of water several millimeters or centimeters deep. This is called *sheet flow.*

Sheet flow moves downslope in response to the pull of gravity, so the sheets of water flow from streets and sidewalks to ditches and street gutters. From the ditches and storm sewers, it flows downhill into small streams. Small streams merge to form larger streams, larger streams merge to form rivers, and rivers flow into lakes and oceans. This entire drainage network, from the smallest *upland* tributaries to larger streams, to the largest river (*main stream* or *main river*), is called a **stream drainage system** (Figure 11.1A).

Stream Drainage Patterns

Stream drainage systems form characteristic patterns of drainage, depending on the relief and geology of the land. The common patterns below are illustrated in Figure 11.2:

- **Dendritic pattern**—resembles the branching of a tree. Water flow is from the branch-like tributaries to the trunk-like main stream or river. This pattern is common where a stream cuts into flat lying layers of rock or sediment. It also develops where a stream cuts into homogeneous rock (crystalline igneous rock) or sediment (sand).

- **Rectangular pattern**—a network of channels with right-angle bends that form a pattern of interconnected rectangles and squares. This pattern often develops over rocks that are fractured or faulted in two main directions that are perpendicular (at nearly right angles) and break the bedrock into rectangular or square blocks. The streams erode channels along the perpendicular fractures and faults.

- **Radial pattern**—channel flow outward from a central area, resembling the spokes of a wheel. Water drains from the inside of the pattern, where

the "spokes" nearly meet, to the outside of the pattern (where the "spokes" are farthest apart). This pattern develops on conical hills, such as volcanoes and some structural domes.

- **Centripetal pattern**—channels converge on a central point, often a lake or playa (dry lake bed), at the center of a closed basin (a basin from which surface water cannot drain because there is no outlet valley).

- **Annular pattern**—a set of incomplete, concentric rings of streams connected by short radial channels. This pattern commonly develops on eroding structural domes and folds that contain alternating folded layers of resistant and nonresistant rock types.

- **Trellis pattern**—resembles a vine or climbing rose bush growing on a trellis, where the main stream is long and intersected at nearly right angles by its tributaries. This pattern commonly develops where alternating layers of resistant and nonresistant rocks have been tilted and eroded to form a series of parallel ridges and valleys. The main stream channel cuts through the ridges, and the main tributaries flow perpendicular to the main stream and along the valleys (parallel to and between the ridges).

- **Deranged pattern**—a random pattern of stream channels that seem to have no relationship to underlying rock types or geologic structures.

Drainage Basins and Divides

The entire area of land that is drained by one stream, or an entire stream drainage system, is called a **drainage basin.** And the linear boundaries that separate one drainage basin from another are called **divides.**

Some divides are easy to recognize on maps as knife-edge ridge crests. However, in regions of lower relief or rolling hills, the divides separate one gentle slope from another and are more difficult to locate precisely (Figure 11.1A, dashed line surrounding the Tributary X drainage basin). For this reason, divides cannot always be mapped as distinct lines. In the absence of detailed elevation data, they must be represented by dashed lines that signify their most probable locations.

You may have heard of something called a *continental divide,* which is a narrow strip of land dividing surface waters that drain in opposite directions across the continent (Figure 11.3). The continental divide in North America is an imaginary line along the crest of the Rocky Mountains. Rainwater that falls

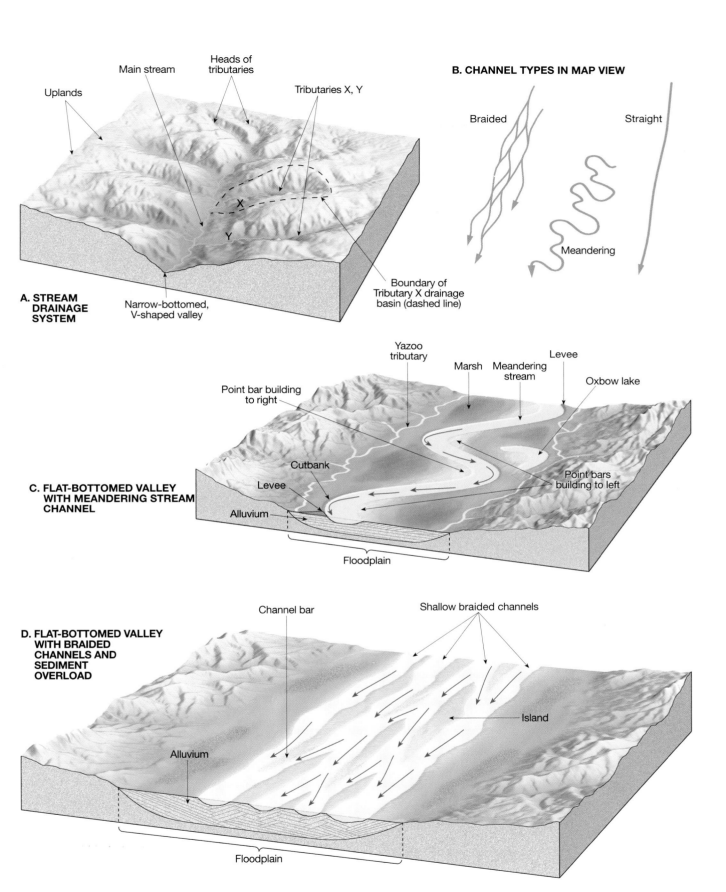

A. STREAM DRAINAGE SYSTEM

Uplands

Main stream

Heads of tributaries

Tributaries X, Y

Narrow-bottomed, V-shaped valley

Boundary of Tributary X drainage basin (dashed line)

X

Y

B. CHANNEL TYPES IN MAP VIEW

Braided

Straight

Meandering

C. FLAT-BOTTOMED VALLEY WITH MEANDERING STREAM CHANNEL

Point bar building to right

Yazoo tributary

Marsh

Meandering stream

Levee

Oxbow lake

Cutbank

Levee

Alluvium

Point bars building to left

Floodplain

D. FLAT-BOTTOMED VALLEY WITH BRAIDED CHANNELS AND SEDIMENT OVERLOAD

Channel bar

Shallow braided channels

Island

Alluvium

Floodplain

FIGURE 11.1 General features of stream drainage basins, streams, and stream channels. Arrows indicate current flow in main stream channels. **A.** Features of a stream drainage basin. **B.** Stream channel types as observed in map view. **C.** Features of a meandering stream valley. **D.** Features of a typical braided stream. Braided streams develop in sediment-choked streams.

STREAM DRAINAGE PATTERNS

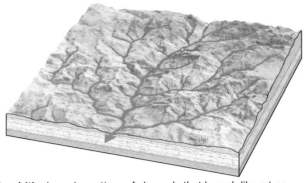

Dendritic: Irregular pattern of channels that branch like a tree. Develops on flat lying or homogeneous rock.

Rectangular: Channels have right-angle bends developed along perpendicular sets of rock fractures or joints.

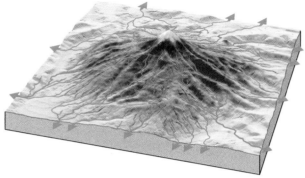

Radial: Channels radiate outward like spokes of a wheel from a high point.

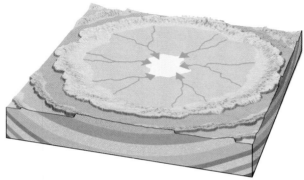

Centripetal: Channels converge on the lowest point in a closed basin from which water cannot drain.

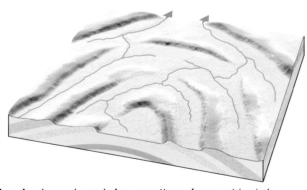

Annular: Long channels form a pattern of concentric circles connected by short radial channels. Develops on eroded domes or folds with resistant and nonresistant rock types.

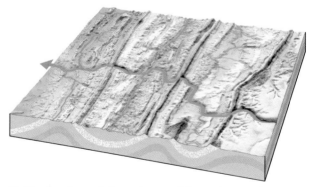

Trellis: A pattern of channels resembling a vine growing on a trellis. Develops where tilted layers of resistant and nonresistant rock form parallel ridges and valleys. The main stream channel cuts through the ridges, and the main tributaries flow along the valleys parallel to the ridges and at right angles to the main stream.

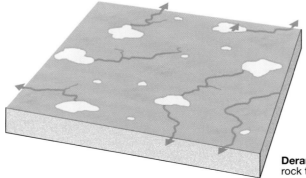

Deranged: Channels flow randomly with no relation to underlying rock types or structures.

FIGURE 11.2 Some stream drainage patterns and their relationship to bedrock geology.

E 11.3: Lake Scott, Kansas

North

.5   1 kilometer

1/4   1/2   1 mile

ur interval = 10 ft.   1:24,000

Drainage basin of the
Mississippi River

Rocky Mts.

Mississippi
River

Rocky Mts.

• Lake Scott, KS

Continental
Divide

Well

PIPELINE

3067

4

3032

3

2985

Gravel
Pit

2850

BM
2837

2968 ×

3068

3061

3046

3006

Garvin
Canyon

29×5

SCOTT STATE GAME
MANAGEMENT AREA

Suicide   Bluf

2900

Battendorf

3050

2900

3065

3066

9

Canyon

10

CANYON

11

Honset

2997

3050

Well

3044

3039

3027

3013

×3031

3000

Epi

TIMBER

×3035

16

3063

15

3040

3026

14

3010

BM 3076

3076

3065

3041

3025

3000

B   E   A   V   E   R

3080

21

3064

22

3021

23

Batt

east of the line drains eastward into the Atlantic Ocean, and rainwater that falls west of the line drains westward into the Pacific Ocean. Therefore, North America's continental divide is sometimes called "The Great Divide."

Stream Weathering, Transportation, and Deposition

Three main processes are at work in every stream. *Weathering* occurs where the stream physically erodes and disintegrates Earth materials and where it chemically decomposes or dissolves Earth materials to form sediment and aqueous chemical solutions. *Transportation* of these weathered materials occurs when they are dragged, bounced, and carried downstream (as suspended grains or chemicals in the water). *Deposition* occurs if the velocity of the stream drops (allowing sediments to settle out of the water) or if parts of the stream evaporate (allowing mineral crystals and oxide residues to form).

The smallest valleys in a drainage basin occur at its highest elevations, called **uplands** (Figure 11.1). In the uplands, a stream's (tributary's) point of origin, or **head,** may be at a spring or at the start of narrow runoff channels developed during rainstorms. Erosion (wearing away rock and sediment) is the dominant process here, and the stream channels deepen and erode their V-shaped channels uphill through time—a process called **headward erosion.** Eroded sediment is transported downstream by the tributaries.

Streams also weather and erode their own valleys along weaknesses in the rocks (fractures, faults), soluble nonresistant layers of rock (salt layers, limestone), and where there is the least resistance to erosion (see Figure 11.2). Rocks comprised of hard, chemically resistant minerals are generally more resistant to erosion and form ridges or other hilltops. Rocks comprised of soft and more easily weathered minerals are generally less resistant to erosion and form valleys. This is commonly called *differential erosion* of rock.

Headward tributary valleys merge into larger stream valleys, and these eventually merge into a larger river valley. Along the way, some new materials are eroded, and deposits (gravel, sand, mud) may form temporarily, but the main processes at work over the years in uplands are erosion (headward erosion and cutting V-shaped valleys) and transportation of sediment.

The end of a river valley is the **mouth** of the river, where it enters a lake, ocean, or dry basin. At this location, the river water is dispersed into a wider area, its velocity decreases, and sediment settles out of suspension to form an alluvial deposit (alluvial fan or

delta). If the river water enters a dry basin, then it will evaporate and precipitate layers of mineral crystals and oxide residues (in a *playa*).

River Valley Forms and Processes

The form or shape of a river valley varies with these main factors:

- **Geology**—the bedrock geology over which the stream flows affects the stream's ability to find or erode its course (Figure 11.2).

- **Gradient**—the steepness of a slope—either the slope of a valley wall or the slope of a stream along a selected length (segment) of its channel (Figure 11.4). Gradient is generally expressed in *feet per mile.* This is determined by dividing the vertical rise or fall between two points on the slope (in feet) by the horizontal distance (run) between them (in miles). For example, if a stream descends 20 feet over a distance of 40 miles, then its gradient is 20 ft/40 mi, or 0.5 ft/mi. You can estimate the gradient of a stream by studying the spacing of contours on a topographic map. Or, you can precisely calculate the exact gradient by measuring how much a stream descends along a measured segment of its course.

- **Base level**—the lowest level to which a stream can theoretically erode. For example, base level is achieved where a stream enters a lake or ocean. At that point, the erosional (cutting) power of the stream is zero and depositional (sediment accumulation) processes occur.

- **Discharge**—the rate of stream flow at a given time and location. Discharge is measured in water volume per unit of time, commonly *cubic feet per second* (ft^3/sec).

- **Load**—the amount of material (mostly alluvium, but also plants, trash, and dissolved material) that is transported by a stream. In the uplands, most streams have relatively steep gradients, so the streams cut narrow, V-shaped valleys. Near their heads, tributaries are quick to transport their load downstream, where it combines with the loads of other tributaries. Therefore, the load of the tributaries is transferred to the larger streams and, eventually, to the main river. The load is eventually deposited at the mouth of the river, where it enters a lake, ocean, or dry basin.

From the headwaters to the mouth of a stream, the gradient decreases, discharge increases, and

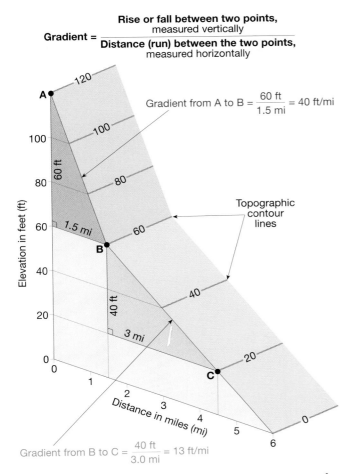

Rise or fall between two points, measured vertically

$$\text{Gradient} = \frac{\text{Rise or fall between two points, measured vertically}}{\text{Distance (run) between the two points, measured horizontally}}$$

$$\text{Gradient from A to B} = \frac{60 \text{ ft}}{1.5 \text{ mi}} = 40 \text{ ft/mi}$$

Topographic contour lines

Elevation in feet (ft)

Distance in miles (mi)

$$\text{Gradient from B to C} = \frac{40 \text{ ft}}{3.0 \text{ mi}} = 13 \text{ ft/mi}$$

FIGURE 11.4 Gradient is a measure of the steepness of a slope. As above, gradient is usually determined by dividing the rise or fall (vertical relief) between two points on the map by the distance (run) between them. It is usually expressed as a fraction in feet per mile (as above) or meters per kilometer.

A second way to determine and express the gradient of a slope is by measuring its steepness in degrees relative to horizontal. Thirdly, gradient can be expressed as a percentage (also called *grade* of a slope). For example, a grade of 10% would mean a grade of 10 units of rise divided by (per) 100 units of distance (i.e., 10 in. per 100 in., 10 m per 100 m).

valleys generally widen. Along the way, the load of the stream may exceed the ability of the water to carry it, so the solid particles accumulate as sedimentary deposits along the river margins, or banks. **Floodplains** develop when alluvium accumulates landward of the river banks, during floods (Figures 11.1C and 11.1D). However, most flooding events do not submerge the entire floodplain. The more abundant minor flooding events deposit sediment only where the water barely overflows the river's banks. Over time, this creates natural **levees** (Figure 11.1C) that are higher than the rest of the floodplain. If a tributary cannot breach a river's levee, then it will become a **yazoo tributary** that flows parallel to the river (Figure 11.1C).

Still farther downstream, the gradient decreases even more as discharge and load increase. The stream valleys develop very wide, flat floodplains with sinuous channels. These channels may become highly sinuous, or **meandering** (see Figures 11.1B and 11.1C). Erosion occurs on the outer edge of meanders, which are called **cutbanks.** At the same time, **point bar** deposits (mostly gravel and sand) accumulate along the inner edge of meanders. Progressive erosion of cutbanks and deposition of point bars makes meanders "migrate" over time.

Channels may cut new paths during floods. This can cut off the outer edge of a meander, abandoning it to become a crescent-shaped **oxbow lake** (see Figure 11.1C). When low gradient/high discharge streams become overloaded with sediment, they may form **braided stream** patterns. These consist of braided channels with linear, underwater sandbars (**channel bars**) and islands (see Figures 11.1B and 11.1D).

Some stream valleys have level surfaces that are higher than the present floodplain. These are remnants of older floodplains that have been dissected (cut by younger streams) and are called **stream terraces.** Sometimes several levels of stream terraces may be developed along a stream, resembling steps.

Where a stream enters a lake, ocean, or dry basin, its velocity decreases dramatically. The stream drops its sediment load, which accumulates as a triangular or fan-shaped deposit. In a lake or ocean, such a deposit is called a **delta.** A similar fan-shaped deposit of stream sediment also occurs where a steep-gradient stream abruptly enters a wide, level plain, creating an **alluvial fan.**

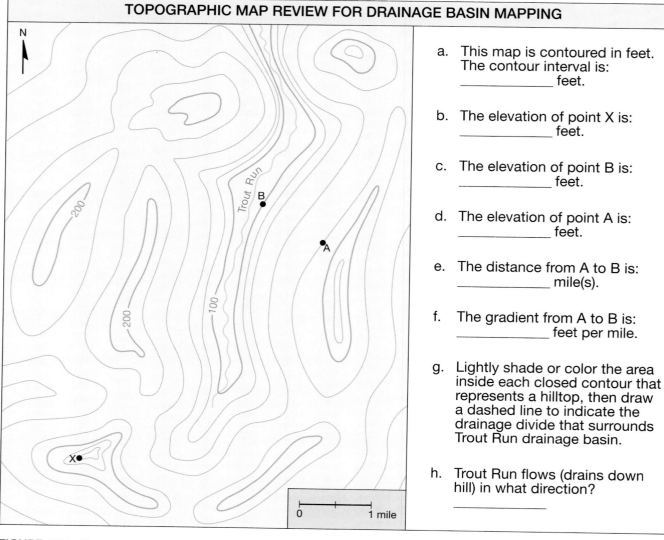

FIGURE 11.5 Topographic map for analysis and interpretation in Question 1.

Questions

1. Refer to Figure 11.5 and complete items **a** through **h**. Refer back to Figures 9.6 and 9.7 as needed.

2. Imagine that drums of oil were emptied (illegally) at location **X** in Figure 11.5. Is it likely that the oil would eventually wash downhill into Trout Run? Explain your reasoning.

3. Refer to the topographic map of the Lake Scott quadrangle, Kansas (Figure 11.3).

   a. Draw and label a dashed line on this map that is the divide between Battendorf Canyon (NE$\frac{1}{4}$ sec. 9) and the smaller canyon that runs from the SW$\frac{1}{4}$ sec. 9 to the "B" in the word "TIMBER" ($\frac{1}{2}$ mile north of the center of sec. 16).

   b. Draw and label a dashed line at the boundary of the Garvin Canyon drainage basin.

4. Notice that the upland surface of the Lake Scott, Kansas, quadrangle (see Figure 11.3) is not horizontal.

   a. In what general direction does water flow over the upland surface?

   b. What is the gradient of this upland surface? (Show your mathematical calculation.)

   c. The mudstone bedrock of these uplands is overlain by alluvium (sand and gravel) that was deposited about 10,000–12,000 years ago by streams draining from melting glaciers (prior to dissection of the surface as it appears today). It is the upper surface of this alluvium that has the

Within the map figure:

TOPOGRAPHIC MAP REVIEW FOR DRAINAGE BASIN MAPPING

a. This map is contoured in feet. The contour interval is: _____ feet.

b. The elevation of point X is: _____ feet.

c. The elevation of point B is: _____ feet.

d. The elevation of point A is: _____ feet.

e. The distance from A to B is: _____ mile(s).

f. The gradient from A to B is: _____ feet per mile.

g. Lightly shade or color the area inside each closed contour that represents a hilltop, then draw a dashed line to indicate the drainage divide that surrounds Trout Run drainage basin.

h. Trout Run flows (drains down hill) in what direction? _____

Trout Run

0   1 mile

attitude described in **a** and **b,** above. What is the probable source area for the water and sediments that formed this alluvial deposit? Explain your reasoning. (*Hint:* Notice at the top of Figure 11.3 that this area is just east of the Great Divide.)

5. Reconsider the uplands and upland alluvium discussed in Question 4. The extensive sheet of upland alluvium was probably deposited by braided streams (see Figures 11.1B and 11.1D) in an older stage of landscape development.

   a. What drainage pattern (shown in Figure 11.2) currently is developed in this area?

   b. What does this modern drainage pattern suggest about the attitude of bedrock layers (the mudstone layers beneath the upland alluvium) in this area (refer to Figure 11.2)? Explain your reasoning.

6. Examine the landscape of the Waldron, Arkansas, quadrangle (Figure 11.6). What drainage pattern is developed in this area, and what does it suggest about the attitude of bedrock layers on this area? Explain your reasoning. (*Hint:* Refer to Figure 11.2. It may also be useful to trace the linear hilltops with a highlighter so you can see their orientation relative to streams.)

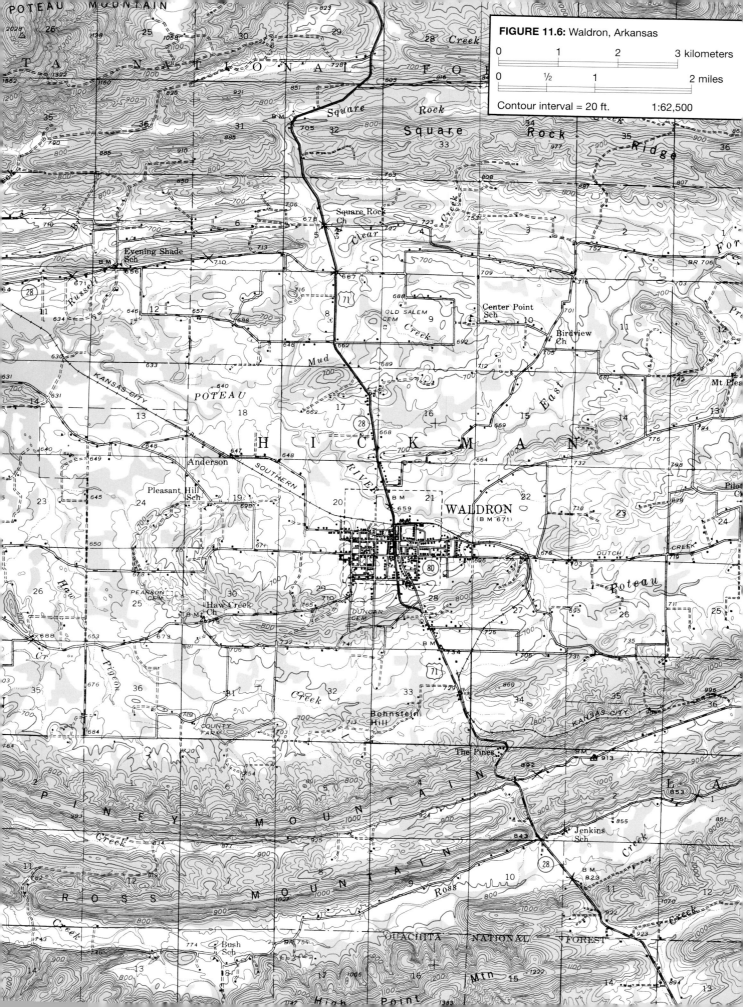

FIGURE 11.6: Waldron, Arkansas

PART 11B: STREAM PROCESSES AND LANDSCAPES NEAR VOLTAIRE, NORTH DAKOTA

Refer to the Voltaire, North Dakota, quadrangle (Figure 11.7) and stereogram (Figure 11.8).

Questions

7. Glaciers (composed of a mixture of ice, gravel, sand, and mud) were present in this region at the end of the Pleistocene Ice Age. When the glaciers melted about 11,000–12,000 years ago, a thick layer of sand and gravel was deposited on top of the bedrock, and streams began forming from the glacial meltwater. Therefore, streams have been eroding and shaping this landscape for about 11,000–12,000 years. Notice how well-developed the meanders and floodplains of the Souris River are. The landscape around Lake Scott, Kansas (Figure 11.3), is about the same age (i.e., 10,000–12,000 years old). Explain why you think there are such differences in forms of the valleys and types of stream channels between the Voltaire region and the Lake Scott region.

8. On Figures 11.7 and 11.8, note the swampy oxbow lakes and depressions (hachured contours on Figure 11.7) in the Souris River floodplain. These show that the river channel has changed course repeatedly. Explain how its course has changed at the oxbow just east of Westgaard Cemetery (see northeast of map center, NE$\frac{1}{4}$ sect. 3, Figure 11.7).

9. Do the hachured contours and other oxbows of the Souris River Valley show that this same process has occurred elsewhere along the valley? If so, then suggest one location.

10. What is one location along the course of the Souris River where the same thing may happen in the future if the course of the channel is not controlled by engineers?

11. Imagine what the topographic profile looks like along **X–X'**. (Refer to the stereogram in Figure 11.8 to help you with this.) Notice the relatively flat areas of the profile, such as those in SW$\frac{1}{4}$ sec. 33 and SE$\frac{1}{4}$ sec. 4 (Figure 11.7).

    a. What are these features called?

    b. How did they form?

    c. How could vegetation be used to map the location of the modern floodplain?

12. In SE$\frac{1}{4}$ sec. 3 (Figure 11.7), a stream trends northeast–southeast. What is the name of this type of stream and how did it probably form?

13. Notice the marsh in sec. 9 (Figure 11.7) and the depression on which it is located. What was this depression before it became a marsh?

14. How might the discharge of the Souris River have changed over the past 12,000 years? Why?

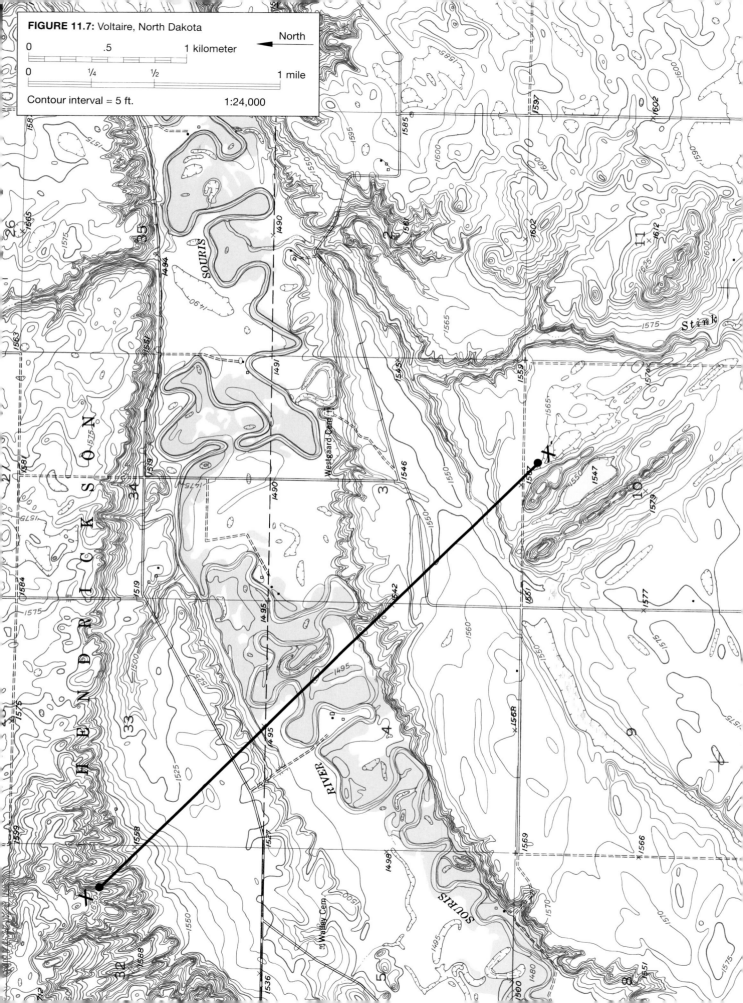

FIGURE 11.7: Voltaire, North Dakota

North

Contour interval = 5 ft.

1:24,000

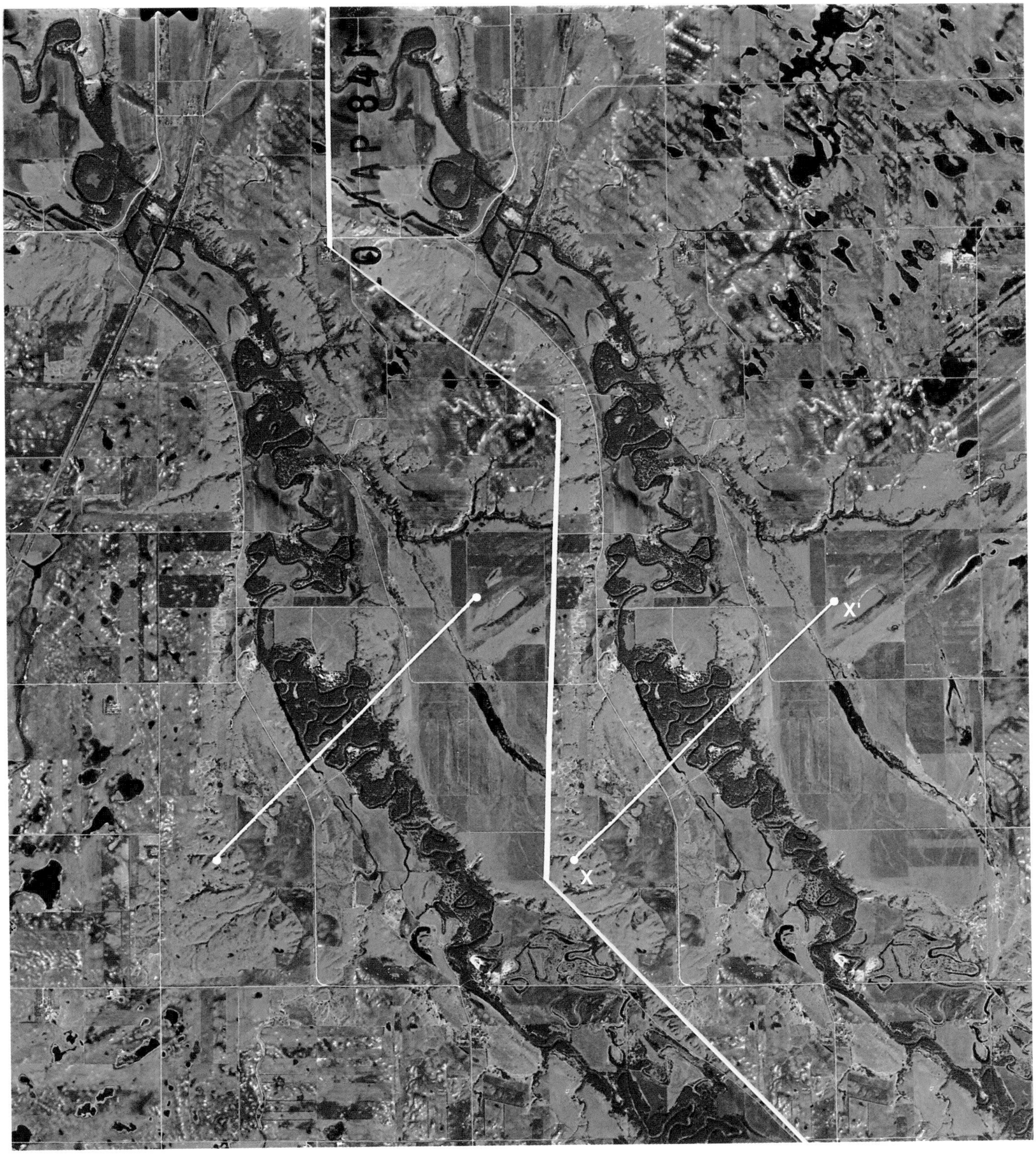

FIGURE 11.8 Color-infrared stereogram of national high-altitude aerial photographs (NHAP) of Voltaire, North Dakota, 1991. Scale 1:58,000. To view in stereo: (a) note that the figure is two images, (b) hold figure at arm's length, (c) cross your eyes until the two images become four images, (d) slightly relax your eyes so the two center images merge in stereo. To view using a pocket stereoscope, refer to Figure 9.23. (Courtesy of U.S. Geological Survey)

PART 11C: STREAM PROCESSES AND LANDSCAPES NEAR ENNIS, MONTANA

Some rivers are subject to large floods, either seasonal or periodic. In mountains, this flooding is due to snow melt. In deserts, it is caused by thunderstorms. During such times, rivers transport exceptionally large volumes of sediment. This causes characteristic features, two of which are braided channels and alluvial fans. Both features are relatively common in arid mountainous regions, such as the Ennis, Montana, area in Figures 11.9 and 11.10. (Both features also can occur wherever conditions are right, even at construction sites!)

Questions

15. What was the source of the sediments that have accumulated on the Cedar Creek Alluvial Fan?

16. What is the approximate stream gradient of:

    a. the main stream in the forested southeastern corner of the map (Figure 11.9) and stereogram (Figure 11.10)?

    b. most streams on the Cedar Creek Alluvial Fan?

    c. the Madison River?

17. What main stream channel types (shown in Figure 11.1B) are present on:

    a. the streams in the forested southeastern corner of the map (Figure 11.9) and stereogram (Figure 11.10)?

    b. the Cedar Creek Alluvial Fan?

    c. the valley of the Madison River (northwestern portion of Figure 11.9)?

18. How are the stream gradients and channel types described above (Questions 16 and 17) related?

19. How did the Cedar Creek Alluvial Fan form?

PART 11D: MEANDER EVOLUTION ON THE RIO GRANDE

Refer to Figure 11.11 showing the meandering Rio Grande, the river that forms the national border between Mexico and the United States. Notice that the position of the river changed in many places between 1936 (red line and leaders by lettered features) and 1992 (blue water bodies and leaders by lettered features). Study the meander terms provided in Figure 11.11, and then proceed to the questions below.

Questions

20. Study the meander cutbanks labeled **A** through **G**. The red leader from each letter points to the cutbank's location in 1936. The blue leader from each letter points to the cutbank's location in 1992. In what two general directions (relative to the meander, relative to the direction of river flow) have these cutbanks moved?

21. Study locations **H** and **I**.

    a. In what country were **H** and **I** located in 1936?

    b. In what country were **H** and **I** located in 1992?

    c. Explain a process that probably caused locations **H** and **I** to change from meanders to oxbow lakes.

22. Based on your answer in Question 21c, predict how the river will change in the future at locations **J** and **K**.

23. What are features **L**, **M**, and **N**, and what do they indicate about the historical path of the Rio Grande?

24. What is the average rate at which meanders like **A** through **G** migrated here (in meters per year) from 1936 to 1992? Explain your reasoning and calculations.

25. Explain in steps how a meander evolves from the earliest stage of its history as a broad slightly-sinuous meander to the stage when an oxbow lake forms.

26. Suggest as many factors as you can think of that could speed up changes in the location of the Rio Grande.

27. Suggest as many factors as you can think of that could slow down changes in the location of the Rio Grande.

FIGURE 11.9: Ennis, Montana

North

Montana

| 0 | 1 | 2 | 3 kilometers |

| 0 | ½ | 1 | 2 miles |

Contour interval = 40 ft.   1:62,500

Quadrangle location

FIGURE 11.10 Color-infrared stereogram of national high-altitude aerial photographs (NHAP) of Ennis, Montana, 1991. Scale 1:58,000. (Courtesy of U.S. Geological Survey)

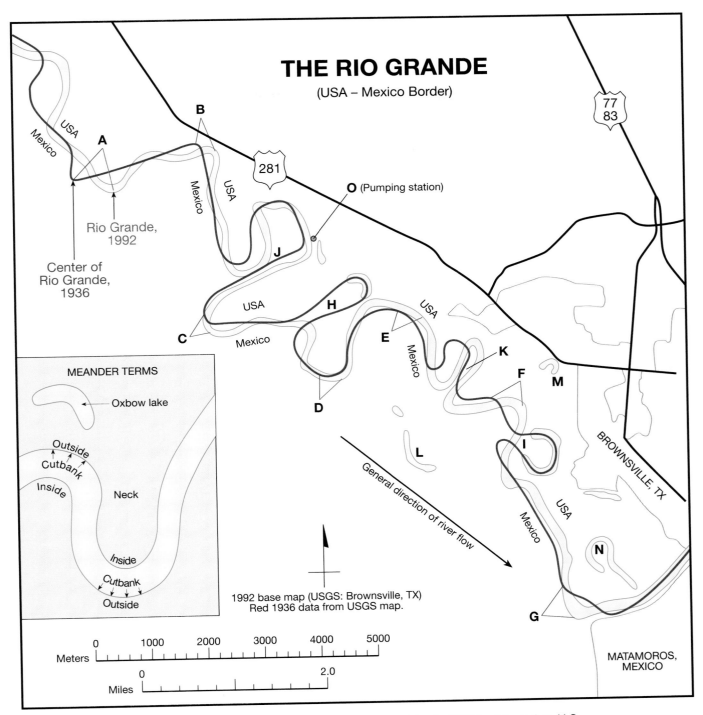

THE RIO GRANDE
(USA – Mexico Border)

MEANDER TERMS

- Oxbow lake
- Outside
- Cutbank
- Inside
- Neck
- Inside
- Cutbank
- Outside

O (Pumping station)

Rio Grande, 1992

Center of Rio Grande, 1936

General direction of river flow

1992 base map (USGS: Brownsville, TX)
Red 1936 data from USGS map.

| Meters | 0 | 1000 | 2000 | 3000 | 4000 | 5000 |

| Miles | 0 | | 2.0 | |

BROWNSVILLE, TX

MATAMOROS, MEXICO

FIGURE 11.11 Map of the Rio Grande in 1992 (blue) and its former position in 1936 (red) based on U.S. Geological Survey topographic maps (Brownsville, Texas, 1992; West Brownsville, Texas, 1936). The river flows east-southeast. Note the inset box of meander terms used to describe features of meandering streams.

PART 11E: STREAM EROSION AND MASS WASTAGE AT NIAGARA FALLS

Mass wastage is the downslope movement of Earth materials such as soil, rock, and other debris. It is common along steep slopes, such as those created where rivers cut into the land. Some mass wastage occurs along the steep slopes of the river valleys. However, mass wastage can also occur in the bed of the river itself, as it does at Niagara Falls.

The Niagara River flows from Lake Erie to Lake Ontario (Figure 11.12). The gorge of the Niagara presents good evidence of the erosion of a caprock falls, Niagara Falls (Figure 11.13). The edge (caprock) of the falls is composed of the resistant Lockport Dolostone. The retreat of the falls is due to undercutting of mudstones that support the Lockport Dolostone. Water cascading from the lip of the falls enters the plunge pool with tremendous force, and the turbulent water easily erodes the soft mudstones. With the erosion of the mudstones, the Lockport Dolostone collapses.

Questions

28. Geologic evidence indicates that the Niagara River began to cut its gorge (Niagara Gorge) about 11,000 years ago as the Laurentide Ice Sheet retreated from the area. The ice started at the Niagara Escarpment shown in Figure 11.12 and receded (melted back) north to form the basin of Lake Ontario. The Niagara Gorge started at the Niagara Escarpment and retreated south to its present location. Based on this geochronology and the length of Niagara Gorge, calculate the average rate of falls retreat in cm/year.

29. Name as many factors as you can that could cause the falls to retreat at a faster rate.

30. Name as many factors as you can that could cause the falls to retreat more slowly.

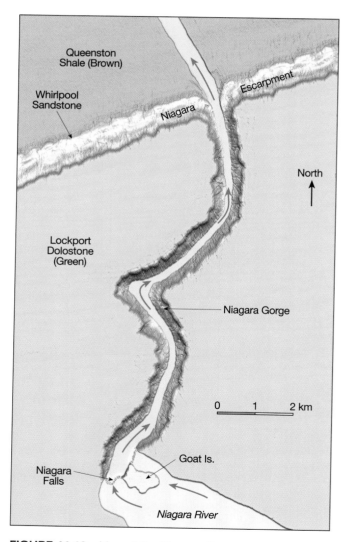

FIGURE 11.12 Map of the Niagara Gorge region of Canada and the United States. The Niagara River flows from Lake Erie north to Lake Ontario. Niagara Falls is located on the Niagara River at the head of Niagara Gorge, about half way between the two lakes.

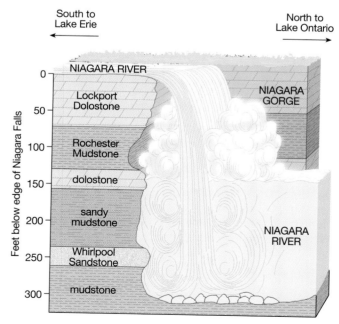

FIGURE 11.13 Cross section of Niagara Falls and geologic units of the Niagara escarpment.

31. Niagara Falls is about 35 km north of Lake Erie, and it is retreating southward. If the falls was to continue its retreat at the average rate calculated in Question 28, then how many years from now would the falls reach Lake Erie?

PART 11F: FLOOD HAZARD MAPPING, ASSESSMENT, AND RISKS

The water level and discharge of a river fluctuates from day to day, week to week, and month to month. These changes are measured at *gaging stations*, with a permanent water-level indicator and recorder. On a typical August day in downtown St. Louis, Missouri, the Mississippi River normally has a discharge of about 130,000 cubic feet of water per second and water levels well below the boat docks and concrete *levees* (retaining walls). However, at the peak of an historic 1993 flood, the river discharged more than a million cubic feet of water per second (8 times the normal amount), swept away docks, and reached water levels at the very edge of the highest levees.

When the water level of a river is below the river's banks, the river is at a **normal stage.** When the water level is even with the banks, the river is at **bankfull stage.** And when the water level exceeds (overflows) the banks, the river is at a **flood stage.** The Federal Emergency Management Agency (FEMA: **www.floodsmart.gov**) notes that there is a 26% chance that your home will be flooded over the 30-year period of a typical home mortgage (compared to a 9% chance of a home fire).

Early in July 1994, Tropical Storm Alberto entered Georgia and remained in a fixed position for several days. More than 20 inches of rain fell in west-central Georgia over those three days and caused severe flooding along the Flint River. Montezuma was one of the towns along the Flint River that was flooded.

This part of the laboratory is designed for you to map and assess the extent of flooding that occurred at Montezuma, based on river gage records and a $7\frac{1}{2}$-minute quadrangle map (Figure 11.14). You will then use river gage records to construct a flood magnitude/frequency graph, determine the probability of floods of specific magnitudes (flood levels), and revise a map of Montezuma to show the area that can be expected to be flooded by a 100-year flood.

Questions

32. On Figure 11.14, locate the gaging station along the east side of the Flint River near the center of the

map, between Montezuma and Oglethorpe. This U.S. Geological Survey (USGS) gaging station is located at an elevation of 255.83 feet above sea level, and the river is considered to be at flood stage when it is 20 feet above this level (275.83 feet). The old flood record was 27.40 feet above the gaging station (1929), but the July 1994 flood established a new record at 35.11 feet above the gaging station, or 289.94 feet above sea level. This corresponds almost exactly to the 290 foot contour line on Figure 11.14. Trace the 290 foot contour line on both sides of the Flint River on Figure 11.14. Label the area within these contours (land areas lower than 290 feet) as the 1994 Flood Hazard Zone.

33. Assess the damage caused by the July 1994 flood.

a. Notice that the gaging station is located adjacent to highway 26-49-90. Was it possible to travel from Montezuma to Oglethorpe on this highway during the 1994 flood? Explain.

b. Notice the railroad tracks parallel to (and south of) highway 26-49-90 between Montezuma and Oglethorpe. Was it possible to travel on these railroads during the 1994 flood? Explain.

34. Notice line **X–Y** near the top center part of the map (Figure 11.14).

a. The map shows the Flint River at its normal stage. What is the width (in km) of the Flint River at its normal stage along line **X–Y**?

b. What was the width of the river (in km) along this line when it was at maximum flood stage (290 feet) during the July 1994 flood?

35. Notice the floodplain of the Flint River along line **X–Y**. It is the relatively flat (as indicated by widely-spaced contour lines) marshy land between the river and the steep (as indicated by more closely spaced contour lines) walls of the valley that are created by erosion during floods.

a. What is the elevation (in feet above sea level) of the floodplain on the west side of the river along line **X–Y**?

b. How deep (in feet above sea level) was the water that covered that floodplain during the 1994 flood? (Explain your reasoning or show your mathematical calculation.)

c. Did the 1994 flood (i.e., the highest river level ever recorded here) stay within the floodplain and its bounding valley slopes? Does this suggest that the 1994 flood was of normal or abnormal magnitude (severity) for this river? Explain your reasoning.

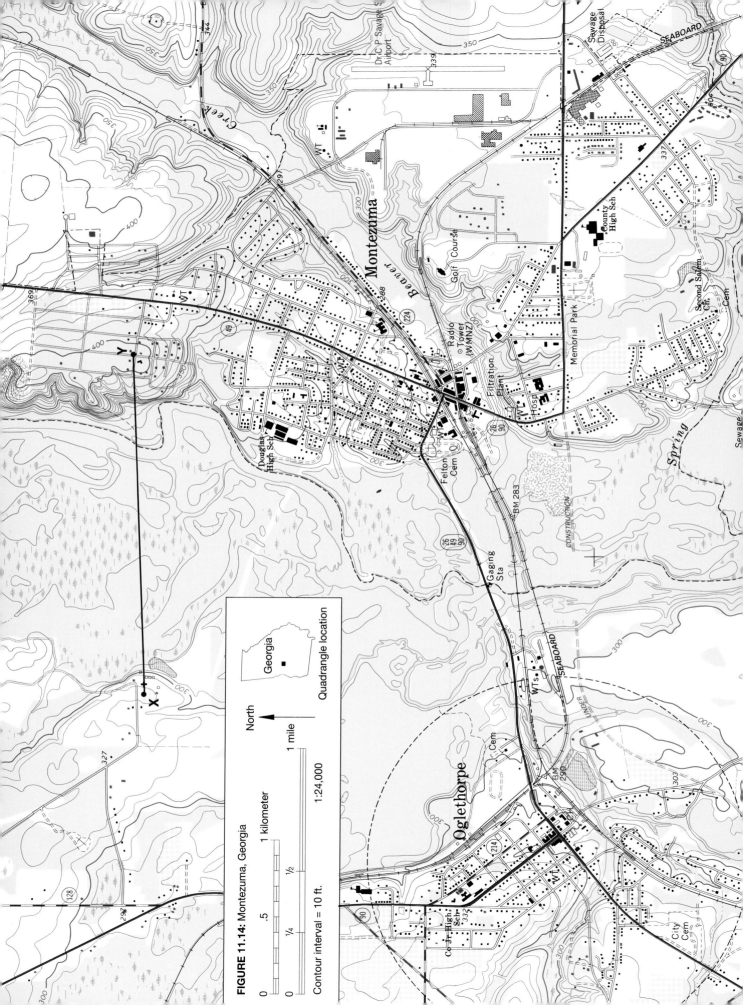

FIGURE 11.14: Montezuma, Georgia

Contour interval = 10 ft.

1:24,000

North

1 mile

1 kilometer

Georgia

Quadrangle location

0   1/4   1/2

0   .5

36. The USGS recorded annual high stages (elevation of water level) of the Flint River at the Montezuma gaging station in Figure 11.14 for 99 years (1897 and 1905–2002). This raw data is available online at http://waterdata.usgs.gov/ga/nwis/inventory/?site_no=02349500. Parts of the data have been summarized in Figure 11.15.

a. The annual highest stages of the Flint River (S) were ranked in severity from S = 1 (highest annual high stage ever recorded; i.e., the 1994 flood) to S = 99 (lowest annual high stage). Data for 14 of these ranked years are provided in the table at the top of Figure 11.15 and can be used to calculate recurrence interval for each magnitude (rank, S). **Recurrence interval** (or **return period**) is the average number of years between occurrences of a flood of a given rank (S) or greater than that given rank. Recurrence interval for a rank of flood can be calculated as: RI = (n + 1)/S. Calculate the recurrence interval for ranks 1 – 5 and write them in the table at the top of Figure 11.15. This has already been done for ranks of 20, 30, 40, 50, 60, 70, 80, 90, and 99.

b. Notice (Figure 11.15) that a recurrence interval of 5.0 means that there is a 1-in-5 probability (or 20% chance) that an event of that magnitude will occur in any given year. This is known as a *5-year flood*. What is a *100-year flood*?

c. Plot (as exactly as you can) points on the flood magnitude/frequency graph (bottom of Figure 11.15) for all 14 ranks of annual high river stage in the table (top of Figure 11.15). Then use a ruler to draw a line through the points (and on to the right edge of the graph) so the number/distance of/to points above and below the line is similar.

d. Your completed flood magnitude/frequency graph can now be used to estimate the probability of future floods of a given magnitude and frequency. A 10-year flood on the Flint River is the point where the line in your graph crosses the flood frequency (RI, return period) of 10 years. What is the probability that a future 10-year flood will occur in any given year, and what will be its magnitude (river elevation in feet above sea level)?

e. What is the probability for any given year that a flood on the Flint River at Montezuma, GA will reach an elevation of 275 feet above sea level?

37. Most homeowners insurance policies do not insure against floods, even though floods cause more damage than any other natural hazard. Homeowners must obtain private or federal flood insurance in addition to their base homeowners policy. The National Flood Insurance Program (NFIP), a Division of the Federal Emergency Management Agency (FEMA) helps communities develop corrective and preventative measures for reducing future flood damage. The program centers on floodplain identification, mapping, and management. In return, members of these communities are eligible for discounts on federal flood insurance. The rates are determined on the basis of a community's FIRM (Flood Insurance Rate Map), an official map of the community on which FEMA has delineated flood *hazard areas* and *risk premium zones* (with discount rates). The hazard areas on a FIRM is defined on a *base flood elevation* (BFE)—the computed elevation to which flood water is estimated to rise during a *base flood*. The regulatory-standard base flood elevation is the 100-year flood elevation. Based on your graph (bottom of Figure 11.15), what is the BFE for Montezuma, GA?

38. The 1996 FEMA FIRM for Montezuma, GA shows hazard areas designated *zone A*. Zone A is the official designation for areas expected to be inundated by 100-year flooding even though no BFEs have been determined. The location of zone A (shaded gray) is shown on a portion of the Montezuma, GA $7\frac{1}{2}$ minute topographic quadrangle map in Figure 11.16. Your work above can be used to revise the flood hazard area. Place a dark line on this map (as carefully as you can) to show the elevation contour of the BFE for this community (your answer in item 37). Your revised map reflects more accurately what area will be inundated by a 100-year flood. In general, how is the BFE line that you have plotted different from the boundary of zone A plotted by FEMA on its 1996 FIRM?

39. Note that the elevation of the 100-year flood is estimated on the basis of historical (existing) data. As in the example of Montezuma, GA, a new flood that sets a new flood record will change the flood magnitude/frequency graph and the estimated BFE. You should always obtain the latest flood hazard map/graph to estimate flood probability for a given location. Determine the flood risk where you live (or another location of your choice) at **www.floodsmart.gov/floodsmart/pages/riskassesment/findpropertyform.jsp**.

40. If you live in an area that is prone to flooding, then what disaster supplies should you keep on hand (**www.fema.gov**)?

Recurrence Intervals for Selected, Ranked, Annual Highest Stages of the Flint River over 99 Years of Observation (1897 and 1905–2002) at Montezuma, Georgia (USGS Station 02349500, data from USGS)

Rank of annual highest river stage (S)	Year (*n = 99)	River elevation above gage, in feet	Gage elevation above sea level, in feet	River elevation above sea level, in feet	Recurrence interval** (RI), in years	Probability of occurring in any given year	Percent chance of occurring in any given year***
1 (highest)	1994	34.11	255.83	289.9		1 in 100	1%
2	1929	27.40	255.83	283.2		1 in 50	2%
3	1990	26.05	255.83	281.9		1 in 33.3	3%
4	1897	26.00	255.83	281.8		1 in 25	4%
5	1949	25.20	255.83	281.0		1 in 20	5%
20	1928	21.30	255.83	277.1	5.0	1 in 5	20%
30	1912	20.60	255.83	276.4	3.4	1 in 3.4	29%
40	1959	19.30	255.83	275.1	2.3	1 in 2.3	43%
50	1960	18.50	255.83	274.3	2.0	1 in 2	50%
60	1934	17.70	255.83	273.5	1.8	1 in 1.8	56%
70	1974	17.25	255.83	273.1	1.5	1 in 1.5	67%
80	1967	14.76	255.83	270.6	1.3	1 in 1.3	77%
90	1907	13.00	255.83	268.8	1.1	1 in 1.1	91%
99 (lowest)	2002	8.99	255.83	264.7	1.0	1 in 1	100%

*n = number of years of annual observations = 99
**Recurrence Interval (RI) = (n + 1) / S = average number of years between occurrences of an event of this magnitude or greater.
***Percent chance of occurrence = 1 / RI x 100.

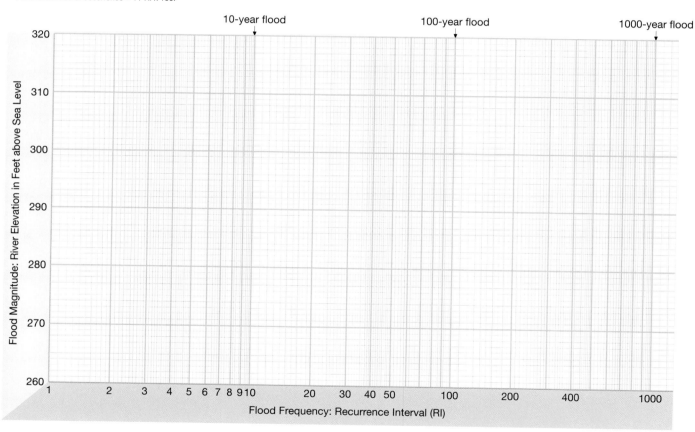

FIGURE 11.15   Flint River, GA historical data (USGS) and flood magnitude/frequency graph.

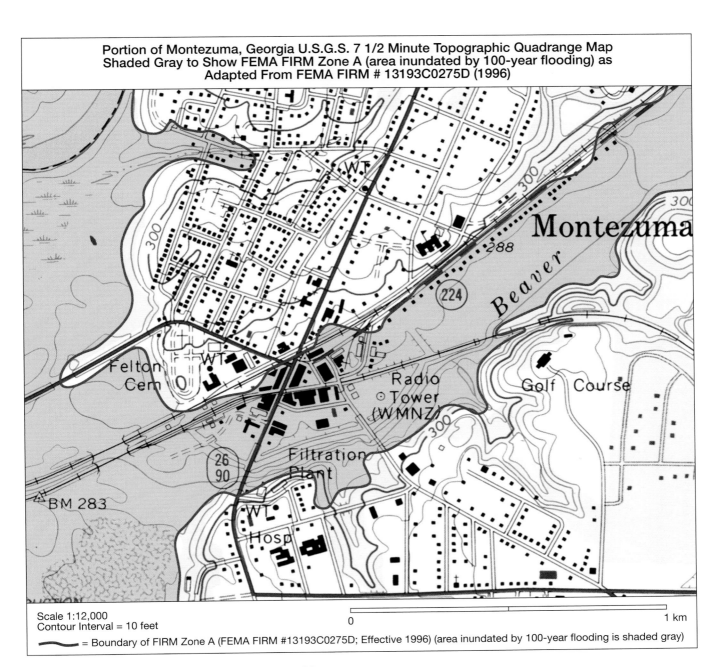

FIGURE 11.16 Flood hazard map of Montezuma, GA.

Groundwater Processes, Resources, and Risks

•CONTRIBUTING AUTHORS•

Gary D. McKenzie • *Ohio State University*

Richard N. Strom • *University of South Florida, Tampa*

James R. Wilson • *Weber State College*

OBJECTIVES

A. Understand the topographic features and groundwater movements associated with *karst topography.*

B. Construct a water-table contour map and determine the rate and direction of groundwater movement.

C. Evaluate how groundwater withdrawal can cause *subsidence* (sinking) of the land.

D. Evaluate hazards and risks associated with the use and contamination of groundwater.

MATERIALS

Pencil, eraser, ruler, and calculator.

INTRODUCTION

Water that seeps into the ground is pulled downward by the force of gravity through spaces in the soil and *bedrock* (rock that is exposed at the land surface or underlies the soil). At first, the water fills just some spaces and air remains in the other spaces. This underground zone with water- and air-filled spaces is called the *zone of aeration* (Figure 12.1; also called the *undersaturated zone* or *vadose zone*). Eventually, the water reaches a zone below the zone of aeration, where all spaces are completely saturated with water.

This water-logged zone is called the *zone of saturation,* and its upper surface is the **water table** (Figure 12.1). Water in the saturated zone is called **groundwater,** which can also be withdrawn from the ground through a **well** (a hole dug or drilled into the ground). Most wells are lined with *casing*, a heavy metal or plastic pipe. The casing is perforated in sections where water is expected to supply the well. Other sections of the casing are left impervious to prevent unwanted rock particles or fluids from entering the well.

Recall the last time that you consumed a drink from a fast-food restaurant (a paper cup containing ice and liquid that you drink using a plastic straw). The mixture of ice and liquid (no air) at the bottom of the cup was a zone of saturation, and your straw was a well. Each time you sucked on the straw, you withdrew liquid from the drink container just as a homeowner withdraws water from a water well. After you drank some of the drink, the cup contained both a zone of saturation (water and ice in the bottom of the cup) and a zone of aeration (ice and mostly air in the upper part of the cup). The boundary between these two zones was a water table. In order to continue drinking the liquid, you had to be sure that the bottom of your straw was within the zone of saturation, below the water table. Otherwise, sucking on the straw produced only a slurping sound, and you obtained mostly air. Natural water wells work the same way. The wells must be drilled or dug to a point below the water table (within the zone of saturation), so water can flow or be pumped out of the ground.

Water Table Contours and Flow Lines

A. Groundwater Zones and the Water Table

$$\text{Hydraulic gradient} = \frac{h_1 - h_2}{d} = 10 \text{ ft/mi}$$

$h_2 = 110$ ft
Well B

$h_1 = 120$ ft
Well A

Flow direction

$h_1 - h_2$

Distance between wells (d)

Elevation above sea level (feet)

0   1 mi

B. Normal Water Table Contours and Flow Lines: Note that flow direction is downhill to streams and the lake

C. Water Table Contours and Flow Lines Changed by a Cone of Depression Developed Around a Pumped Well

Pumped well

Cone of depression around pumped well

120 ——— Water table contour line

⟵——— Flow line (arrow indicates direction of flow)

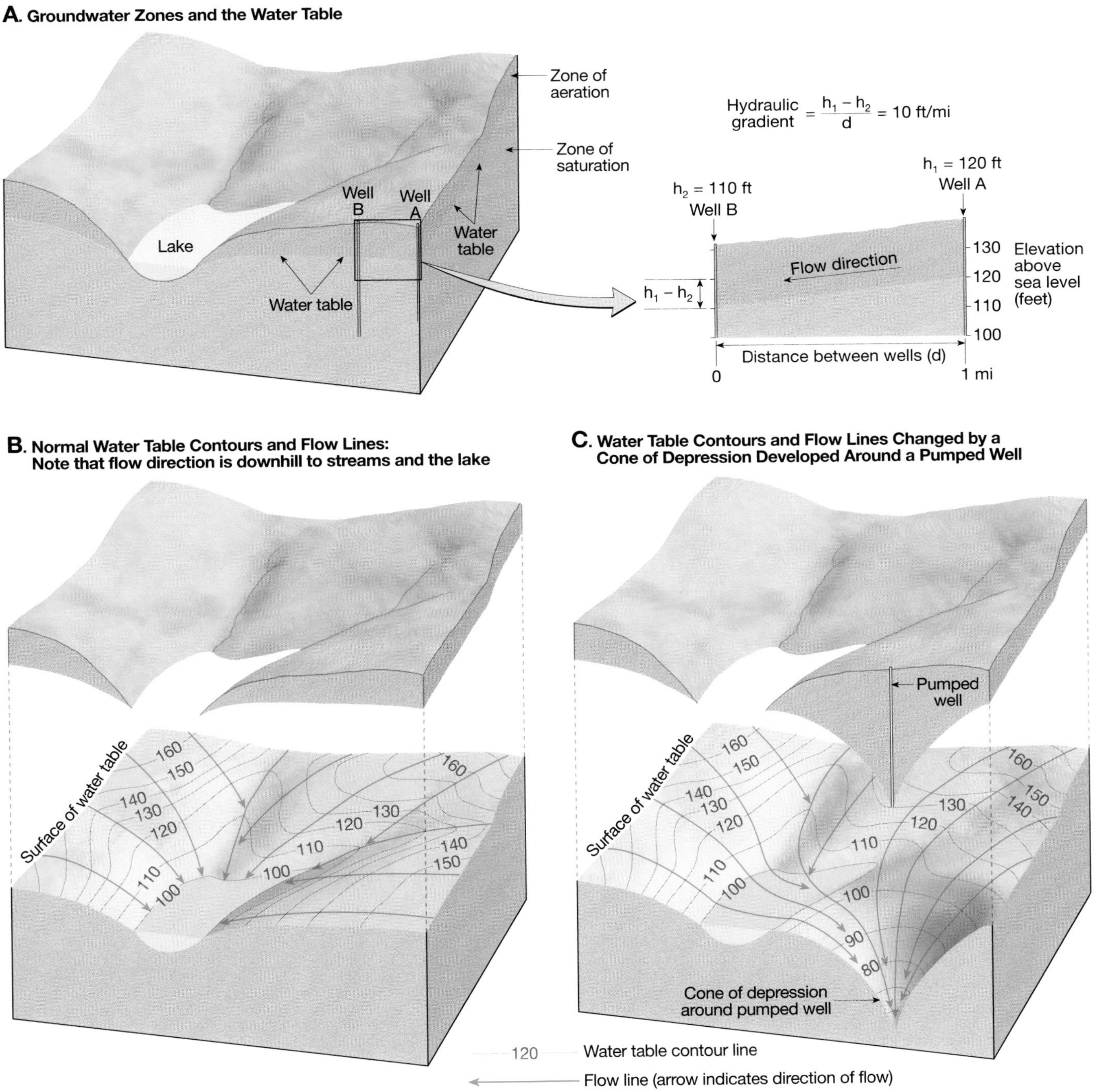

FIGURE 12.1 Water movement through an unconfined aquifer. **A.** Rainwater seeps into the *zone of aeration* (undersaturated zone, vadose zone), where void spaces are filled with air and water. Below it is the *zone of saturation,* where all void spaces are filled with water. Its upper surface is the **water table.** Water in the saturated zone is called **groundwater,** which always flows down the hydraulic gradient in unconfined aquifers. **B.** A water table surface is rarely level. Contour lines (contours) are used to map its topography and identify flow lines—paths traveled by droplets of water from the points where they enter the water table to the points where they enter a lake or stream. **Flow lines** run perpendicular to contour lines, converge or diverge, but never cross. **C.** A pumped well is being used to withdraw water faster than it can be replenished, causing development of a **cone of depression** in the water table and a change in the groundwater flow lines.

The volume of void space (space filled with water or air) in sediment or bedrock is termed *porosity*. The larger the voids, and the greater their number, the higher is the porosity. If void spaces are interconnected, then fluids (water and air) can migrate through them (from space to space), and the rock or sediment is said to be *permeable*. Sponges and paper towels are household items that are permeable, because liquids easily flow into and through them. Plastic and glass are *impermeable* materials, so they are used to contain fluids.

Permeable bedrock materials make good **aquifers,** or rock strata that conduct water. Some examples are sandstones and limestones. Impermeable bedrock materials prevent the flow of water and are called **confining beds.** Some examples are layers of clay, mudstone, shale, or dense igneous and metamorphic rock. But how does groundwater move through aquifers?

When aquifers are sandwiched between confining beds, the groundwater fills them from confining bed to confining bed—like water filling a large, flat pipe. When aquifers are not confined, the groundwater establishes a water table just beneath the surface of the land (Figure 12.1). For this reason, unconfined aquifers are sometimes called *water table aquifers*.

Groundwater in an unconfined aquifer is pulled down by gravity and attempts to spread out through the ground until it forms the water table surface (such as the one in the drink cup described above). You can see the water table where it leaves the ground and becomes the level surface of a lake (Figure 12.1A). However, because groundwater is continuously being replenished (recharged) upslope, and it takes time for the water to flow through the ground, the water table is normally not level. It is normally higher uphill, where water flows into the ground, and lower downhill, where water seeps out of the ground at lakes, streams, or springs. The slope of the water table surface is called the **hydraulic gradient** (Figure 12.1A)—the difference in elevation between two points on the water table (observed in wells or surfaces of lakes and ponds) divided by the distance between those points.

To better understand the topography of the water table in a region, geologists measure its elevation wherever they can find it in wells or where it forms the surfaces of lakes and streams. The elevation data is then contoured to map the **water table contour lines** (Figure 12.1). Since water always flows down the shortest and steepest path (path of highest hydraulic gradient) it can find, a drop of water on the water table surface will flow perpendicular to the slope of the water table contour lines. Geologists use **flow lines** with arrows to show the paths that water droplets will travel from the point where they enter

the water table to the point where they reach a lake, stream, or level water table surface. Notice how flow lines have been plotted on Figure 12.1B and 12.1C. In Figure 12.1C, notice how a pumped well is being used to withdraw water faster than it can be replenished. This has caused a cone-shaped depression in the water table (**cone of depression**) and a change in the regional flow of the groundwater. Thus, water table contour maps are useful for determining:

- paths of water flow (flow lines on a map) along which hydraulic gradients are normally measured

- where the water comes from for a particular well

- paths (flow lines) that contaminants in groundwater will likely follow from their source.

- changes to groundwater flow lines and hydraulic gradients caused by cones of depression at pumped wells.

PART 12A: CAVES AND KARST TOPOGRAPHY

The term **karst** describes a distinctive topography that indicates dissolution of underlying soluble rock, generally limestone (Figure 12.2). The limestone dissolves because rainwater is mildly acidic. The rainwater soaks into the ground to form acidic groundwater, which dissolves the calcite mineral crystals making up the limestone.

Rainwater may contain several acids, but the most common is carbonic acid (H_2CO_3). It forms when water (H_2O) and carbon dioxide (CO_2) combine in the atmosphere. When it rains, the carbonic acid in rainwater dissolves the calcite (and other carbonate minerals) in limestone by this reaction:

$$CaCO_3 \quad + \quad H_2CO_3 \quad = \quad Ca^{+2} \quad + \quad 2\ HCO_3^{-1}$$

| calcite | carbonic acid | calcium ions dissolved in groundwater | bicarbonate ions dissolved in groundwater |

A typical karst topography has these features, which are illustrated in Figure 12.2 and visible on the topographic map in Figure 12.3.

- **Sinkholes**—surface depressions formed by the collapse of caves or other large underground void spaces.

- **Solution valleys**—valley-like depressions formed by a linear series of sinkholes or collapse of the roof of a linear cave.

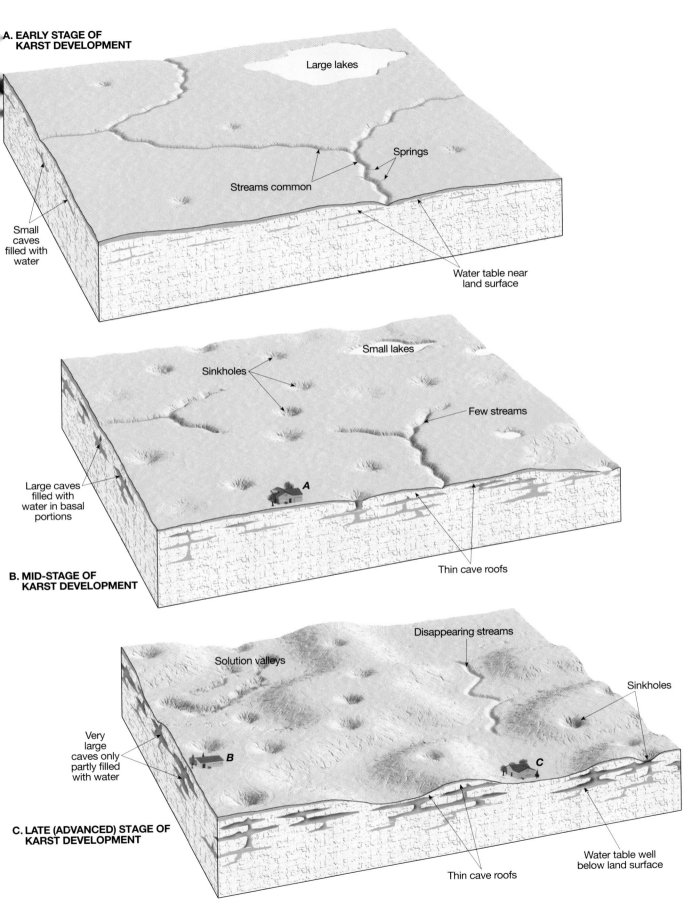

A. EARLY STAGE OF KARST DEVELOPMENT

Large lakes

Springs

Streams common

Small caves filled with water

Water table near land surface

B. MID-STAGE OF KARST DEVELOPMENT

Small lakes

Sinkholes

Few streams

Large caves filled with water in basal portions

A

Thin cave roofs

C. LATE (ADVANCED) STAGE OF KARST DEVELOPMENT

Disappearing streams

Solution valleys

Sinkholes

Very large caves only partly filled with water

B

C

Thin cave roofs

Water table well below land surface

FIGURE 12.2 Stages in the evolution of karst topography, which forms by dissolution of soluble bedrock.

239

FIGURE 12.3: Mammoth Cave, Kentucky

- **Springs**—places where water flows naturally from the ground (from spaces in the bedrock).

- **Disappearing streams**—streams that terminate abruptly by seeping into the ground.

Much of the drainage in karst areas occurs underground rather than by surface runoff. Rainwater seeps into the ground along fractures in the bedrock (Figure 12.4), whereupon the acidic water dissolves the limestone around it. The cracks widen into narrow **caves** (underground cavities large enough for a person to enter), which may eventually widen into huge cave galleries. Sinkholes develop where the ceilings of these galleries collapse, and lakes or ponds form wherever water fills the sinkholes. The systems of fractures and caves that typically develop in limestones are what make limestones good aquifers.

Eventually, the acidic water that was *dissolving* limestone becomes so enriched in calcium and bicarbonate that it turns alkaline (the opposite of acid) and may actually begin *precipitating* calcite (Figure 6.7).

Caves in karst areas often have *stalactites*, icicle-like masses of chemical limestone that hang from cave ceilings (Figure 12.5). They form because calcite precipitates from water droplets as they drip from the cave ceiling. The broken end of a stalactite is shown in Figure 6.7. Water dripping onto the cave floor also can precipitate calcite and form more stout *stalagmites*.

Questions

1. Study Figure 12.4. Notice that there is no soil developed on this limestone bedrock surface, yet abundant plants are growing along linear features in the bedrock. What does this indicate about how water travels through bedrock in this part of Oklahoma?

2. If you had to drill a water well in the area pictured in Figure 12.4, then where would you drill (relative to the linear pattern of plant growth) to find a good supply of water? Explain your reasoning.

3. How is Figure 12.5 related to Figure 12.4?

4. It is common for buildings to sink into newly formed sinkholes as they develop in karst regions. Consider the three new-home construction sites (labeled **A, B,** and **C**) in Figure 12.2, relative to sinkhole hazards.

   a. Which new-home construction site (**A, B,** or **C**) is the **most** hazardous? Why?

   b. Which new-home construction site (**A, B,** or **C**) is the **least** hazardous? Why?

   c. Imagine that you are planning to buy a new-home construction site in the region portrayed in Figure 12.2. What could you do to find out if there is a sinkhole hazard in the location where you are thinking of building your home?

5. Study the portion of the Mammoth Cave (Kentucky) topographic map in Figure 12.3. This area is underlain by limestone, which is capped (overlain) by sandstone in the northern part of the mapped area (north of Park City).

   a. How can you tell the area on this map where limestone crops out at Earth's surface?

   b. Draw and label a line on Figure 12.3 that separates the karst topography from the northern part of the map where mostly forested (green color) sandstone crops out.

   c. Find and label a disappearing stream at the end where it disappears within the karst region.

   d. Find and label a lake that formed by flooding a sinkhole.

   e. Find and label a solution valley anywhere on the map.

   f. Imagine that you are planning to build a home on top of one of the highest hills in the area where mostly sandstone crops out, and that you plan to use well water drawn from the limestone under your property. What is the deepest you would have to drill your well (through the sandstone) to obtain water from the limestone aquifer? Explain your reasoning.

   g. Why are there so few sinkholes developed in the southeast part of this map area (even though limestone crops out there)?

FIGURE 12.4 Looking east toward the Arkansas River from Vap's Pass, Oklahoma (15 miles northeast of Ponca City). The Fort Riley Limestone bedrock *crops out* (is exposed at the surface) here. There is no soil in this location, but plants have grown naturally along linear features in the bedrock.

FIGURE 12.5 Stalactites on part of the ceiling of Cave of the Winds, which has formed in Paleozoic limestones near Manitou Springs, Colorado.

PART 12B: LOCATION AND MOVEMENT OF GROUNDWATER IN THE FLORIDAN LIMESTONE AQUIFER

Figures 12.6–12.8 show karst features developed in the Floridan Limestone Aquifer in the northern part of Tampa, Florida. Notice in Figures 12.6 and 12.7 that most of the lakes occupy sinkholes. They are indicated on Figure 12.6 with hachured contour lines contours with small tick marks that point inward, indicating a depression). These depressions intersect the water table and the subjacent limestone bedrock, as shown in Figure 12.7. By determining and mapping the elevations of water surfaces in the lakes, you can determine the slope of the water table and the direction of flow of the groundwater here (as in Figure 12.1B).

Questions

6. On Figure 12.8, mark the elevations of water levels in the lakes (obtain this information from Figure 12.6). The elevations of Lake Magdalene and some lakes beyond the boundaries of the topographic map already are marked for you.

7. Contour the water-table surface (use a 5-foot contour interval) on Figure 12.8. Draw only contour lines representing whole fives (40, 45, and so on). Do this in the same manner that you contoured land surfaces in Laboratory 9 (topographic maps and aerial photographs).

8. The flow of shallow groundwater in Figure 12.8 is at right angles to the contour lines. The groundwater flows from high elevations to lower elevations, just like a stream. Draw three or four flow lines with arrows on Figure 12.8 to indicate the direction of shallow groundwater flow in this part of Tampa. The southeastern part of Figure 12.6 shows numerous closed depressions but very few lakes. What does this indicate about the level of the water table in this region?

9. Note the Poinsettia Sinks, a pair of sinkholes in the southeast corner of the topographic map (see Figure 12.6). Note their closely spaced hachured contour lines. Next find the cluster of five similar sinkholes, called Blue Sinks, about 1 mile northwest of Poinsettia Sinks (just west of the WHBO radio tower). Use asterisks (*) to mark their locations on Figure 12.8, and label them "Blue Sinks."

   a. Draw a straight arrow (vector) on Figure 12.8 along the shortest path between Blue Sinks and

Poinsettia Sinks. The water level in Blue Sinks is 15 feet above sea level, and the water level in Poinsettia Sinks is 10 feet above sea level. Calculate the hydraulic gradient (in ft/mi) along this arrow and write it next to the arrow on Figure 12.8. (Refer to Figure 12.1 to review hydraulic gradient as needed.)

   b. On Figure 12.6, note the stream and valley north of Blue Sinks. This is a fairly typical disappearing stream. Draw its approximate course onto Figure 12.8. Make an arrowhead on one end of your drawing of the stream to indicate the direction that water flows in this stream. How does this direction compare to the general slope of the water table?

10. In March 1958, fluorescent dye was injected into the northernmost of the Blue Sinks. It was detected 28 hours later in Sulphur Springs, on the Hillsborough River to the south (see Figure 12.8). Use this data to calculate the approximate velocity of flow in this portion of the Floridan Aquifer:

    a. in feet per hour

    b. in miles per hour

    c. in meters per hour

    The velocities you just calculated are quite high, even for the Floridan Aquifer. But this portion of Tampa seems to be riddled with solution channels and caves in the underlying limestone. Sulphur Springs has an average discharge of approximately 44 cubic feet per second (cfs), and its maximum recorded discharge was 165 cfs (it once was a famous spa).

11. During recent years, the discharge at Sulphur Springs has decreased. Water quality has also worsened substantially.

    a. Examine the human-made structures on Figure 12.6. Note especially those in red, the color used to indicate new structures. Why do you think the discharge of Sulphur Springs has decreased in recent years?

    b. Why do you think the water quality has decreased in recent years?

12. Imagine you are selling homeowner's insurance in the portion of the Sulphur Springs quadrangle shown in Figure 12.6. List all the potential groundwater-related hazards to homes and homeowners in the area that you can think of.

FIGURE 12.6: Sulfur Springs, Florida

0   .5   1 kilometer
0   ¼   ½   1 mile
Contour interval = 5 ft.   1:24,000

North

Platt Lake

Long Lake

Lake Gass

Lamps Pond

Lake Magdalene

Bay Lake

Trailer Parks

Trailer Park

Hamner Lookout Tower

FLETCHER

Dorset Lake

Lake Senac

Lake Magdalene Ch

West Lake

Cedar Lake

Noreast Lake

Pine L

Lake Ellen

Lake Carroll Cem

Trailer Park

INGLEWOOD

Drive-in Theater

FOWLER

Radio Tower (WHBO)

Blue Sinks

Lake Carroll

Boat Lake

White Trout Lake

Poinsettia S

BOUGAINVILLEA

LINEBAUGH

Most Holy Redeemer Sch

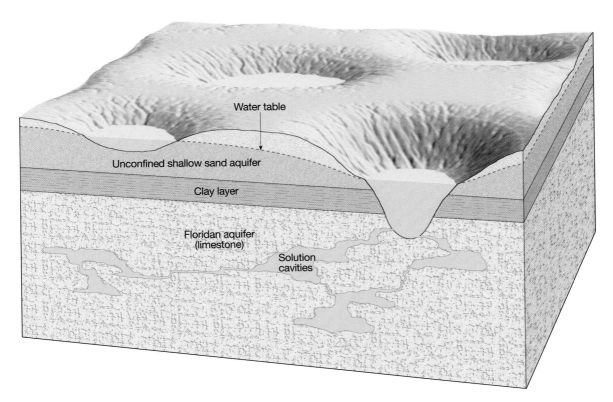

FIGURE 12.7 Geologic cross section showing groundwater distribution in strata underlying the Tampa, Florida, area.

PART 12C: LAND SUBSIDENCE HAZARDS CAUSED BY GROUNDWATER WITHDRAWAL

Land subsidence caused by human withdrawal of groundwater is a serious problem in many places throughout the world. For example, in the heart of Mexico City, the land surface has gradually subsided up to 7.6 m (25 ft). At the northern end of California's Santa Clara Valley, 17 square mi of land have subsided below the highest tide level in San Francisco Bay and now must be protected by earthworks. Other centers of subsidence include Houston, Tokyo, Venice, and Las Vegas. With increasing withdrawal of groundwater and more intensive use of the land surface, we can expect the problem of subsidence to become more widespread.

Subsidence induced by withdrawal of groundwater commonly occurs in areas underlain by stream-deposited (alluvial) sand and gravel that is interbedded with lake-deposited (lacustrine) clays and clayey silts (Figure 12.9A). The sand-and-gravel beds are aquifers, and the clay and clayey silt beds are confining beds.

In Figure 12.10, the water in the lower aquifer ("sand and gravel") is confined between impermeable beds of clay and silt and is under pressure from its own weight. Thus, water in wells **A** and **C** rises to the *potentiometric* (water-pressure) *surface*. Such wells are termed **artesian wells** (water flows naturally from the top of the well) The sand in the water table aquifer (Figure 12.10) contains water that is not confined under pressure, so it is an *unconfined aquifer*. The water in well **B** stands at the level of the water table and must be pumped up to the land surface.

Land subsidence (Figure 12.9B) is related to the compressibility of water-saturated sediments. Withdrawing water from wells not only removes water from the system, it also lowers the potentiometric surface and reduces the water pressure in the confined artesian aquifers. As the water pressure is reduced, the aquifer is gradually compacted and the ground surface above it is gradually lowered. The hydrostatic pressure can be restored by replenishing (or **recharging**) the aquifer with water. But the confining beds, once compacted, will not expand to their earlier thicknesses.

The Santa Clara Valley (Figure 12.11) was one of the first areas in the United States where land subsidence due to groundwater overdraft was recognized. The Santa Clara Valley is a large structural trough

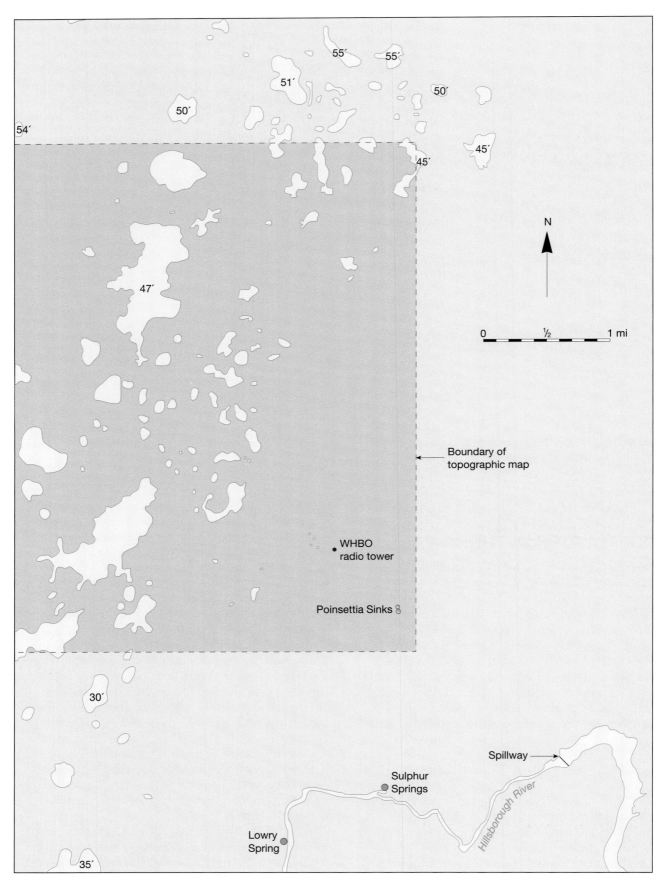

FIGURE 12.8 Sketch map of the area shown in Figure 12.6 (topographic map) and neighboring areas to the north, east, and south.

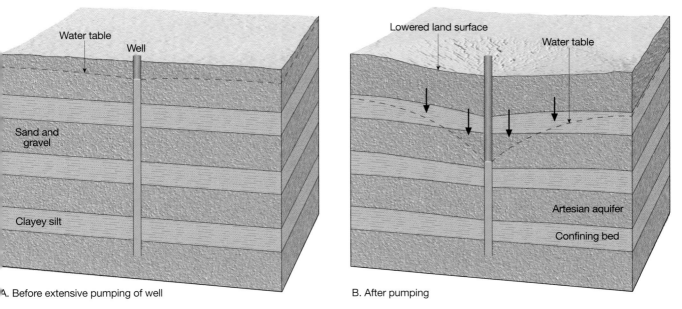

FIGURE 12.9 Before (**A**) and after (**B**) extensive pumping of a well. Note in **B** the lowering of the water-pressure surface, compaction of confining beds between the aquifers, and resulting subsidence of land surface. Arrows indicate the direction of compaction caused by the downward force of gravity, after the opposing water pressure was reduced by excessive withdrawal (discharge) of groundwater from the well.

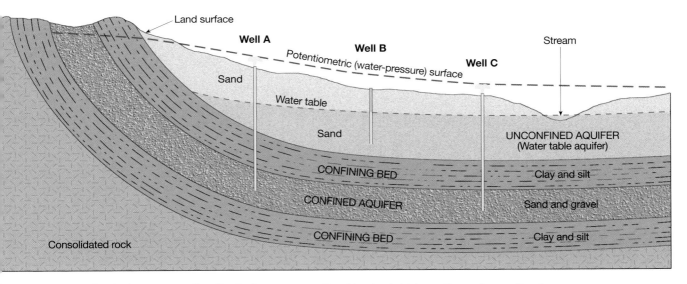

FIGURE 12.10 Geologic cross section illustrating an unconfined (water-table) aquifer and a confined aquifer. Vertical scale is exaggerated.

filled with alluvium (river sediments) more than 460 m (1500 ft) thick. Sand-and-gravel aquifers predominate near the valley margins, but the major part of the alluvium is silt and clay. Below a depth of 60 m (200 ft), the groundwater is confined by layers of clay, except near the margins.

Initially, wells as far south as Santa Clara were artesian, because the water-pressure surface was above the land surface. However, pumping them for irrigation lowered the water-pressure surface 40–60 m (150–200 ft) by 1965. This decline was not continuous; natural recharge of the aquifer occurred between 1938 and 1947, in part because of controlled infiltration from surface reservoirs. As of 1971, the subsidence had been stopped due to a reversal of the water-level decline.

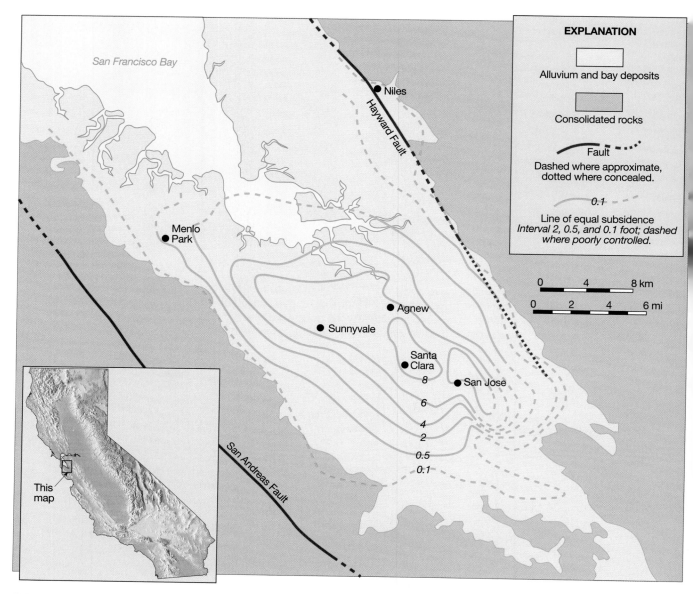

FIGURE 12.11 Land subsidence, 1934–1967, in the Santa Clara Valley, California. (Courtesy of U.S. Geological Survey)

Most wells tapping the artesian system are 150–300 m (500–1000 ft) deep, although a few reach 365 m (1200 ft). Well yields in the valley are 500–1500 gallons per minute (gpm), which is very high.

Questions

13. On Figure 12.12, solid brown contour lines show land surface elevation. Dashed blue lines represent the water-pressure surface (potentiometric surface) of a confined aquifer, as shown

in Figure 12.10. This is the height to which water will rise in a well that is drilled into the aquifer.

a. Find and connect the points on Figure 12.12 where the two sets of contour lines have the same elevation.

b. Shade in the area on this same figure where wells would flow at the land surface without having to be pumped (i.e., where wells would be artesian).

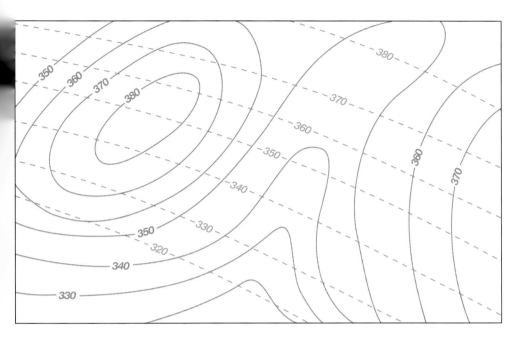

FIGURE 12.12 Graph of water data for Question 13.

14. In Figure 12.11, where are the areas of greatest subsidence in the Santa Clara Valley?

15. What was the total subsidence at San Jose (Figure 12.13) from 1934 to 1967?

16. What was the average annual rate of subsidence for the period 1934 to 1967 in feet per year?

17. At what places in the Santa Clara Valley would subsidence cause the most problems? Explain your reasoning.

18. Would you expect much subsidence to occur in the darker shaded areas of Figure 12.11? Explain.

19. By 1960, the total subsidence at San Jose had reached 9.0 feet (Figure 12.13). What was the average annual rate of subsidence (in feet per year) for the seven-year period from 1960 through 1967?

20. Refer to Figure 12.14. What was the level of the water in the San Jose well in:

    a. 1915?

    b. 1967?

21. During what period would the San Jose well have been a flowing artesian well? Explain.

22. How can you explain the minor fluctuations in the hydrograph (Figure 12.14) like those between 1920 and 1925?

23. In Figure 12.14, the slope of a line joining the level of the land surface in 1915 with subsidence that had occurred by 1967 gives the average rate of subsidence for that period. How did the rate of subsidence occurring between 1935 and 1948 differ from earlier rates?

24. Explain the probable cause of the subsidence rate change noted in Question 23.

25. Subsidence was stopped by 1971. What measures might have been taken to accomplish this?

Year	Total Subsidence (feet) from 1912 level
1912	0.0
1920	0.3
1934	4.6
1935	5.0
1936	5.0
1937	5.2
1940	5.5
1948	5.8
1955	8.0
1960	9.0
1963	11.0
1967	12.7

FIGURE 12.13 Subsidence at benchmark P7 in San Jose, California.

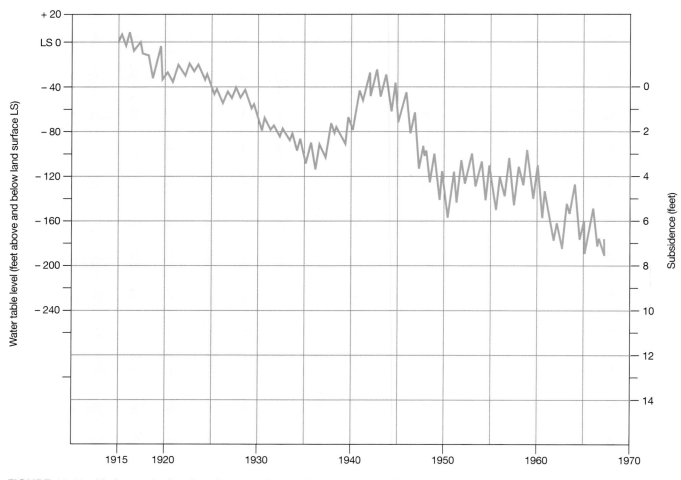

FIGURE 12.14 Hydrograph showing changes of water level in a well at San Jose, California.

PART 12D: HOME SEPTIC SYSTEMS AND GROUNDWATER CONTAMINATION

Many homes are not connected to a public sewage treatment system. Owners of these homes must have on-site sewage disposal systems that are properly located, constructed, and maintained. Otherwise, septic-system failure, groundwater contamination, and health hazards may develop. Such hazards place people's health and lives at risk. Use Internet resources on home sewage treatment and septic systems to complete the questions below. Refer to the *Laboratory Manual in Physical Geology* home page as needed, **www.prenhall.com/agi**

Questions

26. What are the main purposes of a home septic system?

27. Make a sketch of the relationships among the three parts of a typical septic system listed below. Then add a brief description beside each part to describe its function and purpose.

   a. Septic Tank:

   b. Distribution Box:

   c. Absorption Field (or Drain Field):

28. What information about the bedrock and/or soil is used to determine where to install a properly functioning home septic system?

29. What are some of the common contaminants in groundwater that originate in the wastewater of homes, restaurants, and other businesses?

30. What are the procedures for maintaining a properly functioning home septic system?

Glacial Processes, Landforms, and Indicators of Climate Change

---

•CONTRIBUTING AUTHORS•

Sharon Laska • *Acadia University*

Kenton E. Strickland • *Wright State University–Lake Campus*

Nancy A. Van Wagoner • *Acadia University*

OBJECTIVES

A. Understand processes of mountain (alpine) glaciation and the landforms and water bodies it produces.

B. Understand processes of continental glaciation and the landforms and water bodies it produces.

C. Construct and analyze topographic profiles of glaciated valleys and infer ice thicknesses.

D. Analyze glacial features and calculate rates of glacial retreat (ablation) in Glacier National Park.

E. Evaluate the use of Nisqually Glacier as a global thermometer for measuring climate change.

MATERIALS

Pencil, eraser, ruler, and pocket stereoscope (optional).

INTRODUCTION

Glaciers are large ice masses that form on land areas that are cold enough and have enough snowfall to sustain them. They form wherever the winter accumulation of snow and ice exceeds the summer ablation (also called *wastage*). *Ablation* (wastage) is the loss of snow and ice by melting and by *sublimation* to gas (direct change from ice to water vapor, without melting; Figure 1.5). Accumulation commonly occurs in *snowfields*—regions of permanent snow cover (Figure 13.1).

Glaciers can be divided into two zones, accumulation and ablation (Figure 13.1). As snow and ice collect in the **zone of accumulation,** they become compacted and highly recrystallized under their own weight. The ice mass then begins to slide and flow downslope like a very viscous (thick) fluid. If you *slowly* squeeze a small piece of ice in the jaws of a vise or pair of pliers, then you can observe how it flows. In nature, glacial ice formed in the zone of accumulation flows downhill into the **zone of ablation,** where it melts or sublimes (undergoes sublimation) faster than new ice can form. The *snowline* is the boundary between the zones of accumulation and ablation. The bottom end of the glacier is the **terminus.**

It helps to understand a glacier by viewing it as a river of ice. The "headwater" is the zone of accumulation, and the "river mouth" is the terminus. Like a river, glaciers *erode* (wear away) rocks, transport their load (tons of rock debris), and deposit their load "downstream" (down-glacier).

The downslope movement and extreme weight of glaciers cause them to abrade and erode (wear away) rock materials that they encounter. They also *pluck* rock material by freezing around it and ripping it from bedrock. The rock debris is then incorporated into the glacial ice and transported many kilometers by the glacier. The debris also gives glacial ice extra abrasive power. As the heavy rock-filled ice moves over the land, it scrapes surfaces like a giant sheet of sandpaper.

Rock debris falling from valley walls commonly accumulates on the surface of a moving glacier and is

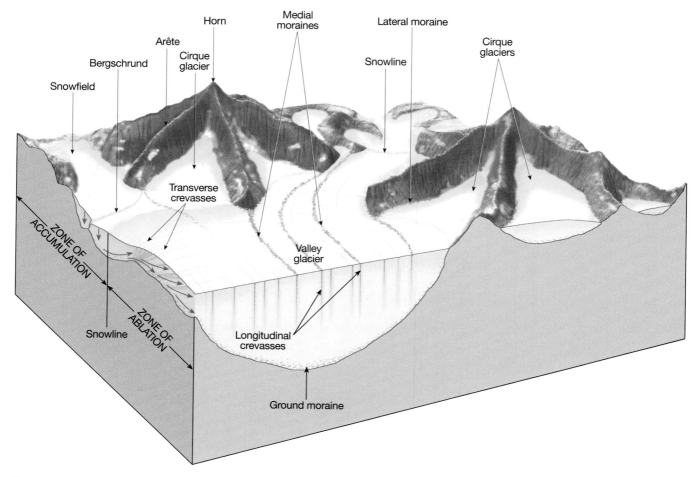

FIGURE 13.1 Active mountain glaciation, in a hypothetical region. Note cutaway view of glacial ice, showing flow lines and direction (blue lines and arrows).

transported downslope. Thus, glaciers transport huge quantities of sediment, not only *in,* but also *on* the ice.

When a glacier melts, it appears to retreat up the valley from which it flowed. This is called **glacial retreat,** even though the ice is simply melting back (rather than moving back up the hill).

As melting occurs, deposits of rocky gravel, sand, silt, and clay accumulate where there once was ice. These deposits collectively are called **drift.** Drift that accumulates directly from the melting ice is unstratified (unsorted by size) and is called **till.** However, drift that is transported by the meltwater becomes sorted by size, layered, and is called **stratified drift.** Wind also can transport the sand, silt, and clay particles from drift. Wind-transported glacial material can form dunes or *loess* deposits (wind-deposited, unstratified accumulations of clayey silt).

There are four main kinds of glaciers based on their size and form.

- **Cirque glaciers**—small, semicircular to triangular glaciers that form on the sides of mountains.

- **Valley glaciers**—long glaciers that flow down stream valleys in the mountains.

- **Piedmont glaciers**—mergers of two or more valley glaciers at the foot (break in slope) of a mountain range.

- **Ice sheet**—a vast, pancake-shaped ice mound that covers a large portion of a continent and flows independent of the topographic features beneath it. The Antarctic Ice Sheet (covering the entire continent of Antarctica) and Greenland Ice Sheet (covering Greenland) are modern examples.

PART 13A: GLACIAL PROCESSES AND LANDFORMS

Cirques, valley glaciers, and piedmont glaciers tend to modify mountainous regions of continents, where climatic conditions are sufficient for them to form. Such regions are said to be under the influence of "mountain glaciation" (Figure 13.1). Ice sheets cover

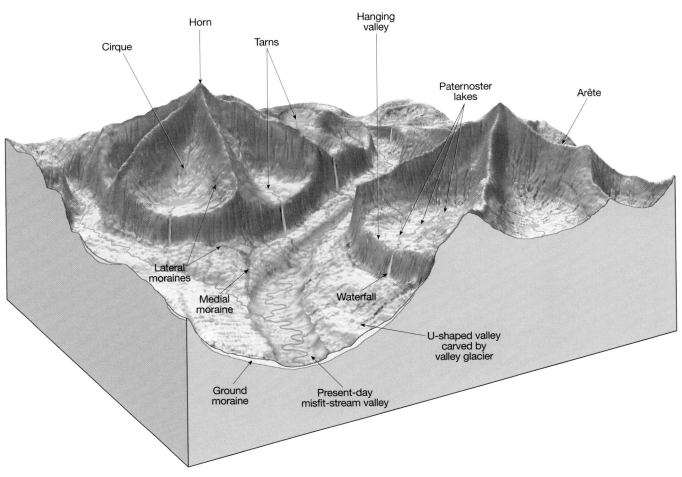

FIGURE 13.2 The same region as Figure 13.1, but showing erosion features remaining after total ablation (melting) of glacial ice.

large parts of continents, or even entire continents, which are then said to be under the influence of "continental glaciation."

Mountain Glaciation

Mountain glaciation is characterized by cirque glaciers, valley glaciers, and piedmont glaciers. Poorly developed mountain glaciation involves only cirques, but the best-developed mountain glaciation involves all three types. In some cases, valley and piedmont glaciers are so well developed that only the highest peaks and ridges extend above the ice. Mountain glaciation also is called *alpine glaciation,* because it is the type seen in Europe's Alps.

Figure 13.1 shows a region with mountain glaciation. Note the extensive *snowfield* in the zone of accumulation. *Snowline* is the elevation above which there is permanent snow cover.

Also note that there are many cracks or fissures in the glacial ice of Figure 13.1. At the upper end of the glacier is the large *bergschrund* (German, "mountain

crack") that separates the flowing ice from the relatively immobile portion of the snowfield. The other cracks are called **crevasses**—open fissures that form when the velocity of ice flow is variable (such as at bends in valleys). **Transverse crevasses** are perpendicular to the flow direction, and **longitudinal crevasses** are aligned with the direction of flow.

Figure 13.2 shows the results of mountain glaciation after the glaciers have completely melted. Notice the characteristic landforms, water bodies, and sedimentary deposits.

For your convenience, distinctive features of glacial lands are summarized in three figures:

- *erosional* features in Figure 13.3
- *depositional* features in Figure 13.4
- *water bodies* in Figure 13.5.

Note that some features are identical in mountain glaciation and continental glaciation, but others are unique to one or the other. Study the descriptions in

EROSIONAL FEATURES OF GLACIATED REGIONS		MOUNTAIN GLACIATION	CONTINENTAL GLACIATION
Cirque	Bowl-shaped depression on a high mountain slope, formed by a cirque glacier	X	
Arête	Sharp, jagged, knife-edge ridge between two cirques or glaciated valleys	X	
Col	Mountain pass formed by the headward erosion of cirques	X	
Horn	Steep-sided, pyramid-shaped peak produced by headward erosion of several cirques	X	
Headwall	Steep slope or rock cliff at the upslope end of a glaciated valley or cirque	X	
Glacial trough	U-shaped, steep-walled, glaciated valley formed by the scouring action of a valley glacier	X	
Hanging valley	Glacial trough of a tributary glacier, elevated above the main trough	X	
Roche moutonnée	Asymmetrical knoll or small hill of bedrock, formed by glacial abrasion on the smooth stoss side (side from which the glacier came) and by plucking (prying and pulling by glacial ice) on the less-smooth lee side (down-glacier side)		X
Glacial striations and grooves	Parallel linear scratches and grooves in bedrock surfaces, resulting from glacial scouring	X	X
Glacial polish	Smooth bedrock surfaces caused by glacial abrasion (sanding action of glaciers analogous to sanding of wood with sandpaper)	X	X

FIGURE 13.3 Erosional features produced by mountain or continental glaciation.

these three figures and compare them with the visuals in Figures 13.1 and 13.2.

Continental Glaciation

During the Pleistocene Epoch, or "Ice Age," that ended approximately 10,000 years ago, thick ice sheets covered most of Canada, large parts of Alaska, and the northern contiguous United States. These continental glaciers produced a variety of characteristic landforms (Figure 13.6, Figure 13.7).

Recognizing and interpreting these landforms is important in conducting work such as regional soil analyses, studies of surface drainage and water supply, and exploration for sources of sand, gravel, and minerals. The thousands of lakes in the Precambrian Shield area of Canada also are a legacy of this continental glaciation, as are the fertile soils of the north-central United States and south-central Canada.

Questions

1. In Figure 13.8, examine the typical stream cobble and typical glacial cobble. Explain how you think the two different physical-abrasion processes (river abrasion versus glacial abrasion) can produce such different-looking cobbles.

DEPOSITIONAL FEATURES OF GLACIATED REGIONS		MOUNTAIN GLACIATION	CONTINENTAL GLACIATION
Ground moraine	Sheetlike layer (blanket) of till left on the landscape by a receding (wasting) glacier.	X	X
Terminal moraine	Ridge of till that formed along the leading edge of the farthest advance of a glacier.	X	X
Recessional moraine	Ridge of till that forms at terminus of a glacier, behind (up-glacier) and generally parallel to the terminal moraine; formed during a temporary halt (stand) in recession of a wasting glacier.	X	X
Lateral moraine	A body of rock fragments at or within the side of a valley glacier where it touches bedrock and scours the rock fragments from the side of the valley. It is visible along the sides of the glacier and on its surface in its ablation zone. When the glacier melts, the lateral moraine will remain as a nerrow ridge of till or boulder train on the side of the valley.	X	
Medial moraine	A long narrow body of rock fragments carried in or upon the middle of a valley glacier and parallel to its sides, usually formed by the merging of lateral moraines from two or more merging valley glaciers. It is visible on the surface of the glacier in its ablation zone. When the glaciers melt, the medial moraine will remain as a narrow ridge of till or boulder train in the middle of the valley.	X	
Drumlin	An elongated mound or ridge of glacial till (unstratified drift) that accumulated under a glacier and was elongated and streamlined by movement (flow) of the glacier. Its long axis is parallel to ice flow. It normally has a blunt end in the direction from which the ice came and long narrow tail in the direction that the ice was flowing.		X
Kame	A low mound, knob, or short irregular ridge of stratified drift (sand and gravel) sorted by and deposited from meltwater flowing a short distance beneath, within, or on top of a glacier. When the ice melted, the kame remained.		X
Esker	Long, narrow, sinuous ridge of stratified drift deposited by meltwater streams flowing under glacial ice or in tunnels within the glacial ice		X
Erratic	Boulder or smaller fragment of rock resting far from its source on bedrock of a different type.	X	X
Boulder train	A line or band of boulders and smaller rock clasts (cobbles, gravel, sand) transported by a glacier (often for many kilometers) and extending from the bedrock source where they originated to the place where the glacier carried them. When deposited on different bedrock, the rocks are called erratics.	X	X
Outwash	Stratified drift (mud, sand and gravel) transported, sorted, and deposited by meltwater streams (usually muddy braided streams) flowing in front of (down-slope from) the terminus of the melting glacier.	X	X
Outwash plain	Plain formed by blanket-like deposition of outwash; usually an outwash braid plain, formed by the coalescence of many braided streams having their origins along a common glacial terminus.	X	X
Valley train	Long, narrow sheet of outwash (outwash braid plain of one braided stream, or floodplain of a meandering stream) that extends far beyond the terminus of a glacier.	X	
Beach line	Landward edge of a shoreline of a lake formed from damming of glacial meltwater, or temporary ponding of glacial meltwater in a topographic depression.		X
Glacial-lake deposits	Layers of sediment in the lake bed, deltas, or beaches of a glacial lake.		X
Loess	Unstratified sheets of clayey silt and silty clay transported beyond the margins of a glacier by wind and/or braided streams; it is compact and able to resist significant erosion when exposed in steep slopes or cliffs.		X

FIGURE 13.4 Depositional features produced by mountain or continental glaciation.

WATER BODIES OF GLACIATED REGIONS		MOUNTAIN GLACIATION	CONTINENTAL GLACIATION
Tarn	Small lake in a cirque (bowl-shaped depression formed by a cirque glacier). A melting cirque glacier may also fill part of the cirque and may be in direct contact with or slightly up-slope from the tarn.	X	
Ice-dammed lake	Lake formed brhind a mass of ice sheets and blocks that have wedged together and blocked the flow of water from a melting glacier and or river. Such natural dams may burst and produce a catastropic flood of water, ice blocks, and sediment.	X	X
Paternoster lakes	Chain of small lakes in a glacial trough.	X	
Finger lake	Long narrow lake in a glacial trough that was cut into bedrock by the scouring action of glacial ice (containing rock particles and acting like sand paper as it flows downhill) and usually dammed by a deposit of glacial gravel (end or recessional moraine).	X	X
Kettle lake or kettle hole	Small lake or water-saturated depression (10s to 1000s of meters wide) in glacial drift, formed by melting of an isolated, detached block of ice left behind by a glacier in retreat (melting back) or buried in outwash from a flood caused by the collapse of an ice-dammed lake.	X	X
Swale	Narrow marsh, swamp, or very shallow lake in a long shallow depression between two moraines.		X
Marginal glacial lake	Lake formed at the margin (edge) of a glacier as a result of accumulating meltwater; the upslope edge of the lake is the melting glacier itself.	X	X
Meltwater stream	Stream of water derived from melting glacial ice, that flows under the ice, on the ice, along the margins of the ice, or beyond the margins of the ice.	X	X
Misfit stream	Stream that is not large enough and powerful enough to have cut the valley it occupies. The valley must have been cut at a time when the stream was larger and had more cutting power or else it was cut by another process such as scouring by glacial ice.	X	X
Marsh or swamp	Saturated, poorly drained areas that are permanently or intermittently covered with water and have grassy vegetation (swamp) or shrubs and trees (marsh).	X	X

FIGURE 13.5 Water bodies produced as a result of mountain or continental glaciation.

Refer to the Siffleur River, Alberta, quadrangle (Figure 13.9) for these questions:

2. What is the name given to features like Marmot Mountain and Conical Peak? How do such features form?

3. The boundary between Improvement Districts 9 and 10 follows a ridge from the Siffleur River Valley to Mount Kentigern. What type of ridge is this, and how did it form?

4. Near the northern edge of the map, what type of valley is located above the falls west of the Siffleur River, and how did it form?

5. What type of lake is at the headwaters of the stream that forms these falls?

6. What other features produced by mountain glaciation can you see on this map?

Refer to Figure 13.10, a portion of the Anchorage (B-2), Alaska, quadrangle, for the following questions. In the southwestern corner, note the Harvard Arm of Prince William Sound. The famous *Exxon Valdez* oil spill occurred just south of this area (it did not affect Harvard Arm).

7. Lateral and medial moraines in/on the ablation zone of Harvard Glacier are indicated by the

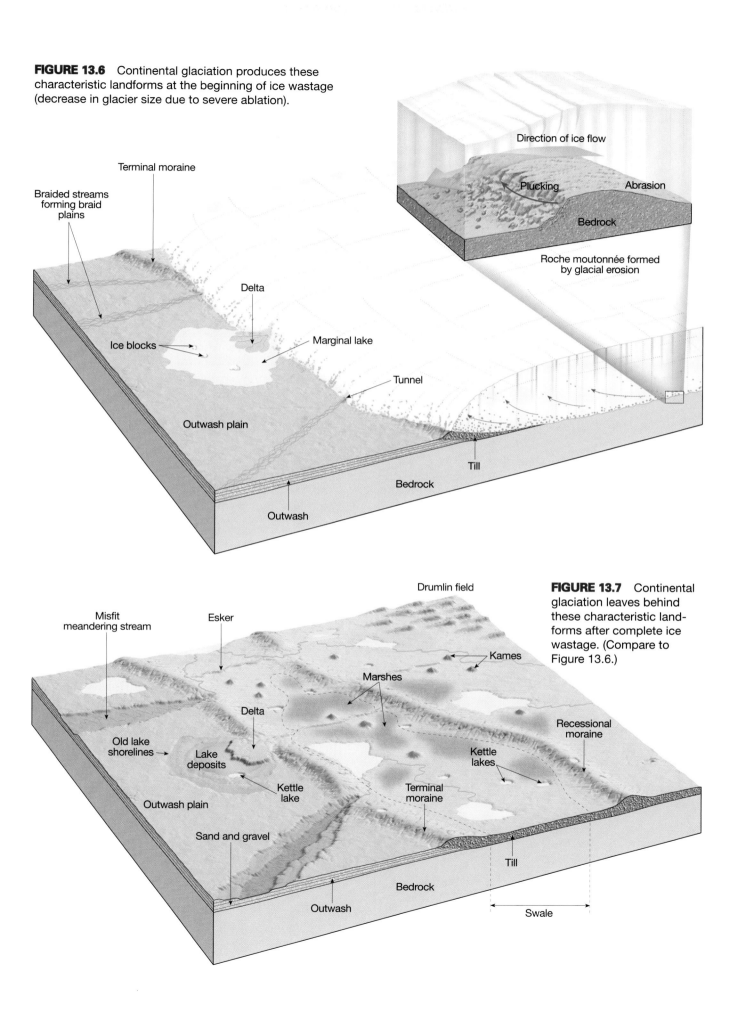

FIGURE 13.6 Continental glaciation produces these characteristic landforms at the beginning of ice wastage (decrease in glacier size due to severe ablation).

Direction of ice flow

Plucking

Abrasion

Bedrock

Roche moutonnée formed by glacial erosion

Terminal moraine

Braided streams forming braid plains

Delta

Ice blocks

Marginal lake

Tunnel

Outwash plain

Till

Bedrock

Outwash

Drumlin field

FIGURE 13.7 Continental glaciation leaves behind these characteristic landforms after complete ice wastage. (Compare to Figure 13.6.)

Misfit meandering stream

Esker

Kames

Marshes

Delta

Old lake shorelines

Lake deposits

Recessional moraine

Kettle lake

Kettle lakes

Terminal moraine

Outwash plain

Sand and gravel

Till

Bedrock

Outwash

Swale

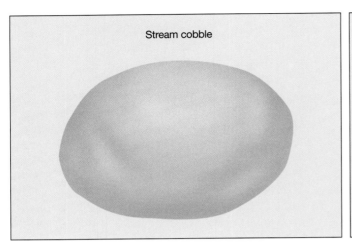

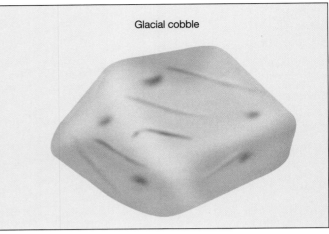

FIGURE 13.8 Note the differences between a stream cobble and a glacial cobble. Stream cobbles are rounded to well-rounded and have smooth surfaces. Glacial cobbles are angular or faceted and have many scratch marks. (A cobble is a clast between a pebble and a boulder in size, 64–256 mm diameter.)

brown stippled (finely dotted) pattern. If a hiker found gold in rock fragments on the glacier at location **C,** then would you look for gold near location **X, Y,** or **Z?** Explain your reasoning.

8. Notice the crevasses within a mile of Harvard Glacier's terminus. What specific kind of crevasses are they, and why do you think they formed only on this part of the glacier?

9. Between the Harvard and Yale glaciers, notice how the Dora Keen Range has been shaped by these two main glaciers and thins to the southwest. How could you use this information to infer how ice has flowed in regions where glaciers are no longer present?

Refer to Figure 13.11, part of the Peterborough, Ontario, quadrangle, for the following questions. This area lies north of Lake Ontario.

10. Study the size and shape of the short, oblong rounded hills. Fieldwork has revealed that they are made of till. What type of feature are they and how did they form?

11. Using a ruler, draw a line on ten of the oblong hills of question 10 to indicate their long axis. In what direction did glaciers move (advance toward) in this area? Explain.

12. Find the red highway, Route 7, that crosses the northern part of the map. About 1 mile south of the circled number 7 that identifies Route 7, there is the long, narrow, sinuous hill that runs northeast from the pale blue number 36. What would you call this feature, and how do you think it formed?

13. Imagine that you could visit the feature in question 12 and dig through it with a bulldozer. What would you expect to find inside of the feature (be as exact as you can)?

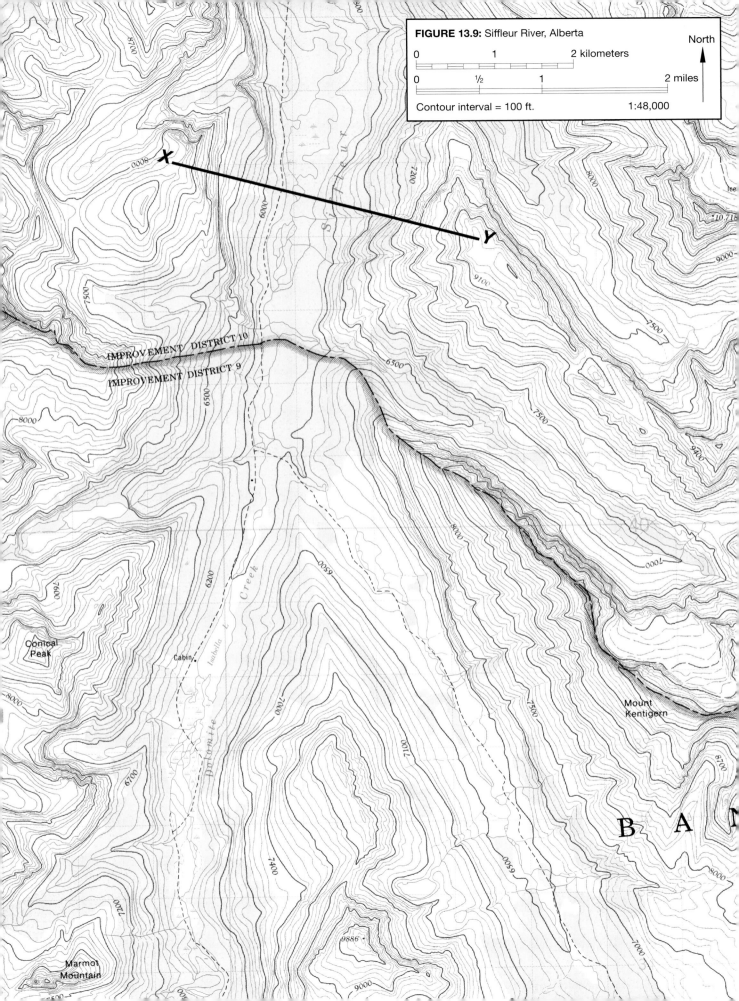

FIGURE 13.9: Siffleur River, Alberta

North

0   1   2 kilometers

0   ½   1   2 miles

Contour interval = 100 ft.   1:48,000

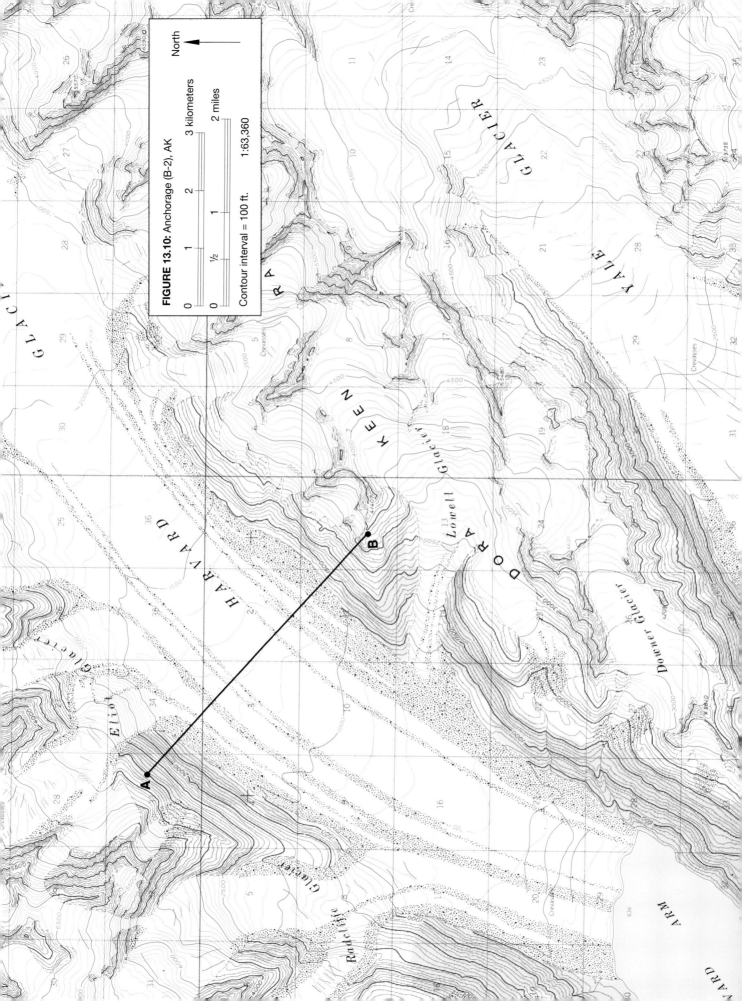

FIGURE 13.10: Anchorage (B-2), AK

North ⟵

0   1   2   3 kilometers

0 ½ 1   2 miles

1:63,360

Contour interval = 100 ft.

FIGURE 13.11: Peterborough, Ontario

1   2 kilometers

½   1   2 miles

Contour interval = 25 ft.   1:48,000

North

PART 13B: GLACIATION IN WISCONSIN

The most recent glaciation of Earth is called the *Wisconsinan glaciation,* named for the state of Wisconsin, where it left behind many erosional and depositional features. The Wisconsinan glaciation reached its maximum development about 18,000 years ago, when a *Laurentide ice sheet* covered central and eastern Canada, the Great Lakes Region, and the northeastern United States. It ended by about 10,000 years ago, at the start of the recent interglaciation (Recent or Holocene Epoch).

Refer to Figure 13.12, a portion of the Whitewater, Wisconsin, quadrangle and Figure 13.13, the accompanying stereogram, for these questions:

Questions

14. List the features of glaciated regions from Figures 13.3 and 13.4 that are present in this region.

15. Based on your answer to Question 14 and the information provided above, what kind of glaciation (mountain versus continental) has shaped this landscape?

16. Describe what direction the ice flowed over this region. Cite evidence for your inference.

17. What kinds of lakes are present in this region, and how did they form? (Refer to Figure 13.5.)

18. In the southeastern corner of the map, the forested area is probably what kind of feature?

19. Note the swampy and marshy area running from the west-central edge of the map to the northeastern corner. Describe the probable origin of this feature (more than one answer is possible).

PART 13C: COMPARING TOPOGRAPHIC PROFILES OF GLACIATED VALLEYS

Questions

20. Complete the topographic profile on the left-hand side of Figure 13.14 for line **X–Y,** *across the Siffleur River Valley.* Refer to Figure 9.18 (Topographic Profile Construction), if needed.

   a. What is the vertical exaggeration of this topographic profile?

   b. Is this a normal profile for a river valley? Why?

   c. Why does the Siffleur River Valley have this shape?

21. On the right-hand side of Figure 13.14, complete the topographic profile for line *A–B* across the *Harvard Glacier.* Refer to Figure 9.18 (Topographic Profile Construction), if needed.

   a. What is the vertical exaggeration of this topographic profile?

   b. Label the part of the profile that is the top surface of the glacier.

   c. Using a dashed line draw where you think the rock bottom of the valley is located under the Harvard Glacier. (Your drawing may extend slightly below the figure.)

22. Based on your work in Questions 20 and 21, what is the maximum thickness of Harvard Glacier at line *A–B*? Explain your reasoning.

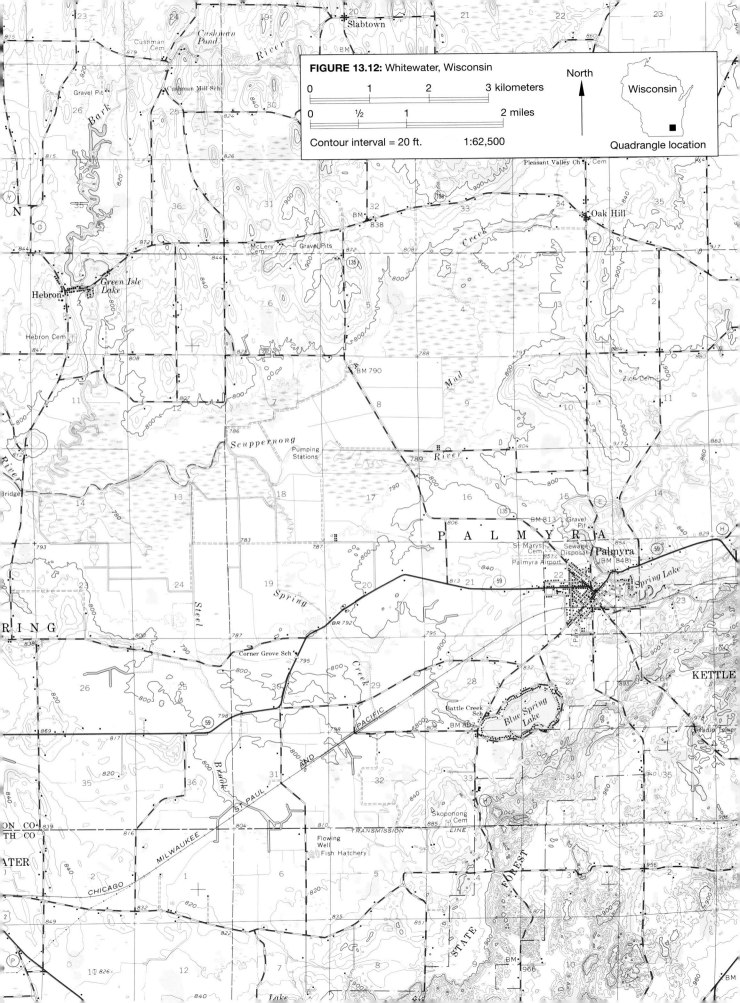

FIGURE 13.12: Whitewater, Wisconsin

North

Wisconsin

Quadrangle location

Contour interval = 20 ft.   1:62,500

0   1   2   3 kilometers

0   ½   1   2 miles

FIGURE 13.13 National high-altitude photograph (NHAP, color-infrared) stereogram of the Whitewater, Wisconsin, region. View the stereogram using a pocket stereoscope and compare it to Figure 13.12. (Courtesy of U.S. Geological Survey)

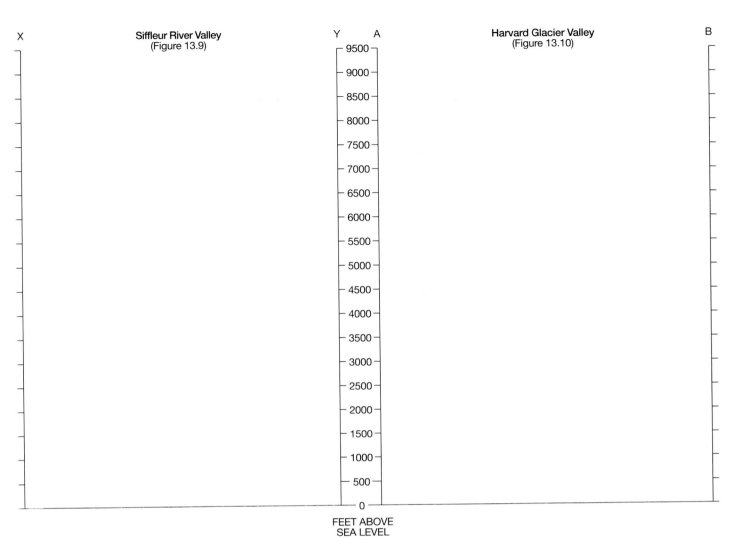

X — Siffleur River Valley (Figure 13.9) — Y   A — Harvard Glacier Valley (Figure 13.10) — B

9500
9000
8500
8000
7500
7000
6500
6000
5500
5000
4500
4000
3500
3000
2500
2000
1500
1000
500
0

FEET ABOVE
SEA LEVEL

FIGURE 13.14 Graphs for completing topographic profiles of glaciated valleys (Part 13C).

PART 13D: GLACIER NATIONAL PARK, MONTANA

Glacier National Park is located on the northern edge of Montana, across the border from Alberta and British Columbia, Canada. Most of the erosional features formed by glaciation in the park were formed during the Wisconsinan glaciation that ended about 10,000 years ago. Today, only small cirque glaciers exist in the park. Thirty-seven of them are named, and nine of those can be observed on the topographic map of part of the park in Figure 13.15. Use the map and chart of glacier data in Figure 13.15 to answer the following questions.

Questions

23. List the features of glaciated regions from Figures 13.3 and 13.4 that are present in this region.

24. Locate Quartz Lake and Middle Quartz Lake in the southwest part of the map. Notice the Patrol Cabin located between these lakes. Describe the chain of geologic/glacial events (steps) that led to formation of Quartz Lake, the valley of Quartz Lake, the small piece of land on which the Patrol Cabin is located, and the cirque in which Rainbow Glacier is located today.

25. Based on your answers to Questions 23 and 24, what kind of glaciation (mountain versus continental) has shaped this landscape?

26. Locate the Continental Divide and think of ways that it may be related to weather and climate in the region. Recall that weather systems generally move across the United States from west to east.

   a. Describe how modern glaciers of this region are distributed in relation to the Continental Divide.

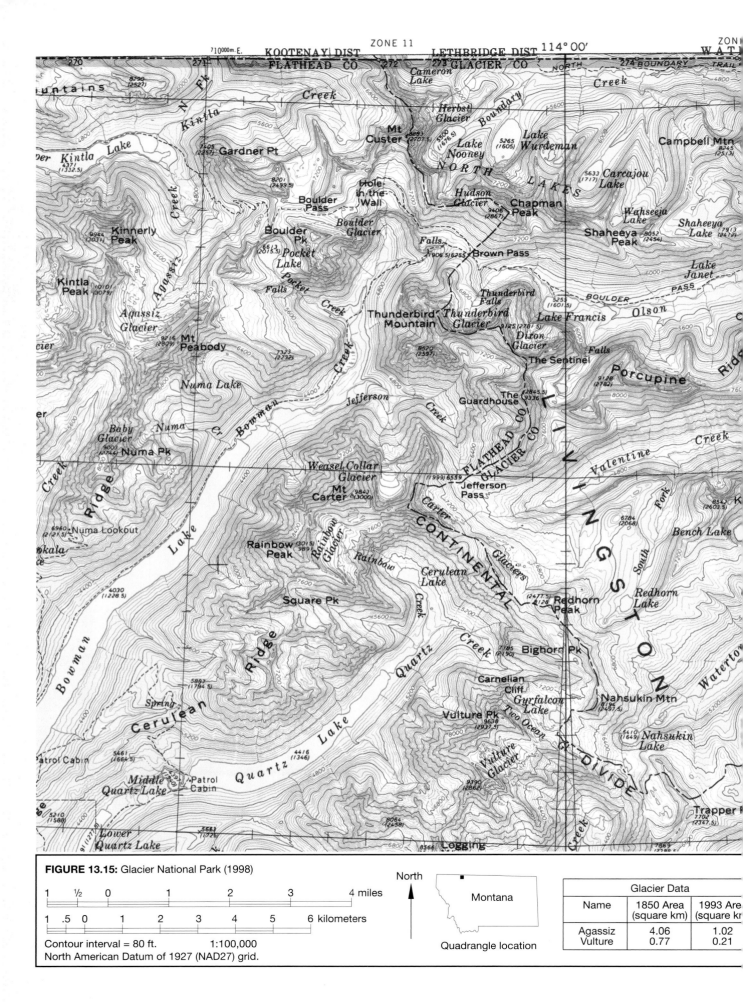

FIGURE 13.15: Glacier National Park (1998)

| 1 | ½ | 0 | 1 | 2 | 3 | 4 miles |

| 1 | .5 | 0 | 1 | 2 | 3 | 4 | 5 | 6 kilometers |

Contour interval = 80 ft.   1:100,000
North American Datum of 1927 (NAD27) grid.

North

Montana

Quadrangle location

Glacier Data		
Name	1850 Area (square km)	1993 Are (square kr
Agassiz	4.06	1.02
Vulture	0.77	0.21

b. Based on the distribution you observed, describe the weather/climate conditions that may exist on opposite sides of the Continental Divide in this region.

27. By what percentage did each of the glaciers below decrease in size between 1850 and 1993?

a. Agassiz Glacier:

b. Vulture Glacier:

28. What was the rate, in km^2/yr, that each of the glaciers receded between 1850 and 1993?

a. Agassiz Glacier:

b. Vulture Glacier:

29. Based on the rates you calculated in Question 28, calculate the year in which each of these glaciers will be completely melted.

a. Agassiz Glacier:

b. Vulture Glacier:

30. The largest glacier in Figure 13.15, and one of the largest in Glacier National Park, is Rainbow Glacier. Predict what year Rainbow Glacier will be completely melted (based on your calculations in Question 28 and assuming no significant change in climate). Explain your reasoning and calculation.

PART 13E: NISQUALLY GLACIER— A GLOBAL THERMOMETER?

Nisqually Glacier is one of many active valley glaciers that occupy the radial drainage of Mt. Rainier—an active volcano located near Seattle, Washington, in the Cascade Range of the western United States. Nisqually Glacier occurs on the southern side of Mt. Rainier and flows south toward the Nisqually River Bridge in Figure 13.16. The position of the glacier's terminus (downhill end) was first recorded in 1857, and it has been measured and mapped by numerous geologists since that time. The map in Figure 13.16 was prepared by the U.S. Geological Survey in 1976 and shows where the terminus of Nisqually Glacier was located at various times from 1840 to 1997. (The 1994 and 1997 positions were added for this laboratory, based on NHAP aerial photographs and satellite imagery.) Notice how the glacier has more or less retreated up the valley since 1840. Measure, chart, graph, and analyze this retreat below.

Questions

31. Fill in the Nisqually Glacier Data Chart on the left side of Figure 13.17. To do this, use a ruler and the map's bar scale to measure the distance in kilometers from Nisqually River Bridge to the position of the glacier's terminus (red dot) for each year of the chart. Be sure to record your distance measurements to two decimal points (hundredths of kilometers).

32. Plot your data from Question 31 (Nisqually Glacier Data Chart) in the graph on the right side of Figure 13.17. After plotting each point of data, connect the dots with a smooth, light pencil line. Notice that the glacier terminus retreated up-valley at some times, but advanced back down the valley at other times. Summarize these changes in a chart or paragraph, relative to specific years of the data.

33. Notice the blue and red graph of climatic data at the bottom of Figure 13.17 provided by the NOAA National Climatic Data Center (NCDC). NCDC's global mean temperatures are mean temperatures for Earth calculated by processing data from thousands of observation sites throughout the world (from 1880 to 2005). The temperature data were corrected for factors such as increase in temperature around urban centers and decrease in temperature with elevation. Although NCDC collects and processes data on land and sea, this graph only shows the variation in annually averaged global land surface temperature since 1880.

a. Describe the long-term trend in this graph—how averaged global land surface temperature changed from 1880 to 2005.

b. Lightly in pencil, trace any shorter-term pattern of cyclic climate change that you can identify in the graph. Describe this cyclic shorter-term trend.

34. Describe how the changes in position of the terminus of Nisqually Glacier compare to variations in annually averaged global land surface temperature. Be as specific as you can.

35. Based on all of your work above, do you think Nisqually Glacier can be used as a global thermometer for measuring climate change? Explain.

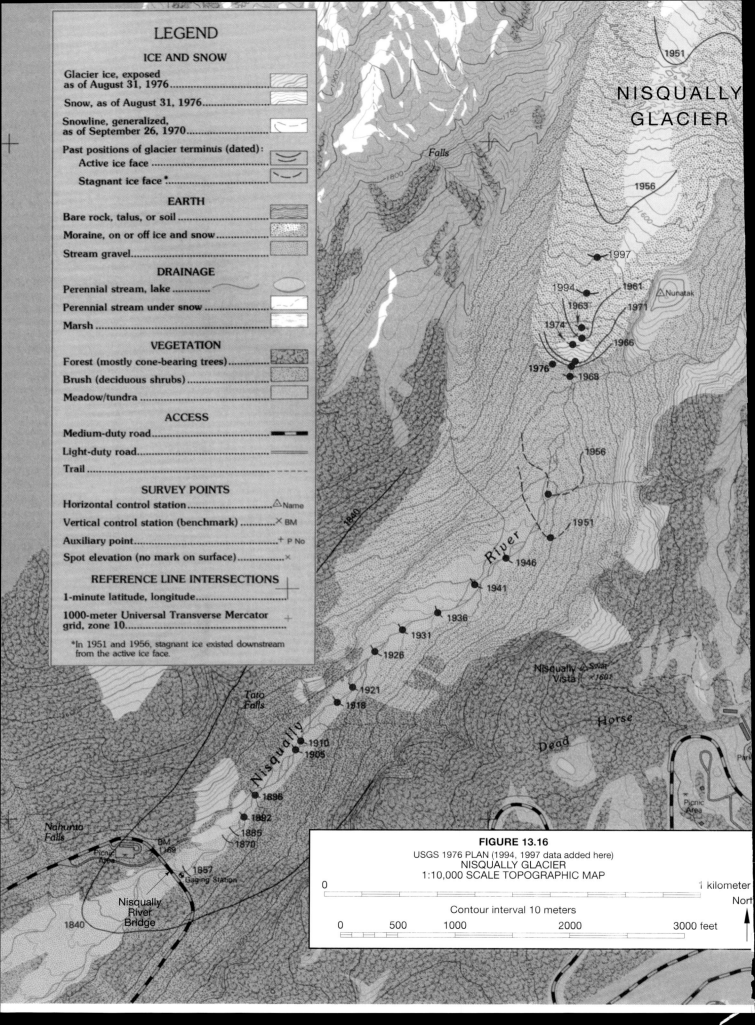

LEGEND

ICE AND SNOW

Glacier ice, exposed
as of August 31, 1976.............................

Snow, as of August 31, 1976....................

Snowline, generalized,
as of September 26, 1970.........................

Past positions of glacier terminus (dated):

Active ice face ....

Stagnant ice face *...................................

EARTH

Bare rock, talus, or soil ....

Moraine, on or off ice and snow...............

Stream gravel...

DRAINAGE

Perennial stream, lake ....

Perennial stream under snow ....

Marsh ....

VEGETATION

Forest (mostly cone-bearing trees)..............

Brush (deciduous shrubs) ....

Meadow/tundra ....

ACCESS

Medium-duty road.....................................

Light-duty road..

Trail ....

SURVEY POINTS

Horizontal control station............................△Name

Vertical control station (benchmark) ....× BM

Auxiliary point..+ P No

Spot elevation (no mark on surface)................×

REFERENCE LINE INTERSECTIONS

1-minute latitude, longitude............................

1000-meter Universal Transverse Mercator
grid, zone 10..+

*In 1951 and 1956, stagnant ice existed downstream
from the active ice face.

NISQUALLY
GLACIER

1951

Falls

1956

1997

1994   1961   △Nunatak
1963   1971
1974
1966
1976
1968

1956

1951

River   1946

1941

1936

1931

1926   Nisqually   △Sugar
                         Vista   × 1601

1921
1918   Horse

Tato
Falls   Dead

1910
1905

1898

1892
1885
1870

BM
1169

1857
Gaging Station

Nahunta
Falls   Picnic
Area

Nisqually
River
Bridge

1840

Park

Picnic
Area

FIGURE 13.16

USGS 1976 PLAN (1994, 1997 data added here)
NISQUALLY GLACIER
1:10,000 SCALE TOPOGRAPHIC MAP

0   1 kilometer

Contour interval 10 meters

North

0   500   1000   2000   3000 feet

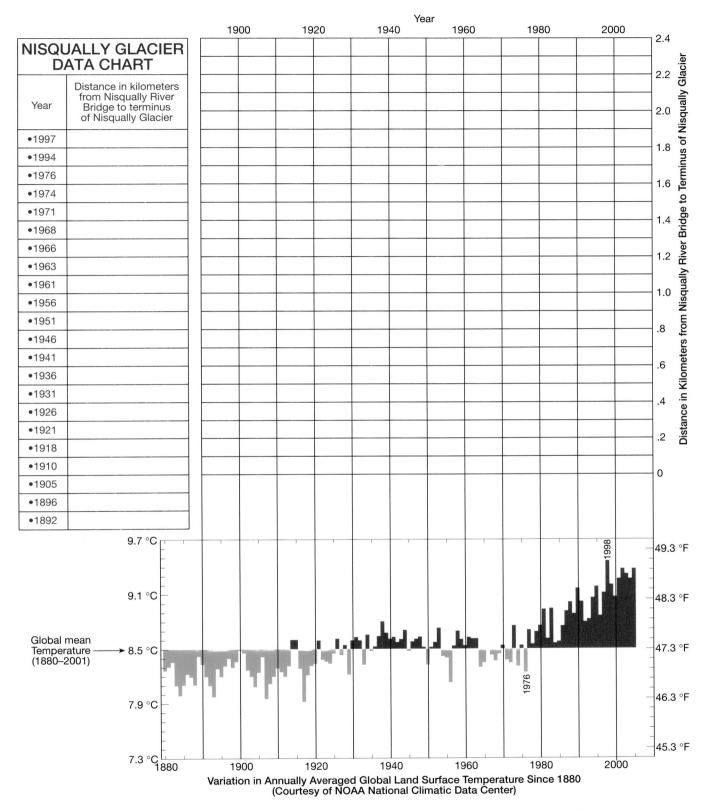

NISQUALLY GLACIER
DATA CHART

Year	Distance in kilometers from Nisqually River Bridge to terminus of Nisqually Glacier
•1997	
•1994	
•1976	
•1974	
•1971	
•1968	
•1966	
•1963	
•1961	
•1956	
•1951	
•1946	
•1941	
•1936	
•1931	
•1926	
•1921	
•1918	
•1910	
•1905	
•1896	
•1892	

Year

Distance in Kilometers from Nisqually River Bridge to Terminus of Nisqually Glacier

Global mean Temperature (1880–2001)

1998

1976

Variation in Annually Averaged Global Land Surface Temperature Since 1880
(Courtesy of NOAA National Climatic Data Center)

FIGURE 13.17   Graph of changes in position of the terminus of Nisqually Glacier compared to the variation in annually averaged global land surface temperature since 1880 (Part 13E).

Dryland Landforms, Hazards, and Risks

·CONTRIBUTING AUTHORS·

Charles G. Oviatt • *Kansas State University*

James B. Swinehart • *Institute of Agriculture & Natural Resources, University of Nebraska*

James R. Wilson • *Weber State University*

OBJECTIVES

A. Understand basic eolian (having to do with wind) processes that lead to formation of specific kinds of sand dunes and other eolian deposits.

B. Identify and understand the formation of landforms typically developed in drylands (lands in arid, semi-arid, and dry subhumid climates) of Arizona, California, and the midwestern United States.

C. Analyze the Utah desert to evaluate the history of Lake Bonneville.

D. Understand the nature and magnitude of desertification hazards.

E. Analyze drylands and evaluate their risk of desertification.

MATERIALS

Pencil, eraser, ruler, calculator, pocket stereoscope, and set of colored pencils.

INTRODUCTION

Drylands are lands in arid, semi-arid, and dry subhumid climates. The United Nations Environment Programme (UNEP) estimates that they comprise 36% of all land on Earth and that they support one-sixth of the world's human population. Sixteen percent of all existing drylands (about 6% of all land areas on Earth) are so dry that their biological productivity is too poor to support any type of agriculture. These regions are true **deserts.**

When people rely on land for farming or ranching, they must assess the potential for land **degradation**—a state of declining agricultural productivity due to natural and/or human causes. Humid lands (lands in humid climates) may undergo degradation from factors such as soil *erosion* (wearing away), farming without crop rotation or fertilization, overgrazing, or dramatic increases or decreases in soil moisture. However, degraded humid lands always retain the capability of some level of agricultural production. This is not true in drylands, where degradation may cause the land to become true desert with no agricultural value. This type of degradation is called **desertification** (the process of land degradation toward drier, true desert conditions).

UNEP estimates that 70% of all existing drylands (about 25% of all land on Earth) are now experiencing the hazard of desertification from factors related to human population growth, climate change, poor groundwater use policies, overgrazing, and other poor land management practices. More than 100 nations now risk degradation of their productive cropland and grazeland to useless desert. For this reason, the Third World Academy of Sciences declared the 1990s as the "Decade of the Desert," and the United Nations General Assembly declared 2006 the International Year of Deserts and Desertification.

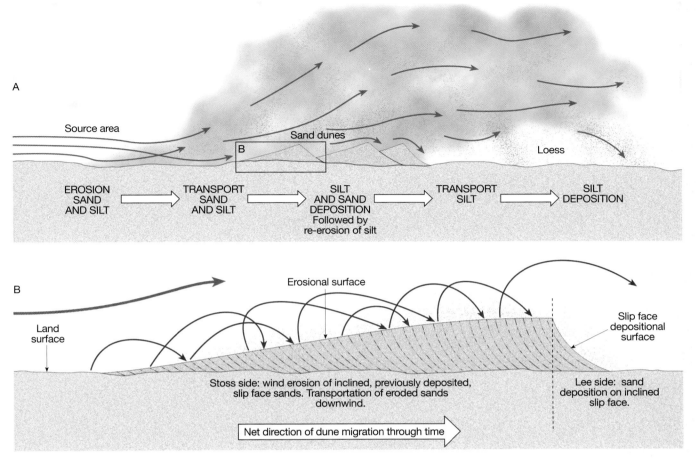

FIGURE 14.1 Eolian (wind-related) erosion, transportation, and deposition. **A.** Strong winds erode sand and silt from a source area and transport them to new areas. As the wind velocity decreases, the sand accumulates first (closest to the source) and the silt (loess) is carried further downwind. **B.** Hypothetical cross section through a sand dune. Wind erodes and transports sand up the *stoss side* (upwind side) of the dune. Sand rolls down the slip face, forming the *lee side* (downwind side) of the dune. This continuing process of wind erosion and transportation of sand on the stoss side of the dune, and simultaneous deposition of sand on the slip face of the dune, results in net downwind migration of the dune.

PART 14A: EOLIAN PROCESSES, DRYLAND LANDFORMS, AND DESERTIFICATION

Many drylands have specific landforms that result primarily from processes associated with degradation, erosion by streams and flash floods developed after infrequent rains (fluvial processes), or erosion and deposition associated with wind (eolian processes). Fluvial and eolian processes erode dryland landscapes, transport Earth materials, and deposit sediments. Rocky surfaces, sparsely vegetated surfaces, sand dunes, and arroyos (steep-walled canyons with gravel floors) are typical of dryland landscapes. Therefore, humans living in drylands must adapt to their landforms, conditions and

processes that have created the landforms, and the prospect of land degradation or even desertification.

Eolian Processes and Landforms

Water and ice are capable of moving large particles of sediment. The wind can move only smaller particles (Figures 14.1 and 14.2). For this reason, the eolian (wind-related) landforms may be subtle or even invisible on a topographic map. (However, they may be more evident on aerial photographs that have a higher resolution.) They may be superimposed on **fluvial** (stream-related) or *glacial* (ice- or glacier-related) features, particularly where recently exposed and unvegetated sediment occurs.

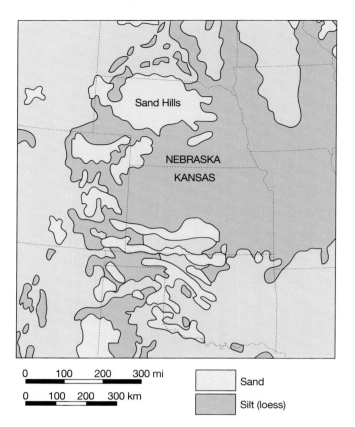

FIGURE 14.2 Map of part of the midwestern United States showing the location of Nebraska's Sand Hills, sand deposits, and silt (loess) deposits.

A lack of a dense vegetation cover is a prerequisite for significant wind erosion. This lack of vegetation can occur:

- On recently deposited sediment, such as floodplains,

- In areas where vegetation has been destroyed by fire, overgrazing, or human activity, or

- In true deserts where the lack of water precludes substantial growth of vegetation.

When examining a topographic map, keep in mind that the green overprint represents only trees and shrubs. There could be an important soil-protecting grass cover present that is not indicated on the map. Your evaluation of the present climate of a topographic map area should consider surface water features, groundwater features, and the geographic location of the area.

The most common wind-eroded landform visible on a topographic map is usually a **blowout**—a shallow

depression developed where wind has eroded and blown out the soil and fragmented rock (Figure 14.3C). Blowouts may resemble sinkholes (depressions formed where caves have collapsed), or kettles (depressions formed where sediment-covered blocks of glacial ice have melted), but you can distinguish these different types of depressions in the context of other features observable on the map. Unlike sinkholes and kettles, blowouts usually have an adjacent sand dune or dunes that formed where sand-sized grains were deposited after being removed from the blowout. Blowouts also range in size from a few meters to a few kilometers in diameter.

Sand transported by wind also erodes many rock surfaces by sandblasting them. **Ventifacts** are rocks that have flat or scoop-shaped surfaces that were abraded in this manner (Figure 14.4).

When the wind transports and deposits sand, it creates sand dunes and silty deposits called *loess* (see Figure 14.2). The process of dune and loess formation is shown in Figure 14.1. Four major types of dunes are illustrated in Figure 14.3 and are described below:

- **Barchan dunes** are crescent shaped. They occur where sand supply is limited and wind direction is fairly constant. Barchans generally form around shrubs or large rocks, which serve as minor barriers to sand transportation. The *horns* (tips) of barchans point downwind.

- **Transverse dunes** occur where sand supply is greater. They form as ridges perpendicular to the prevailing wind direction. The crests of transverse dunes generally are sinuous to very sinuous.

- **Parabolic dunes** somewhat resemble barchans. However, their horns point in the opposite direction—upwind. Parabolic dunes always form adjacent to **blowouts,** oval depressions from which come the sandy sediments that form the parabolic dunes.

- **Longitudinal (linear) dunes** occur in some modern deserts where sand is abundant and cross winds merge to form these high, elongated dunes. They can be quite large, up to 200 km long and up to 100 m high. The crests of longitudinal dunes generally are straight to slightly sinuous.

Dunes tend to migrate slowly in the direction of the prevailing wind (Figures 14.1 and 14.3). However, revegetation of exposed areas, due to changes in climate or mitigation, may stabilize them.

On topographic maps, large areas of dunes may be marked with a brown pattern (see topographic

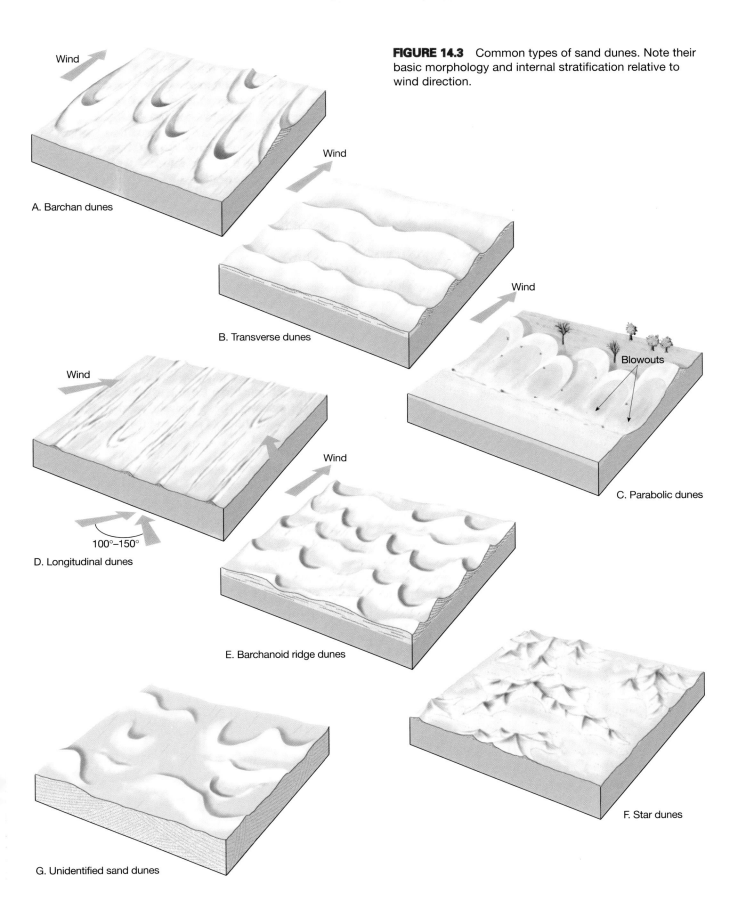

FIGURE 14.3 Common types of sand dunes. Note their basic morphology and internal stratification relative to wind direction.

Wind

A. Barchan dunes

Wind

B. Transverse dunes

Wind

Blowouts

C. Parabolic dunes

Wind

100°–150°

D. Longitudinal dunes

Wind

E. Barchanoid ridge dunes

F. Star dunes

G. Unidentified sand dunes

FIGURE 14.4 Photograph (actual size) of a sandblasted rock, or *ventifact*. Note the three flat surfaces, which were abraded by windblown sand.

map symbols, Figure 9.2). Small groups of stabilized dunes may be indicated by contour lines, without a pattern. In that case, you can recognize dunes by their distinctive shape and overall pattern, or by linearity of the contour lines.

Dryland Landforms

Two characteristics of dryland precipitation combine to create some of the most characteristic dryland (including desert) landforms other than blowouts and dunes. First, rainfall in drylands is minimal. Second, when rainfall does occur, it generally is in the form of violent thunderstorms. The high volume of water falling from such storms causes flash floods over dry ground. These floods develop suddenly, have high discharge, and last briefly. They carve steep-walled canyons, often floored with gravel that is deposited as the flow decreases and ends. Such steep-walled canyons with gravel floors commonly are called **arroyos** (or **wadis,** or **dry washes**).

Flash flooding in arid regions also erodes vertical cliffs along the edges of hills. When bedrock lies roughly horizontal, such erosion creates broad, flat-topped **mesas** bounded by cliffs. In time, the mesas can erode to small, stout, barrel-like rock columns, called **buttes.**

Figures 14.5 and 14.6 are portions of topographic maps of drylands in the southwest part of the United

States. You should be able to identify arroyos, mesas, or buttes on these maps.

In regions where Earth's crust has been lengthened by tensional forces (pulled apart), mountain ranges and basins develop by **block faulting**—a type of regional rock deformation where Earth's crust is broken into fault-bounded blocks of different elevations (Figure 14.7). The higher blocks are called *horsts* and the lower blocks are *grabens* (see Figure 14.7). The horst mountain ranges are eroded by running water and fluvial processes, which also transport the rock debris to adjacent graben *basins* (depressions where water and sediment accumulate). In a humid climate, these basins might collect water in permanent lakes. In a desert, however, precipitation usually is insufficient to fill and maintain permanent lakes.

Another factor is that climate is not constant through geologic time. Regions that now are deserts may have been more humid in the past. Landforms produced in the past, when the climate was different, still may be preserved on the present landscape. These landforms are valuable clues to understanding environmental changes.

These dryland landforms are illustrated in Figure 14.7 and present on the topographic maps in Figures 14.5 and 14.6:

- **Alluvial fan**—a fan-shaped, delta-like deposit of alluvium made at the mouth of a stream or arroyo, where it enters a graben, level plain, or basin.

- **Bajada**—(Spanish, "slope") a continuous apron of coalescing alluvial fans below a mountain front.

- **Mountain front**—the sharp-angled intersection where the steep lower slope of a mountain range meets a pediment or alluvial fan and the slope of the land (and spacing of topographic contours) changes.

- **Pediment**—a gently inclined erosion surface in the upper part of the piedmont slope. It is carved into bedrock and generally has a thin veneer of alluvium.

- **Playa**—the shallow, almost flat, central part of a desert basin, in which water gathers after a rain and evaporates to leave behind silt, clay, and evaporites (salts).

- **Sand dune**—a small hill, mound, or ridge (linear or sinuous) of windblown sand.

- **Bolson**—a basin into which water drains from the surrounding mountains to a central playa; a closed basin—one that has no outlet.

FIGURE 14.5: Antelope Peak, Arizona

Contour interval = 25 ft. 1:62,500

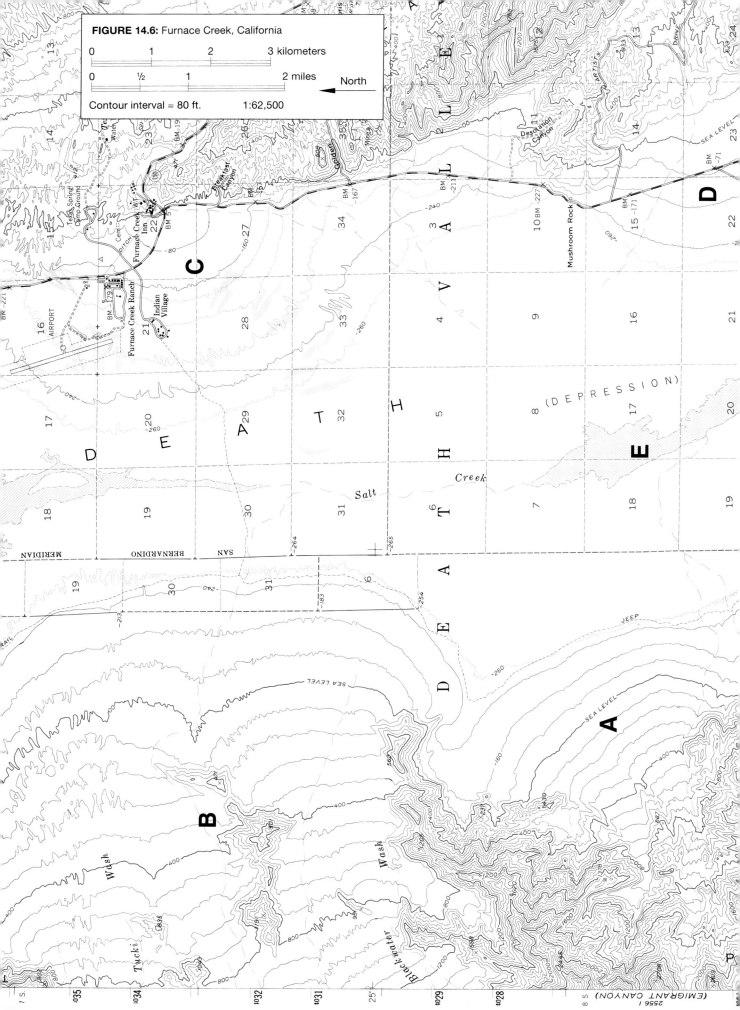

FIGURE 14.6: Furnace Creek, California

0   1   2   3 kilometers

0   ½   1   2 miles

North

Contour interval = 80 ft.   1:62,500

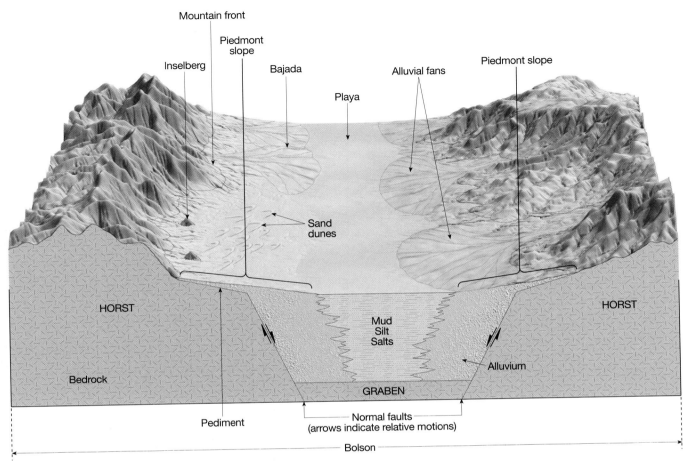

FIGURE 14.7 Typical landforms of arid mountainous deserts in regions where Earth's crust has been lengthened by tensional forces (pulled apart). Mountain ranges and basins develop by **block faulting—** a type of regional rock deformation where Earth's crust is broken into fault-bounded blocks of different elevations. The higher blocks form mountains called *horsts* and the lower blocks form valleys called *grabens.* Note that the boundaries between horsts and grabens are typically normal faults. Sediment eroded from the horsts is transported into the grabens by wind and water. **Alluvial fans** develop from the mountain fronts to the valley floors. They may surround outlying portions of the **mountain fronts** to create **inselbergs** (island-mountains). The fans may also coalesce to form a bajada. In cases where there is no drainage outlet from the valley, the valley is a closed basin or **bolson.**

Questions

1. Refer to the ventifact in Figure 14.4. This is one of many such rocks in a local area, all of which were worn this way. In this area, from which direction does the strongest sustained wind likely blow? Explain your reasoning.

2. Examine the **star dunes** in Figure 14.3F. What kind of wind pattern would cause this kind of dune form? Explain your reasoning.

3. Examine the unidentified dunes in Figure 14.3G. What kind of wind pattern would cause this kind of dune form? Explain your reasoning.

Study Figure 14.5 of part of the Antelope Peak, Arizona, quadrangle. See if you can identify any of the features listed above and then complete Questions 4–8.

4. What is the hilly feature labeled **A** in Figure 14.5? How did it form?

5. How do you think the Table Top Mountains got their name?

6. What is the name of the linear landform labeled **B** in Figure 14.5? How did it form?

7. How are landforms **C** and **D** similar?

8. List the stages of development of landform **D.**

PART 14B: DEATH VALLEY, CALIFORNIA

Study Figure 14.6 of part of the Furnace Creek, California, quadrangle. Then complete Questions 9–12. The large *graben* valley in the middle of the map is Death Valley, California, which is the lowest valley in the United States. The mountains on each side of the valley are *horsts*. Faults separate the horsts from the graben valley, as portrayed in Figure 14.7.

Questions

9. Obtain your set of colored pencils and do the following.

   a. Color alluvial fan **A** yellow, including the two arroyos at the top (upslope end) of the fan.

   b. Color the inselbergs red in the vicinity of location **B**.

   c. Color alluvial fan **C** yellow.

   d. Color alluvial fan **D** yellow, to the point upslope where its drainage touches the 800-foot topographic contour line.

   e. Color the 00 (sea level) topographic contours blue on both sides of the valley.

   f. Make a green line along the downhill edge of the *mountain front* (on Figure 14.6) on both sides of the valley and label it "mountain front."

10. Notice the intermittent stream that drains from the upstream end of the alluvial fan/arroyo system **A** (that you have already colored yellow) to the playa at **E**.

    a. What is the *average* gradient of this stream, from points **X** to **Y**, in ft/mi? (Refer to Figure 11.4: gradient.)

    b. How would the grain size of the sediments along this stream change as you walk downslope from the high arroyo to **E**?

11. Draw dark dashed lines (with a normal pencil or a black colored pencil) to indicate where you think the main faults are present on each side of the graben.

12. Notice that people chose to build a ranch on alluvial fan **C**, even though this entire region is dryland. What do you think was the single most important reason why those people chose alluvial fan **C** for their ranch instead of one of the other fans?

PART 14C: DRYLAND LAKES

The amount of rain that falls on a particular dryland normally fluctuates over periods of several months, years, decades, centuries, or even millennia. Therefore, a dryland may actually switch back and forth between arid and semi-arid conditions, semi-arid and dry-subhumid conditions, arid and dry-subhumid conditions, and so on. Where lakes persist in the midst of drylands, their water levels fluctuate up and down in relation to such periodic changes in precipitation and climate. Periods of higher rainfall (or snow that eventually melts) and reduced aridity and evaporation create lakes that dry up during intervening periods of less rain and greater aridity and evaporation. The Great Salt Lake, Utah, is an example.

Great Salt Lake is a closed basin, so water can escape from the lake only by evaporation. When it rains, or when snow melts in the surrounding hills, water raises the level of the lake. Therefore, the level of Great Salt Lake has varied significantly in historic times over periods of months, years, and decades. During one dry period of many years, people ignored the dryland hazard of fluctuating lake levels and constructed homes, roads, farms, and even a 2.5-million-dollar resort, the Saltair, near the shores of Great Salt Lake. When a wet period occurred from 1982–87, many of these structures (including the resort) were submerged. The State of Utah installed huge pumps in 1987 to pump lake water into another valley, but the pumps were left high and dry during a brief dry period that lasted for two years (1988–89) after they were installed.

Geologic studies now suggest that the historic fluctuations of Great Salt Lake are minor in comparison to those that have occurred over millennia. Great Salt Lake is actually all that remains of a much larger lake that covered 20,000 square miles of Utah—Lake Bonneville. Lake Bonneville reached its maximum depth and geographic extent about 17,000 years ago as glaciers were melting near the end of the last Ice Age. One arm of the lake at that time extended into Wah Wah Valley, Utah, which is now a dryland (Figures 14.8 and 14.9).

Questions

13. What specific type of feature is the Wah Wah Valley Hardpan?

14. On Figure 14.8, identify and label the bajada, the alluvial fans, and the mountain front east of the Wah Wah Valley Hardpan.

15. A dashed blue line surrounds the Wah Wah Valley Hardpan. What does it represent? (It is not a stream.)

16. If the Wah Wah Valley Hardpan were to fill with water, how deep could the lake become before it overflows to the northeast?

17. On the stereogram (Figure 14.9), what evidence can you identify for a former deeper lake (an arm of Lake Bonneville) in Wah Wah Valley (deeper than your answer in Question 16)?

18. What is the age of the former deep lake (Question 17) relative to the age of the alluvial fans and bajada? Explain your reasoning. (*Hint*: Study the stereogram.)

19. On Figure 14.9, note the steplike terraces on the slope just to the north (right) of **B**. How did these terraces form?

20. Also on Figure 14.9, study the coastal landforms upslope and downslope from letter **B**. What are these depositional landforms called? How did they form in a line from upslope to downslope?

21. The patch of white at point **A** (see Figure 14.9) is modern tufa, a kind of porous limestone similar to travertine. Infer why it may have formed here.

22. On Figure 14.8, use a blue colored pencil to show the position (line) of the shoreline of ancient Lake Bonneville where it reached its highest elevation. Then shade (blue) the area that was submerged (i.e., color in the area that was the lake).

23. Abundant studies by geologists of the Utah Geologic Survey and U.S. Geological Survey indicate that ancient Lake Bonneville stabilized in elevation at least three times before present. About 23,000 years ago, the lake stabilized at an elevation of about 4400 ft above sea level. About 17,000 years ago, the lake stabilized at an elevation of about 5100 ft above sea level. About 16,000 years ago, the lake stabilized at an elevation of about 4800 ft above sea level. About 12,000 years ago, the lake stabilized at an elevation of about 4300 ft above sea level.

   a. What is the elevation of the lake level that you identified in Question 22?

   b. What is the age of the lake level that you identified in Question 22?

   c. Modern Great Salt Lake has an elevation of about 4200 ft and is 30 ft deep. How deep was the Great Salt Lake location at the time identified in **a** and **b**? Explain your calculations.

24. Study Figure 14.10.

   a. What percentage (by visual estimation) of modern Utah is submerged by the Great Salt Lake?

   b. Based on your answers in Question 23, use a blue colored pencil to shade in Figure 14.10 so it shows how Utah was covered by lake water about 17,000 years ago. What percentage (by visual estimation) of Utah was submerged by Lake Bonneville at that time?

PART 14D: DRYLAND HAZARDS AND RISKS IN NEBRASKA'S SAND HILLS

There are many deposits of eolian sand and silt (called *loess*) in the midwestern United States that were deposited during Pleistocene and Holocene times. The distribution of these eolian deposits is illustrated in Figure 14.2, which also shows the location of Nebraska's Sand Hills region.

The Sand Hills of Nebraska covers about 50,000 square kilometers of land and is the largest sand-dune area—or **sand sea**—in the Western Hemisphere. This sand sea was undoubtedly present in Late Pleistocene time, but the largest of the modern sand dunes were active as late as 8,000 years ago. This was determined by dating the radioactive carbon of organic materials that have been covered up by the large dunes. The large dunes are now covered with grass (short-grass prairie) that is suitable for limited ranching. About 17,000 people (mostly ranchers) now live in the Sand Hills.

Questions

25. The large hills in Figure 14.11 are either large barchan dunes (called *megabarchans*) or barchanoid ridges (compare outlined examples on Figure 14.11 with art in Figure 14.3). Using colored pencils, color another isolated megabarchan yellow (not the one that is already outlined) and another one of the barchanoid ridges green (not the one that is already outlined).

   a. What is the relief, in feet, of the megabarchan that you colored?

   b. What is the length of the megabarchan that you colored (measured from stoss edge to lee edge of the dune; refer also to Figure 14.1)?

   c. According to the orientations of the megabarchans and the barchanoid ridges, the winds that made these dunes were coming *from* what direction? Explain your reasoning.

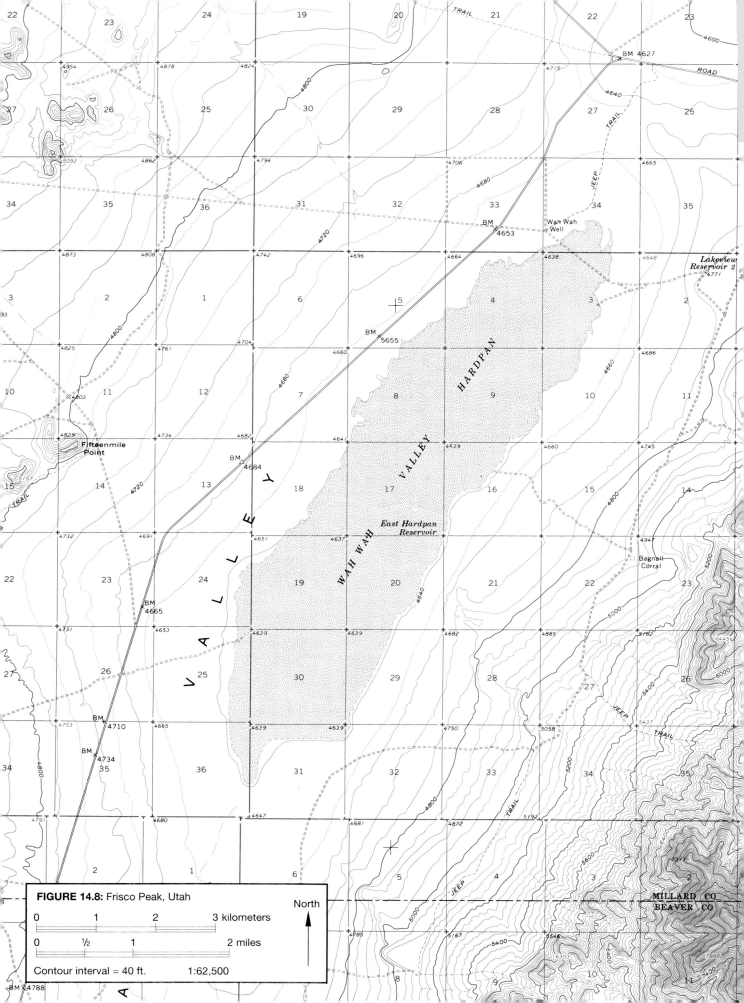

FIGURE 14.8: Frisco Peak, Utah

North

| 0 | | 1 | | 2 | | 3 kilometers |

| 0 | ½ | 1 | | 2 miles |

Contour interval = 40 ft.   1:62,500

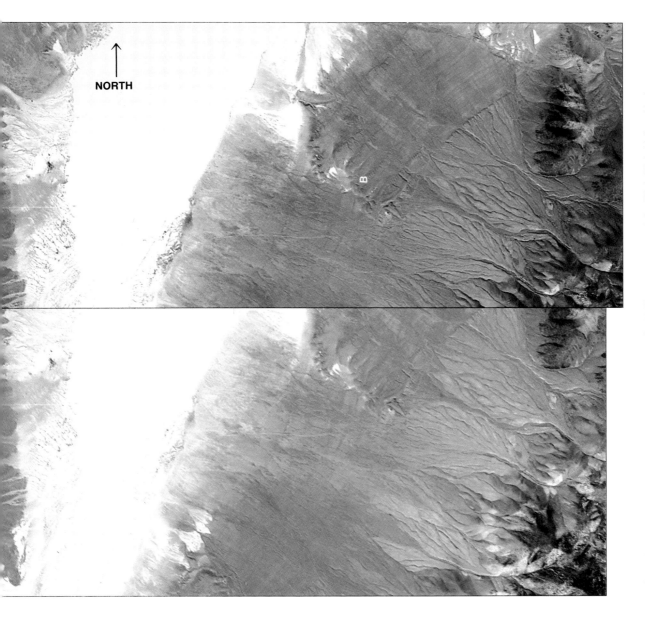

FIGURE 14.9 National high-altitude photograph (NHAP) stereogram of the Wah Wah Valley, Utah, area: 1:58,000 scale. 1991 (Courtesy of U.S. Geological Survey)

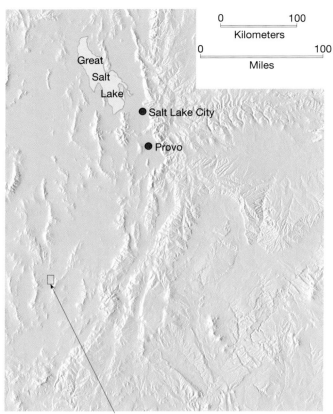

Frisco Peak Quadrangle
Figure 14.8

FIGURE 14.10   Relief map of Utah showing location of Frisco Peak Quadrangle and the Great Salt Lake.

d. What evidence is there in Figure 14.2 that the source of sand for the Sand Hills was located northwest of the Sand Hills? Explain your reasoning.

26. Notice that there are small closed depressions on the surfaces of most of the megabarchans and barchanoid ridges. Fresh sand is exposed in these depressions.

a. What are these small depressions called?

b. What kind of small dunes are associated with these depressions (*Hint:* Refer to Figure 14.3)?

c. What does this suggest about the risk of desertification here?

27. Locate Star Ranch in the northwest corner of the map in Figure 14.11 (section 17, T26N, R43W). This ranch was present in the 1930s during the famous Dust Bowl days, when dry windy conditions persisted for years and dust storm after dust storm scoured the land here. Star Ranch was not covered by any advancing dunes during the Dust Bowl.

a. What does this suggest about whether the megabarchans and barchanoid ridges are active or inactive at this time?

b. What does this suggest about how much desertification must occur here in order for the large dunes of the Sand Hills to once again become an active sand sea?

28. This area now receives an average of 580 mm (about 2 in.) of rain per year, and large active dunes tend to occur in regions that receive less than 250 mm of rain per year (less than an inch). Winds capable of moving sand already occur in this region, but the rainfall is presently enough to sustain grasses that hold the sand in place. If you were a rancher in the Sand Hills, what grazing practices would you follow to decrease the risk of desertification there?

29. The lakes of this region are important sources of water for hayfields cultivated on the lake margins (so the hay can be fed to cattle), water for human use, and water for large populations of migratory birds like ducks and geese. The large number of lakes represent a water table that is above the land surface in the basins between large dunes. Contour the water-table surface of the lakes on Figure 14.11 by using the elevations shown on the lakes and drawing directly on the map. Make contours for levels of 3880, 3885, 3890, 3895, 3900, and 3905 feet.

a. What do you think is the direction of groundwater flow? Why?

b. Many cities in central and eastern Nebraska rely on groundwater for consumption, industry, and pleasure. As these cities continue to grow, and their use of groundwater increases, what effect might this have on the environments and people of the Sand Hills?

FIGURE 14.11: Lakeside, Nebraska

R 43 W

0   1   2   3 kilometers

0   ½   1   2 miles

Contour interval = 20 ft.   1:62,500

North

T 26 N
T 25 N

Coastal Processes, Landforms, Hazards, and Risks

•CONTRIBUTING AUTHORS•

James G. Titus • *U.S. Environmental Protection Agency*

Donald W. Watson • *Slippery Rock University*

OBJECTIVES

A. Identify and interpret natural shoreline landforms.

B. Distinguish between emergent and submergent shorelines.

C. Know the common types of artificial structures that are used to modify shorelines and understand their effects on coastal environments.

D. Be aware of the probability of global sea-level rise and the coastal hazards and increased risks that this sea-level rise may cause.

MATERIALS

Pencils, eraser, ruler, set of colored pencils, and pocket stereoscope.

INTRODUCTION

The shorelines of lakes and oceans are among the most rapidly changing parts of the Earth's surface. All coastlines are subject to *erosion* (wearing away) by waves. A coastline comprised of loose sediment can be eroded easily and rapidly. A coastline composed of dense bedrock or plastic-like mud erodes much more slowly.

Several factors determine the characteristic landforms of shorelines. They include the shape of the shoreline, the materials that comprise the shoreline (rock, plastic mud, loose sediment, concrete), the source and supply of sediments, the direction that currents move along the shoreline, and the effects of major storms.

Most coastlines also are affected by changes in mean (average) sea level:

- A *rising* sea level creates a **submergent coastline**—one that is flooding and receding (*retrogradational*). Sea level rise is caused either by the water level actually rising (called *transgression*), or by the land getting lower (called *subsidence*).

- A *falling* sea level creates an **emergent coastline**—one that is being elevated above sea level and building out into the water (*progradational*). Sea level fall is caused either by the water level actually falling (called *regression*), or by rising of the land (called *uplift*).

Submergent coastlines may display some emergent features, and vice versa. For example, the Louisiana coastline is submergent, enough so that dikes and levees have been built to keep the ocean from flooding New Orleans. However, the Mississippi Delta is progradational—building out into the water—a feature of most emergent coastlines. It is progradational because of the vast supply of sediment being carried there and deposited by the Mississippi River.

Thus, *sediment supply* is a major factor in determining whether a coastline is progradational or retrogradational, regardless of vertical changes of land level or water level.

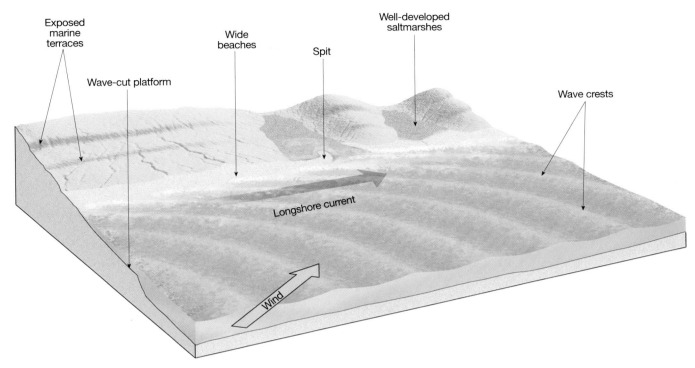

FIGURE 15.1 *Emergent* coastline features. An emergent coastline is caused by sea-level lowering, the land rising, or both. Emergence causes tidal flats and coastal wetlands to expand, wave-cut terraces are exposed to view, deltas prograde at faster rates, and wide stable beaches develop.

Sediment transport and the effects of major storms also are very important agents of shoreline change. A single storm can completely change the form of a coastline.

Figures 15.1 and 15.2 illustrate some features of *emergent* and *submergent* shorelines. Study these features and their definitions below.

- **Barrier island**—a long, narrow island that parallels the mainland coastline and is separated from the mainland by a lagoon, tidal flat, or salt marsh.

- **Beach**—a gently sloping deposit of sand or gravel along the edge of a shoreline.

- **Berm crest**—the highest part of a beach; it separates the *foreshore* (seaward part of the shoreline) from the *backshore* (landward part of the shoreline).

- **Washover fan**—a fan-shaped deposit of sand or gravel transported and deposited landward of the beach during a storm or very high tide.

- **Estuary**—a river valley flooded by a rise in the level of an ocean or lake. (A flooded glacial valley is called a *fjord*.)

- **Longshore current**—a water current in the *surf zone* (zone where waves break). It flows slowly parallel to shoreline, driven by waves that were caused by wind.

- **Delta**—a sediment deposit at the mouth of a river where it enters an ocean or lake.

- **Headland**—projection of land that extends into an ocean or lake and generally has cliffs along its water boundary.

- **Spit**—a sand bar extending from the end of a beach into the mouth of an adjacent bay.

- **Tidal flat**—muddy or sandy area that is covered with water at high tide and exposed at low tide.

- **Saltmarsh**—a marsh that is flooded by ocean water at high tide.

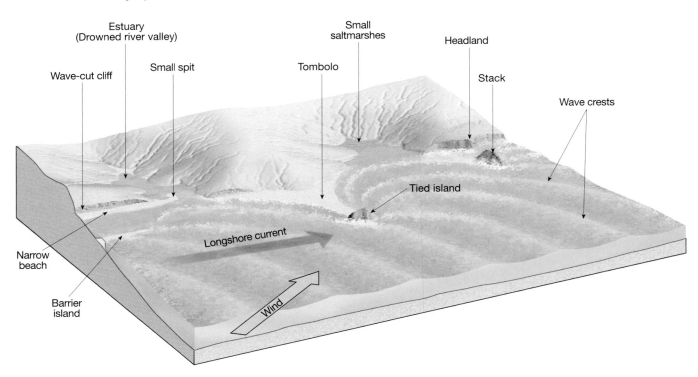

FIGURE 15.2 *Submergent* (drowning) coastline features. A submergent coastline is caused by sea-level rising (transgression), sinking of the land, or both. As the land is flooded, the waves cut cliffs, valleys are flooded to form estuaries, wetlands are submerged, deep bays develop, beaches narrow, and islands are created.

- **Wave-cut cliff** (or *sea cliff*)—seaward-facing cliff along a steep shoreline, produced by wave erosion.

- **Wave-cut platform**—a bench or shelf at sea level (or lake level) along a steep shore, and formed by wave erosion.

- **Marine terrace**—an elevated platform that is bounded on its seaward side by a cliff or steep slope (and formed when a wave-cut platform is elevated by uplift or regression).

- **Stack**—an isolated rocky island near a headland cliff.

- **Tombolo**—a sand bar that connects an island with the mainland or another island.

- **Tied island**—an island connected to the mainland or another island by a tombolo.

Humans build several common types of coastal structures in order to protect harbors, build up sandy beaches, or extend the shoreline. Study these four kinds of structures and their effects both in Figure 15.3 and below.

- **Sea wall**—an embankment of boulders, reinforced concrete, or other material constructed against a shoreline to prevent erosion by waves and currents.

- **Breakwater**—an offshore wall constructed parallel to a shoreline to break waves. The longshore current is halted behind such walls, so the sand accumulates there and the beach widens. Where the breakwater is used to protect a harbor from currents and waves, sand often collects behind the breakwater and may have to be dredged.

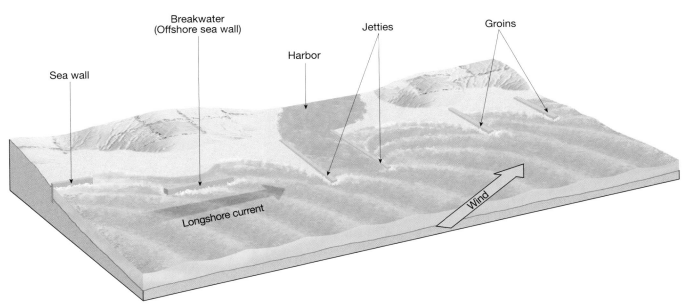

FIGURE 15.3 Coastal structures—sea walls, breakwaters, groins, and jetties. **Sea walls** are constructed along the shore to stop erosion of the shore or extend the shoreline (as sediment is used to fill in behind them). **Breakwaters** are a type of offshore sea wall constructed parallel to shoreline. The breakwaters stop waves from reaching the beach, so the longshore drift is broken and sand accumulates behind them (instead of being carried down shore with the longshore current). **Groins** are short walls constructed perpendicular to shore. They trap sand on the side from which the longshore current is carrying sand against them. **Jetties** are long walls constructed at entrances to harbors to keep waves from entering the harbors. However, they also trap sand just like groins.

- **Groin** (or *groyne*)—a short wall constructed perpendicular to shoreline in order to trap sand and make or build up a beach. Sand accumulates on the up-current side of the groin in relation to the longshore current.

- **Jetties**—long walls extending from shore at the mouths of harbors and used to protect the harbor entrance from filling with sand or being eroded by waves and currents. Jetties are usually constructed of boulders and in pairs (one on each side of a harbor or inlet).

PART 15A: DYNAMIC NATURAL COASTLINES

Refer to the Space Shuttle photograph of the Po Delta, Italy (Figure 15.4). The city of Adria, on the Po River in northern Italy, was a thriving seaport during Etruscan times (600 B.C.). Adria had such fame as to give its name to the Adriatic Sea, the gulf into which the Po River flows. Over the years, the Po River has deposited sediment at its mouth in the Po Delta. Because of the Po Delta's progradation, Adria is no longer located on the shoreline of the Adriatic Sea. The modern shoreline is far downstream from Adria.

Questions

1. What has been the average annual rate of Po Delta progradation in centimeters per year (cm/yr) since Adria was a thriving seaport on the coastline of the Adriatic Sea?

2. Based on the average annual rate calculated in Question 1, how many centimeters would the Po Delta prograde during the lifetime of someone who lived to be 60 years old?

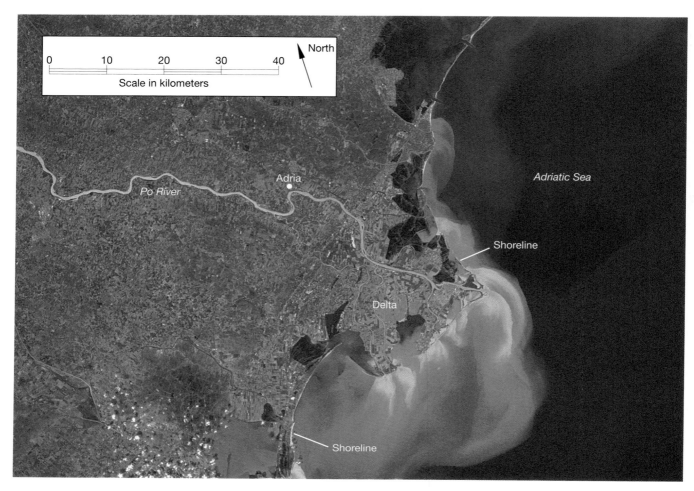

FIGURE 15.4 Space Shuttle photograph of the Po Delta region, northern Italy. (Courtesy of NASA)

Refer to the Oceanside, California, quadrangle (Figure 15.5) and complete Questions 3 and 4.

3. If you climb inland from the Pacific Ocean at South Oceanside to Fire Mountain, you will cross a series of relatively flat surfaces located at successively higher elevations and separated by steep hills or cliffs. All together, they resemble a sort of giant staircase.

   a. About how many of these coastal features are there?

   b. What are the approximate elevations of the flat surfaces, from lowest to highest?

   c. What are these coastal features called, and what is their probable origin?

4. Is this a coastline of emergence or of submergence? Why?

Refer to the Point Reyes, California, quadrangle (Figure 15.6). Point Reyes is a subtriangular landmass bounded on the west by the Pacific Ocean, on the south by Drakes Bay, and on the east by the San Andreas Fault. The fault runs along Sir Francis Drake Road in the northeast corner of the map.

5. Which area is *more* resistant to wave erosion: Point Reyes or Point Reyes Beach? Why?

6. How did Drakes Estero (Spanish: "estuary") form?

7. What is the direction of longshore drift in Drakes Bay? How can you tell?

8. If a groin was constructed from Limantour Spit, at the "n" in Limantour, then on what side of the groin would sand accumulate (east or west)? Why?

9. Is this a coastline of emergence or submergence? Explain.

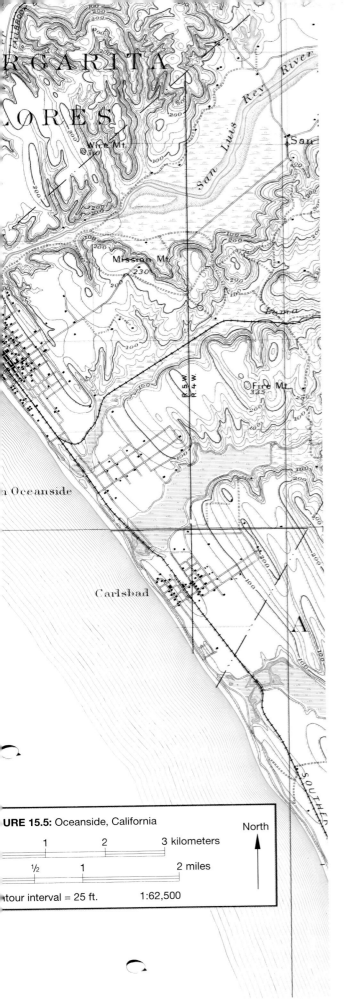

Refer to the map and photographs of Saint Catherines Island, Georgia (Figure 15.7). Note that on the east-central portion of the island there is a large area of salt-marsh mud. Living saltmarsh plants are present there, as shown on the right (west) in Figures 15.7A and B. Also, note the linear sandy beach in Figures 15.7A and B, bounded on its seaward side (left) by another strip of saltmarsh mud. However, all of the living, surficial saltmarsh plants and animals have been stripped from this area. This is called **relict** saltmarsh mud (mud remaining from an ancient saltmarsh).

10. What type of sediment is probably present beneath the beach sands in Figures 15.7A and B?

11. Explain how you think the beach sands became located landward of the relict saltmarsh mud.

12. Portions of the living saltmarsh (wetland) in Figure 15.7C recently have been buried by bodies of white sand that was deposited from storm waves that crashed over the beach and sand dunes. What is the name given to such sand bodies?

13. Photograph 15.7C was taken from a landform called Aaron's Hill. It is the headland of this part of the island. What will eventually happen to Aaron's Hill? Why?

14. Based upon your answer in Question 13, would Aaron's Hill be a good location for a resort hotel?

15. Based upon your inferences, observations, and explanations in Questions 11, 12, and 13, what will eventually happen to the living saltmarsh in Figures 15.7B and C?

16. What can you infer about global sea level, based on your answers to Questions 4, 9, and 15?

URE 15.5: Oceanside, California

North

1   2   3 kilometers

½   1   2 miles

ntour interval = 25 ft.   1:62,500

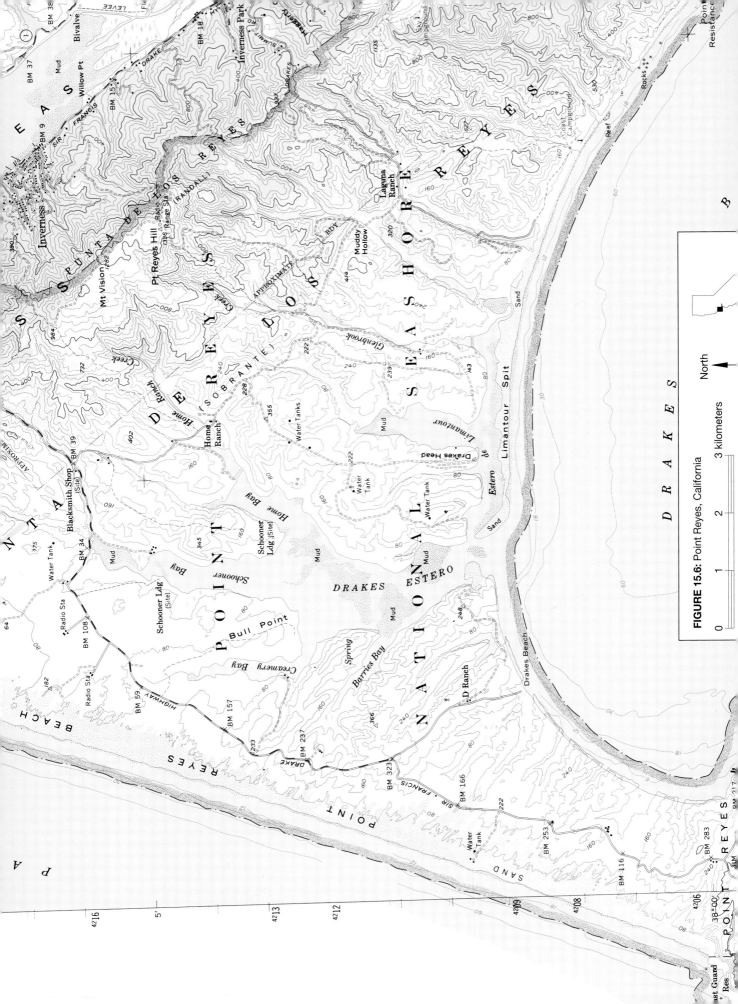

FIGURE 15.6: Point Reyes, California

North

0   1   2   3 kilometers

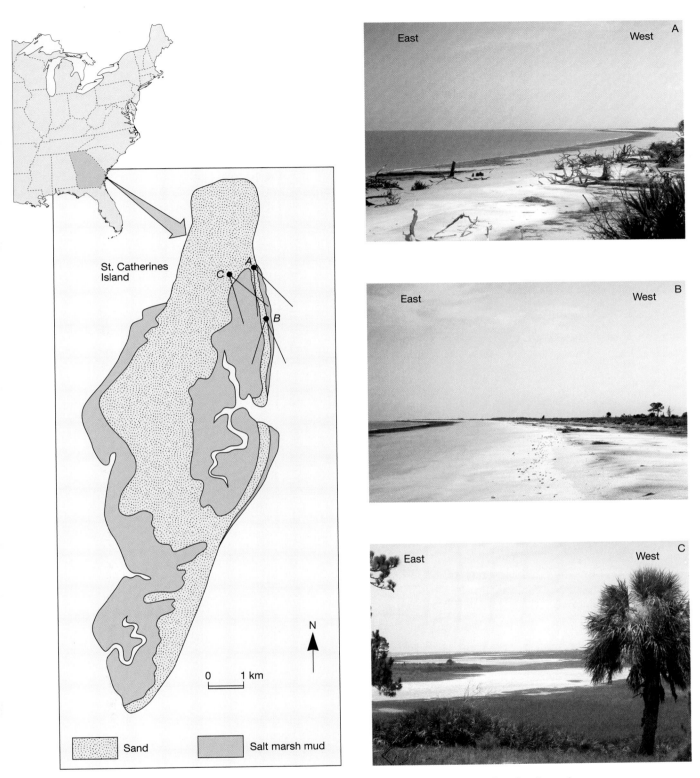

FIGURE 15.7 Saint Catherines Island, Georgia: coastal features and distribution of sand and saltmarsh mud. **A.** View south–southeast from point **A** on map, at low tide. Dark-brown "ribbon" adjacent to ocean is saltmarsh mud. Light-colored area is sand. **B.** View south from point **B** on map at low tide. **C.** View southeast from point **C** (Aaron's Hill) on map. (Photos by R. Busch)

PART 15B: HUMAN MODIFICATION OF SHORELINES

Examine the portion of the Ocean City, Maryland, topographic quadrangle map provided in Figure 15.8. Purple features show changes made in 1972 to a 1964 map, so you can see how the coastline changed from 1964–1972. Also note the outline of the barrier island as it appeared in 1849 according to the U.S. Geological Survey.

Ocean City is located on a long, narrow barrier island called Fenwick Island. During a severe hurricane in 1933, the island was breached by tidal currents that formed Ocean City Inlet and split the barrier island in two. Ocean City is still located on what remains of Fenwick Island. The city is a popular vacation resort that has undergone much property development over the past 50 years. The island south of Ocean City Inlet is called Assateague Island. It has remained undeveloped, as a state and national seashore.

Questions

17. After the 1933 hurricane carved out Ocean City Inlet, the Army Corps of Engineers constructed a pair of jetties on each side of the inlet to keep it open. The southern jetty is labeled "seawall" on the map. Sand filled in behind the northern jetty, so it is now a sea wall forming the straight southern edge of Ocean City on Fenwick Island (a straight black line on the map). Based on this information, would you say that the longshore current is traveling north to south, or south to north? Explain.

18. Notice that Assateague Island has migrated landward (west), relative to its 1849 position. This migration began in 1933.

    a. Why did Assateague Island migrate landward?

    b. Field inspection of the west side of Assateague Island reveals that muds of the lagoon (Sinepuxent Bay) are being covered up by the westward-advancing island. What is the rate of Assateague Island's westward migration in feet/year and meters/year?

    c. Based on your last answer (Question 18b), predict the approximate year in which the west side of Assateague Island will merge with saltmarshes around Ocean City Harbor. What natural processes and human activities could prevent this?

19. Notice the groins (short black lines) that have been constructed on the east side of Fenwick Island (Ocean City) in the northeast corner of the map.

    a. Why do you think these groins have been constructed there?

    b. What effect could these groins have on the beaches around Ocean City's Municipal Pier? Why?

20. Hurricanes normally approach Ocean City from the south–southeast. In 1995, one of the largest hurricanes ever recorded (Hurricane Felix) approached Ocean City but miraculously turned back out into the Atlantic Ocean. How does the westward migration of Assateague Island increase the risk of hurricane damage to Ocean City?

21. The westward migration of Assateague Island could be halted and probably reversed if all of the groins, jetties, and sea walls around Ocean City were removed. How would removal of all of these structures place properties in Ocean City at greater risk to environmental damage than they now face?

FIGURE 15.8: Ocean City, Maryland (1964)
(Photorevised, 1972–purple areas)

0   .5   1 kilometer

0   ¼   ½   1 mile

Contour interval = 5 ft.   1:24,000

North

Maryland

Quadrangle location

PART 15C: THE THREAT OF RISING SEAS

All of the topographic maps in this laboratory manual rely on a zero reference datum of *mean sea level*. Sea level actually fluctuates both above and below mean sea level during daily tidal cycles and storm surges. A **storm surge** is a bulge of water pushed landward by abnormally high winds and/or low atmospheric pressure associated with storms. Storm surges cause the ocean to rise by about 2–24 feet, depending on the magnitude of the storm. However, except for hurricanes, most storm surges are in the range of 2–3 feet.

Given the fact that daily tides cause sea level to fluctuate 2–3 feet above and below mean sea level, and most storm surges are in the range of 2–3 feet, it might be wise to generally not build dwellings and businesses on elevations less than 6 feet near marine coastlines.

Notice that Ocean City, Maryland (Figure 15.8) has not followed this rule of thumb. Dense construction (pink areas on Figure 15.8) has occurred in many areas less than 5 feet above mean sea level. Therefore, Ocean City is at a high risk of flooding from rising sea levels even during normal winter storms. One of these storms flooded most of the city in 1962, and a hurricane could submerge the entire city (because the city's highest elevation is only 10 feet above mean sea level).

A more long-term hazard to coastal cities is the threat that mean sea level may rise significantly over the coming decades. A report on *The Probability of Sea Level Rise* was issued by the U.S. Environmental Protection Agency (EPA) in 1996, and was based on data from dozens of the most respected researchers in this field throughout the world. The report demonstrates that mean sea level is already rising at rates of 2.5–3.0 mm/yr (10–12 inches per century) along U.S. coastlines, and these rates are expected to increase. According to this comprehensive study, there is a 50% probability that sea level will rise 34 cm (about 13 inches) over the next century. A 50% probability is the same probability that you will get heads if you flip a coin. The EPA study also suggests that there is only a 1% probability (1 chance in 100) that sea level will rise 104 cm (well over 3 feet) over the next century.

In planning for safe and economical coastal development, planning commissions and real estate developers could "play it safe" and assume that sea level could rise about 1 meter (about 3 feet) in the next century.

Questions

22. From our discussion on storm surges and the threat of actual sea-level rise, it seems logical that there are two main rules of planning for safe and economical coastal development in relation to the threat of property damage from coastal flooding. Planners should account for the probability that storm surges will normally cause sea level to rise approximately 6 feet above mean sea level. Planners should also account for the long-term probability that mean sea level will actually rise approximately 3 feet over the next century. Given the fact that most existing topographic maps of coastal areas have contour intervals of 5 feet, what would you suggest as the contour line below which construction should not occur along coastlines? Explain.

23. Let us assume that dwellings constructed at elevations less than 10 feet above mean sea level are at increasingly high risk to flooding over the next century. Using a blue colored pencil, color in all of the land areas in Figure 15.9 (Charleston, South Carolina) that are now at elevations less than and equal to 10 feet above mean sea level.

    a. What amount of the new buildings (purple buildings in Figure 15.9) in Charleston have been built in this high-risk, 10-foot zone?

    b. Why do you think that so much new construction has occurred in the high-risk, 10-foot zone around Charleston?

    c. What effect would a 10-foot sea level rise have on the abundant saltmarshes (wetlands) in the Charleston region?

24. Study the map of Miami, Florida (Figure 15.10), and list some of the significant properties that are now located within the high-risk, 10-foot zone where flood hazards will increase over the next century.

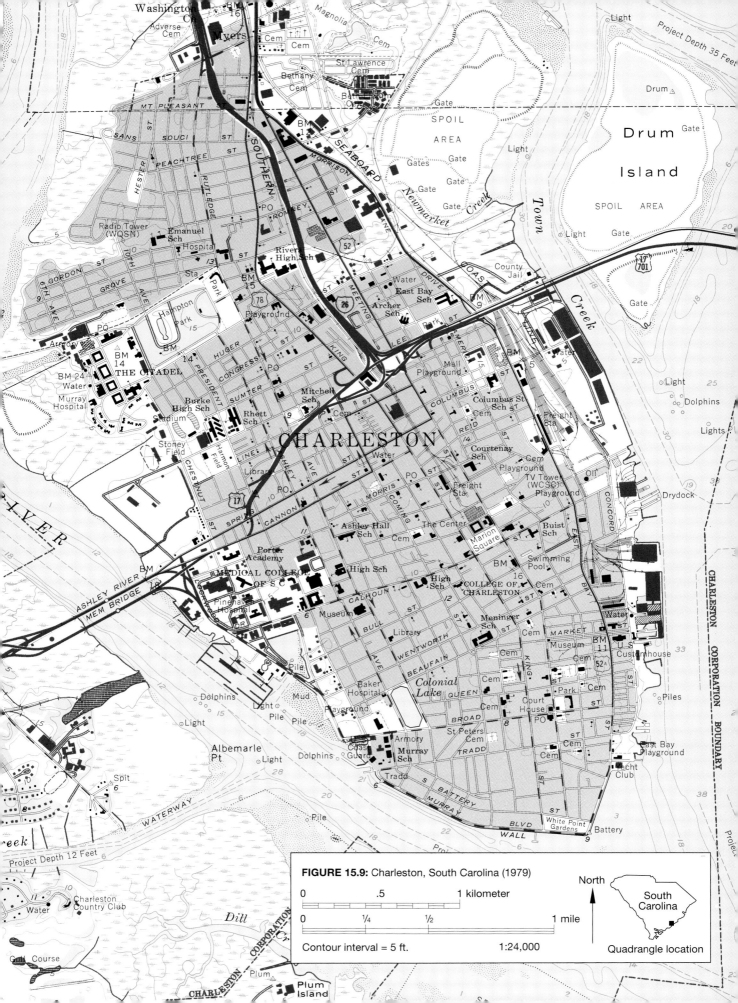

FIGURE 15.9: Charleston, South Carolina (1979)

0 .5 1 kilometer

0 1/4 1/2 1 mile

Contour interval = 5 ft.   1:24,000

North

South
Carolina

Quadrangle location

FIGURE 15.10: Miami, Florida (1988)

0   .5   1 kilometer

Earthquake Hazards and Human Risks

·CONTRIBUTING AUTHORS·

Thomas H. Anderson • *University of Pittsburgh*

David N. Lumsden • *University of Memphis*

Pamela J.W. Gore • *Georgia Perimeter College*

OBJECTIVES

A. Experiment with models to determine how earthquake damage to buildings is related to the Earth materials on which they are constructed. Apply your experimental results to evaluate earthquake hazards and human risks in San Francisco.

B. Graph seismic data to construct and evaluate travel time curves for P-waves, S-waves, and L-waves. Use seismograms and your travel time curves to locate the epicenter of an earthquake.

C. Analyze and evaluate active faults using remote sensing and geologic maps.

D. Interpret seismograms to infer relative movements along the New Madrid Fault System within the North American Plate.

E. Explore real-time earthquake data, hazards, and impacts on humans using resources from the Internet.

MATERIALS

Pencil, eraser, laboratory notebook, ruler, calculator, drafting compass, several coins, a small plastic or paper cup containing dry sediment (fine sand, sugar, or salt), and a wash bottle.

INTRODUCTION

Earthquakes are shaking motions and vibrations of the Earth caused by large releases of energy that accompany volcanic eruptions, explosions, and movements of Earth's bedrock along fault lines. News reports usually describe an earthquake's **epicenter,** which is the point on Earth's surface (location on a map) directly above the **focus** (underground origin of the earthquake, in bedrock). The episodic releases of energy that occur along fault lines strain the bedrock like a person jumping on a diving board. This strain produces elastic waves of vibration and shaking called **seismic waves** (earthquake waves). Seismic waves originate at the earthquake's focus and travel in all directions through the rock body of Earth and along Earth's surface. The surface seismic waves travel in all directions from the epicenter, like the rings of ripples (small waves) that form when a stone is cast into a pond. In fact, people who have experienced strong surface seismic waves report that they saw and felt wave after wave of elastic motion passing by like the above-mentioned ripples on a pond. These waves are strongest near the epicenter and grow weaker with distance from the epicenter. For example, when a strong earthquake struck Mexico City in 1985, it caused massive property damage and 9500 deaths in a circular area radiating about 400 km (250 mi) in every direction from the city. By the time these same surface seismic waves of energy had traveled 3200 km to Pennsylvania, they were so weak that people could not even feel them passing beneath their feet. They did, however, cause water levels in wells and swimming pools to fluctuate by as much as 12 cm. They also were recorded by earthquake-detecting instruments called *seismographs*. Therefore, although most damage from an earthquake usually occurs close to its epicenter, seismographs can detect the earthquake's waves of energy even when they travel through Earth's rocky body or along Earth's surface to locations thousands of kilometers away from the epicenter.

Fault motions (movements of Earth's crust along breaks in the rocks) are the most common source of earthquakes felt by people. These motions can occur along faults that do not break the Earth's surface or along faults that do break the Earth's surface. Fault motions at Earth's surface can directly cause *hazards* such as the destruction of buildings, breakage of pipes and electric lines, development of open fissures in the soil, change in the course of streams, and generation of tsunamis (destructive ocean waves, generally 1–10 m high, that devastate coastal environments). However, *all* earthquakes cause some degree of vibration and shaking of the Earth, which can also cause most of the above-mentioned hazards.

Therefore people who live where strong earthquakes occur are at *risk* for experiencing personal injury, property damage, and disruption of their livelihoods and daily routines. Geologists study seismic waves, map active faults, determine the nature of earthquake-induced hazards, assess human risk where such hazards occur, and assist in the development of government policies related to public safety in earthquake-prone regions.

PART 16A: SIMULATE EARTHQUAKE HAZARDS TO ESTIMATE RISKS

Geoscientists and engineers commonly simulate earthquakes in the laboratory to observe their effects on models of construction sites, buildings, bridges, and so on. Now is your turn to give it a try. Start by making simple models of buildings constructed in dry, uncompacted sediment (Model 1) and moist, compacted sediment (Model 2). Then simulate earthquakes and observe what happens to them.

Questions

Obtain a small plastic or paper cup. Fill it three-quarters full with a dry sediment like sand, dirt, salt, or sugar. Place several coins in the sediment so they resemble vertical walls of buildings constructed on a substrate of uncompacted sediment (as in Figure 16.1). This is Model 1. Observe what happens to Model 1 when you *simulate an earthquake* by tapping the cup on a table top while you also rotate it counterclockwise.

1. What happened to the vertically positioned coins in the uncompacted sediment of Model 1 when you simulated an earthquake?

FIGURE 16.1 Photograph of Model 1 being subjected to a simulated earthquake.

Now make Model 2. Remove the coins from Model 1, and add a small bit of water to the sediment in the cup so that it is moist (but not soupy). Press down on the sediment in the cup so that it is well compacted, and then place the coins into this compacted sediment just as you placed them in Model 1 earlier. *Simulate an earthquake* as you did for Model 1, and then answer Questions 2 and 3.

2. What happened to the vertically positioned coins in the compacted sediment of Model 2 when you simulated an earthquake?

3. Based on your experimental Models 1 and 2, which kind of Earth material is more hazardous to build on in earthquake-prone regions: compacted sediment or uncompacted sediment? (Justify your answer by citing evidence from your experimental models.)

4. Consider the moist, compacted sediment in Model 2. Do you think this material would become *more* hazardous to build on, or *less* hazardous to build on, if it became totally saturated with water during a rainy season? (To find out and justify your answer, design and conduct another experimental model of your own. Call it Model 3.)

5. Write a statement that summarizes how water in a sandy substrate beneath a home can be beneficial or hazardous. Justify your reasoning with reference to your experimental models.

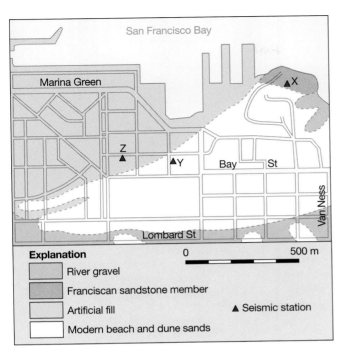

FIGURE 16.2 Map of the nature and distribution of Earth materials on which buildings and roads have been constructed for a portion of San Francisco, California. (Courtesy of U.S. Geological Survey)

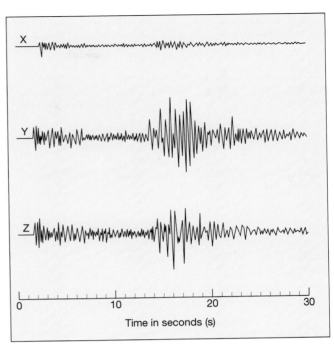

FIGURE 16.3 Seismograms recorded at Stations **X, Y,** and **Z,** for a strong (Richter Magnitude 4.6) aftershock of the Loma Prieta, California, earthquake. During the earthquake, little damage occurred at **X,** but significant damage to houses occurred at **Y** and **Z.** (Courtesy of U.S. Geological Survey)

San Francisco is located in a tectonically active region, so it occasionally experiences strong earthquakes. Figure 16.2 is a map showing the kinds of Earth materials upon which buildings have been constructed in a portion of San Francisco. These materials include hard compact Franciscan Sandstone, uncompacted beach and dune sands, river gravel, and artificial fill. The artificial fill is mostly debris from buildings destroyed in the great 1906 earthquake that reduced large portions of the city to blocks of rubble. Also note that three locations have been labeled **X, Y,** and **Z** on Figure 16.2. Imagine that you have been hired by an insurance company to assess what risk there may be in buying newly constructed apartment buildings located at **X, Y,** and **Z** on Figure 16.2. Your job is to infer whether the risk of property damage during strong earthquakes is **low** (little or no damage expected) or **high** (damage can be expected). All that you have as a basis for reasoning is Figure 16.2 and knowledge of your experiments with models in Questions 1–4.

6. What is the risk at location **X?** Why?

7. What is the risk at location **Y?** Why?

8. What is the risk at location **Z?** Why?

On October 17, 1989, just as Game 3 of the World Series was about to start in San Francisco, a strong earthquake occurred at Loma Prieta, California, and shook the entire San Francisco Bay area. Seismographs at locations **X, Y,** and **Z** (see Figure 16.2) recorded the shaking, and the resulting seismograms are shown in Figure 16.3. Earthquakes are recorded on the seismograms as deviations (vertical zigzags) from a flat, horizontal line. Thus, notice that much more shaking occurred at locations **Y** and **Z** than at location **X.**

9. The Loma Prieta earthquake caused no significant damage at location **X,** but there was moderate damage to buildings at location **Y** and severe damage at location **Z.** Explain how this damage report compares to your predictions of risk in Questions 6, 7, and 8.

10. The Loma Prieta earthquake shook all of the San Francisco Bay region. Yet Figure 16.3 is evidence that the earthquake had very different effects on properties located only 600 m apart. Explain how the kind of substrate (uncompacted vs. firm and compacted) on which buildings are constructed influences how much the buildings are shaken and damaged in an earthquake.

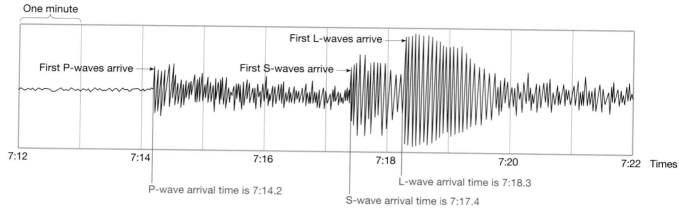

FIGURE 16.4 Seismogram of a New Guinea earthquake recorded at a location in Australia. Most of the seismogram shows only minor background deviations (short zigzags) from a horizontal line, such as the interval recorded between 7:12 and 7:14. Large vertical deviations indicate motions caused by the arrival of P-waves, S-waves, and L-waves of the earthquake (note arrows with labels). By making detailed measurements with a ruler, you can determine that the arrival time of the P-waves was 7:14.2 (14.2 minutes past 7 o'clock), the arrival time of the S-waves was 7:17.4, and the arrival time of the L-waves was 7:18.3.

11. Imagine that you are a member of the San Francisco City Council. What actions could you propose to **mitigate** (decrease the probability of) future earthquake hazards such as the damage that occurred at locations **Y** and **Z** in the Loma Prieta earthquake?

PART 16B: GRAPHING SEISMIC DATA AND LOCATING THE EPICENTER OF AN EARTHQUAKE

An earthquake produces three main types of seismic waves that radiate from its focus/epicenter at different rates. Seismographs are instruments used to detect these seismic waves and produce a **seismogram**—a record of seismic wave motions obtained at a specific recording station (Figure 16.4).

Seismograms can detect and record several types of *body waves*, which are seismic waves that travel through Earth's interior (rather than along its surface) and radiate in all directions from the focus. Two of these body waves are used to locate earthquake epicenters:

- **P-waves:** *P* for primary, because they travel fastest and arrive at seismographs first. (They are compressional, or "push-pull" waves.)

- **S-waves:** *S* for secondary, because they travel more slowly and arrive at seismographs after the P-waves. (They are perpendicular, shear, or "side-to-side" waves.)

Seismographs also detect the surface seismic waves, called **L-waves** or *Love waves* (named for A. E. H. Love, who discovered them). L-waves travel along Earth's surface (a longer route than the body waves) and thus are recorded after the S-waves and P-waves arrive at the seismograph.

Figure 16.4 is a seismogram recorded at a station located in Australia. Seismic waves arrived there from an earthquake epicenter located 1800 kilometers (1125 miles) away in New Guinea. Notice that the seismic waves were recorded as deviations (vertical zigzags) from the nearly horizontal line of normal background vibrations. Thus, the first pulse of seismic waves was P-waves, which had an **arrival time** of 7:14.2 (i.e., 14.2 minutes after 7:00). The second pulse of seismic waves was the slower S-waves, which had an arrival time of 7:17.4. The final pulse of seismic waves was the L-waves that traveled along Earth's surface, so they did not begin to arrive until 7:18.3. The earthquake actually occurred at the New Guinea epicenter at 7:10:23 (10 minutes and 23 seconds after 7:00) Greenwich Mean Time, which can be written as 7:10.4. Therefore the **travel time of the main seismic waves** (to go 1800 km) was 3.8 minutes for P-waves (7:14.2 minus 7:10.4), 7.0 minutes for S-waves (7:17.4 minus 7:10.4), and 7.9 minutes for L-waves (7:18.3 minus 7:10.4).

Notice the seismic data provided in Figure 16.5 for 11 recording stations where seismograms were recorded after the same New Guinea earthquake (at 3° North latitude and 140° East longitude). The **distance from epicenter** (surface distance between the recording station and the epicenter) and travel time of main seismic waves are provided for each

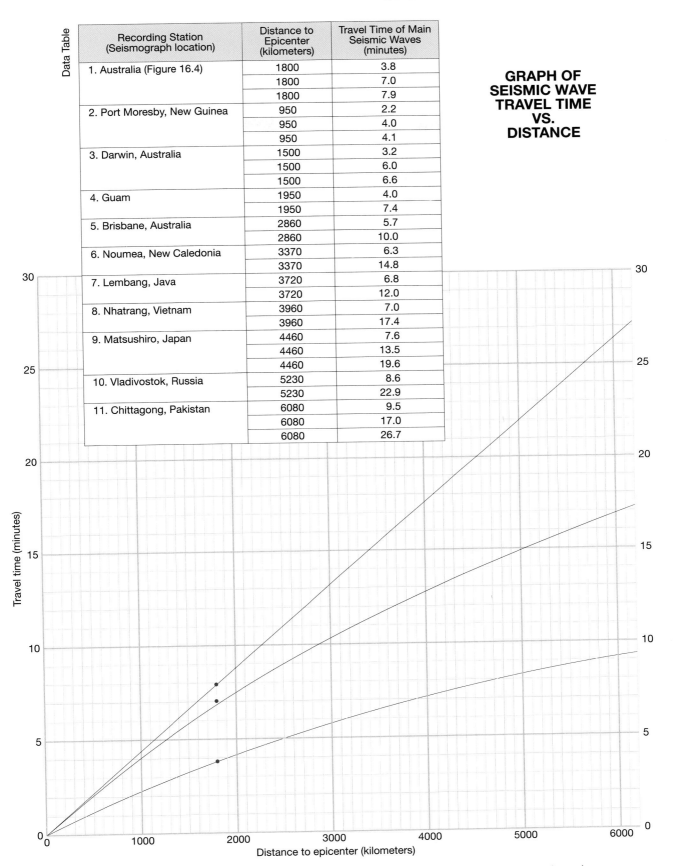

Data Table	Recording Station (Seismograph location)	Distance to Epicenter (kilometers)	Travel Time of Main Seismic Waves (minutes)
	1. Australia (Figure 16.4)	1800	3.8
		1800	7.0
		1800	7.9
	2. Port Moresby, New Guinea	950	2.2
		950	4.0
		950	4.1
	3. Darwin, Australia	1500	3.2
		1500	6.0
		1500	6.6
	4. Guam	1950	4.0
		1950	7.4
	5. Brisbane, Australia	2860	5.7
		2860	10.0
	6. Noumea, New Caledonia	3370	6.3
		3370	14.8
	7. Lembang, Java	3720	6.8
		3720	12.0
	8. Nhatrang, Vietnam	3960	7.0
		3960	17.4
	9. Matsushiro, Japan	4460	7.6
		4460	13.5
		4460	19.6
	10. Vladivostok, Russia	5230	8.6
		5230	22.9
	11. Chittagong, Pakistan	6080	9.5
		6080	17.0
		6080	26.7

GRAPH OF SEISMIC WAVE TRAVEL TIME VS. DISTANCE

FIGURE 16.5 Seismic wave data for an earthquake that occurred in New Guinea (at 3° North latitude and 140° East longitude) at Greenwich Mean Time of 7 hours, 10 minutes, 23 seconds (7:10.4). The travel time of a main seismic wave is the time interval between when the earthquake occurred in New Guinea and when that wave first arrived at a recording location. The surface distance is the distance between the recording location and the earthquake epicenter. Graph is for plotting points that represent the travel time of each main seismic wave at each location versus the surface distance that it traveled.

recording station. Notice that the data from most of the recording stations includes travel times for all three main kinds of seismic waves (P-waves, S-waves, and L-waves). However, instruments at some locations recorded only one or two kinds of waves. Location 1 is the Australian recording station where the seismogram in Figure 16.4 was obtained.

Questions

12. On the graph paper provided at the base of Figure 16.5, *plot points in pencil to show the travel time of each main seismic wave in relation to its distance from the epicenter* (when recorded on the seismogram at the recording station). For example, the data for location 1 (obtained from Figure 16.4) have already been plotted as red points on Figure 16.5. Recording station 1 was located 1800 km from the earthquake epicenter and the main waves had travel times of 3.8 minutes, 7.0 minutes, and 7.9 minutes. Plot points in pencil for data from all of the remaining recording stations, and then examine the graph.

    Notice that your points do not produce a *random pattern*. They fall in *discrete paths* close to the three narrow black lines (or curves) already drawn on the graph. These black lines (or curves) were formed by plotting many thousands of points from hundreds of earthquakes, exactly as you just plotted your points. Explain why you think that your points, and all of the points from other earthquakes, occur along three discrete lines (or curves).

13. Study the three discrete, narrow black lines (or curves) of points in Figure 16.5. Label the line (curve) of points that represents travel times of the P-waves. Label the line or curve that connects the points representing travel times of the S-waves. Label the line or curve that connects the points representing travel times of the L-waves. Why is the S-wave curve steeper than the P-wave curve?

14. Why do the L-wave data points form a straight line whereas data points for P-waves and S-waves form curves? (*Hint:* The curved lines are evidence of how the physical environments and rocks deep inside Earth are different from the physical environments and rocks just beneath Earth's surface.)

15. Notice that the origin on your graph (travel time of zero and distance of zero) represents the location of the earthquake epicenter and the start of the seismic waves. The time interval between first arrival of P-waves and first arrival of S-waves at the same recording station is called the **S-minus-P time interval.** How does the S-minus-P time interval change with distance from the epicenter?

16. Imagine that an earthquake occurred this morning. The first P-waves of the earthquake were recorded at a recording station in Houston at 6:12.6 a.m. and the first S-waves arrived at the same Houston station at 6:17.1 a.m. Use Figure 16.5 to determine an answer for each question below.

    a. What is the S-minus-P time interval of the earthquake?

    b. How far from the earthquake's epicenter is the Houston recording station located?

    c. You have determined the distance (radius of a circle on a map) between Houston and the earthquake epicenter. What additional data would you require to determine the location of the earthquake's epicenter (point on a map), and how would you use the data to locate the epicenter?

Locate the Epicenter of an Earthquake

See if you can use the travel time curves in Figure 16.5 to locate the epicenter (point on a map) of the earthquake that produced the seismograms in Figure 16.6. These seismograms were recorded at stations in Alaska, North Carolina, and Hawaii.

Questions

17. Estimate, to the nearest tenth of a minute, the times that P-waves and S-waves first arrived at each recording station (seismograph location) in Figure 16.6. Then, subtract P from S to get the S-minus-P time interval:

	First P arrival	First S arrival	S-minus-P
Sitka, AK	___	___	___
Charlotte, NC	___	___	___
Honolulu, HI	___	___	___

18. Using the S-minus-P time intervals and Figure 16.5, determine the distance from epicenter (in kilometers) for each recording station.

    Sitka, AK ___ kilometers

    Charlotte, NC ___ kilometers

    Honolulu, HI ___ kilometers

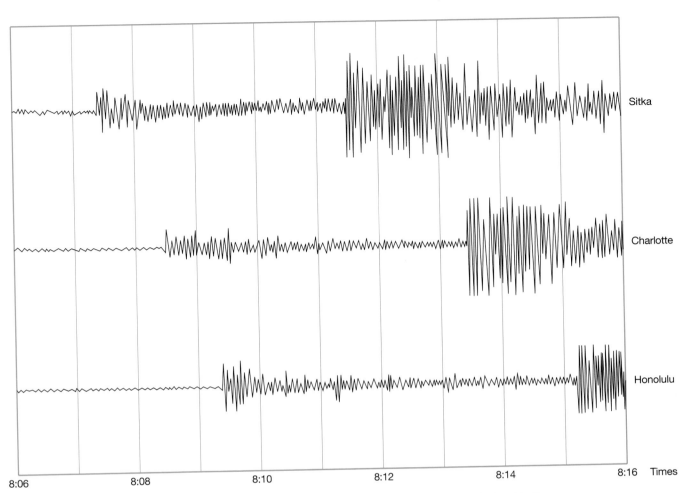

FIGURE 16.6 Seismograms for an earthquake recorded at three different locations in Alaska, North Carolina, and Hawaii. Times have been standardized to Charlotte, North Carolina, to simplify comparison.

19. Next, find the earthquake's epicenter using the distances just obtained.

a. First use the geographic coordinates below to locate and mark the three recording stations on the world map in Figure 16.7.

Sitka, AK: 57°N latitude, 135°W longitude

Charlotte, NC: 35°N latitude, 81°W longitude

Honolulu, HI: 21°N latitude, 158°W longitude

b. Use a drafting compass to draw a circle around each recording station. Make the radius of each circle equal to the *distance from epicenter* determined

for the station in Question 18. (Use the scale on Figure 16.7 to set this radius on your drafting compass.) The circles you draw should intersect approximately at one point on the map. This point is the epicenter. (If the three circles do not quite intersect at a single point, then find a point that is equidistant from the three edges of the circles, and use this as the epicenter.) Record the location of the earthquake epicenter:

N Latitude _____ W Longitude _____

20. What is the name of a major fault that occurs near this epicenter?

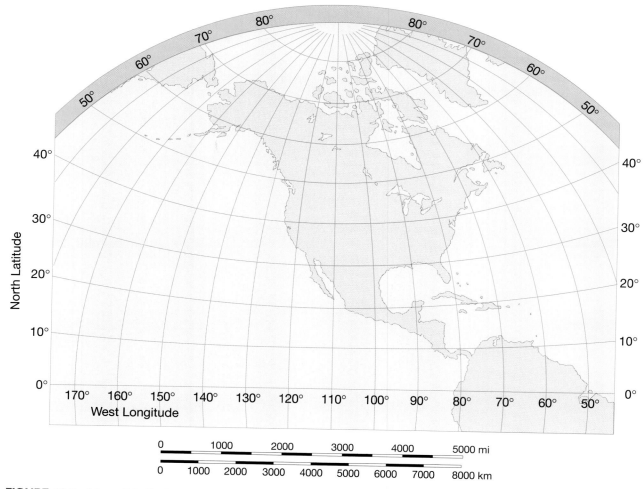

FIGURE 16.7   Map of Earth, for use in plotting data and locating the earthquake's epicenter.

PART 16C: ANALYSIS OF ACTIVE FAULTS USING AERIAL PHOTOGRAPHS

There are many faults that can be imaged, photographed, mapped, and studied where they break Earth's surface. Some of these faults are **active faults,** meaning that they can move and generate earthquakes at the present time.

Examine the aerial photograph of a portion of southern California (Figure 16.8) for evidence of faults and fault motions. Notice the roads, small streams, and fine features of the landscape. Also notice that the figure shows a portion of the San Andreas Fault, which is a tectonic plate boundary separating the Pacific Plate from the North American Plate.

Questions

21. Geologists have inferred that the San Andreas Fault is an active fault and that the blocks of rock on either side of the fault are moving in the directions indicated with half-arrows.

    a. What evidence, visible in this photograph, could you use to suggest that this fault is both active and moving relative to the arrows? Explain your reasoning.

    b. How much has the San Andreas Fault offset the present-day channel of Wallace Creek?

    c. Is the San Andreas Fault a left-lateral fault or a right-lateral fault? Explain.

22. How wide is the San Andreas Fault (tectonic plate boundary) here?

FIGURE 16.8 Aerial photograph of a portion of the San Andreas Fault (a tectonic plate boundary) at Wallace Creek, Carrizo Plain, southern California. (Photo by Randall Marrett, University of Texas, Austin)

23. Notice the small dry valley in the lower-left part of the photograph. Infer how this valley may have formed.

PART 16D: DETERMINING RELATIVE MOTIONS ALONG THE NEW MADRID FAULT ZONE

The relative motions of blocks of rock on either side of a fault zone can be determined by mapping the way the pen on a seismograph moved (up or down on the seismogram) when P-waves first arrived at various seismic stations adjacent to the fault. This pen motion is called **first motion** and represents the reaction of the P-wave to dilation (pulling rocks apart) or compression (squeezing rocks together) as observed on seismograms (see Figure 16.9, left).

If the first movement of the P-wave was up on a seismogram, then that recording station (where the seismogram was obtained) experienced compression during the earthquake. If the first movement of the P-wave was down on a seismogram, then that recording station was dilational during the earthquake. What was the first motion at all of the seismic stations in Figure 16.3? (Answer: The first movement of the pen was up for each P-wave, so the first motion at all three sites was compressional.)

By plotting the first motions observed at recording stations on both sides of a fault that has experienced an earthquake, a picture of the relative motions of the fault emerges. For example, notice that the first motions observed at seismic stations on either side of a hypothetical fault are plotted in relation to the fault in Figure 16.9 (right side). The half-arrows indicate how motion proceeded away from seismic stations where dilation was recorded and toward seismic stations where compression was recorded (for each side of the fault). So the picture of relative motion along this fault is that Block **X** is moving southeast and Block **Y** is moving northwest. Now study a real example using Figures 16.10 and 16.11.

The New Madrid Fault System is located within the *Mississippi Embayment*, a basin filled with Mesozoic and Cenozoic rocks that rest unconformably on (and are surrounded by) Paleozoic and Precambrian rocks (see Figure 16.11). Faults of the New Madrid System are not visible on satellite images and photographs, because

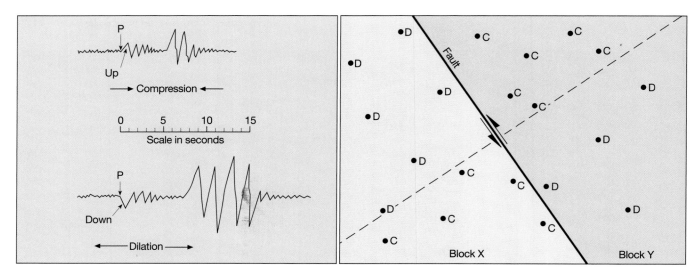

FIGURE 16.9 **Left**—Sketch of typical seismograms for compressional first motion (first P-wave motion is up) compared with dilational first motion (first P-wave motion is down). **Right**—Map of a hypothetical region showing a fault along which an earthquake has occurred, and the P-wave first motions (C = compressional, D = dilational) observed for the earthquake at seismic stations adjacent to the fault. Stress moves away from the field of dilation and toward the field of compression on each side of the fault (large open arrows), so the relative motion of the fault is as indicated by the smaller half-arrows.

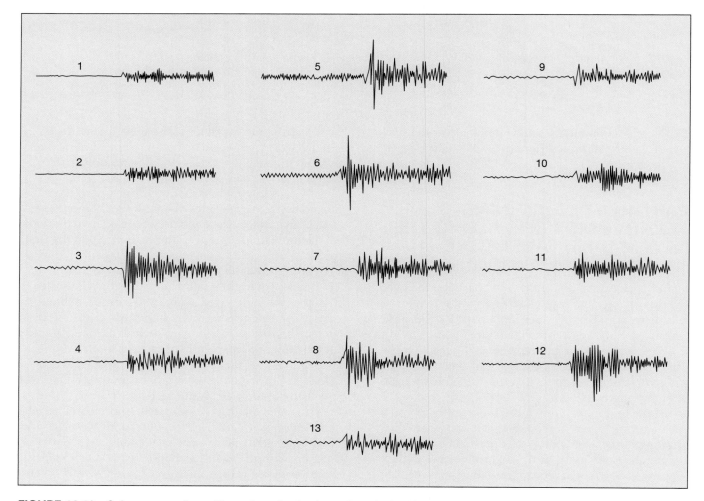

FIGURE 16.10 Seismograms from 13 numbered seismic stations in the Mississippi Embayment after an earthquake that occurred in the New Madrid Fault System. Numbers in this figure correspond to the numbered sites on the map in Figure 16.11.

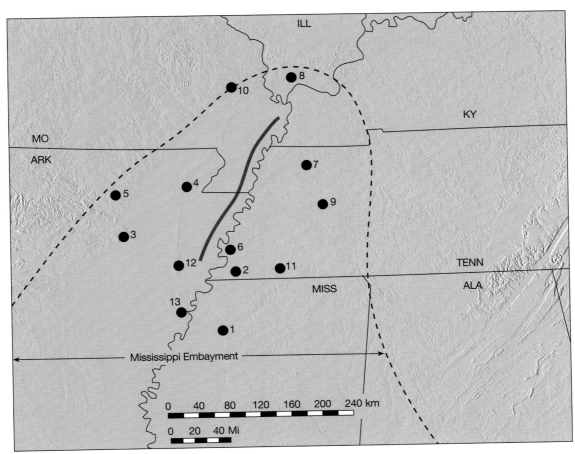

FIGURE 16.11 Map of a portion of the Mississippi Embayment showing the generalized surface geology, location (in red) of the main fault of the New Madrid (Blind) Fault System, numbered seismic stations (as in Figure 16.10), and state boundaries.

they are **blind faults** (faults that do not break Earth's surface). These blind faults occur in the Paleozoic and Precambrian rocks that are buried beneath approximately a kilometer of Mesozoic and Cenozoic rocks.

The main fault of the New Madrid System is plotted in red on Figure 16.11. It is well known, because a series of strong earthquakes occurred along it in 1811 and 1812. One of these earthquakes was the strongest earthquake ever recorded in North America, and the potential for more strong earthquakes here is a lingering hazard. The locations of 13 seismic stations are also plotted on Figure 16.11. Seismograms obtained at these stations (after an earthquake along the New Madrid Fault System) are provided in Figure 16.10.

Question

24. Analyze the seismograms in Figure 16.10 to determine if their P-wave first motions indicate

compression or dilation (refer to Figure 16.9 as needed). Plot this information on Figure 16.11 by writing a *C* beside the stations where compression occurred and a *D* beside the stations where dilation occurred. When you have finished plotting these letters, draw half-arrows on Figure 16.11 to indicate the relative motions of the blocks of rock on either side of the main fault. Does the main fault have a right-lateral motion or a left-lateral motion? Explain.

Unlike most major active fault zones that occur at plate boundaries (such as the San Andreas Fault), the New Madrid Fault System is an active and hazardous system of faults that occurs *within* the North American Plate. Intraplate stresses are apparently causing adjustments along these blind faults and the potential for more earthquakes that place humans at risk.

PART 16E: TRACKING EARTHQUAKE HAZARDS IN REAL TIME AND ASSESSING THEIR IMPACT ON RISK TAKERS

Find and explore Internet sites that contain real-time information about earthquake hazards and risks (go to **http://www.prenhall.com/agi**). Record the date and time that you conducted this exploration, and proceed to the items below.

Questions

25. How many earthquakes of Richter Magnitude 2.5 or greater have occurred in each of the following areas in the past week?

   a. Southern California (from motions along a plate boundary):

   b. Hawaii (from tectonism at the world's most active hot spot):

   c. Along the New Madrid Fault Zone (from intraplate stresses):

26. From your answers in Question 25, which plate tectonic setting seems to generate:

   a. the most earthquakes? Why?

   b. the fewest earthquakes? Why?

27. Search the Federal Emergency Management Agency server (**http://www.fema.gov/**) to find out about some of the damage done to properties and lives of people after a strong earthquake. What should *you* do:

   a. to *prepare* for a strong earthquake?

   b. to *survive during and shortly after* a strong earthquake?

9
8
7
6
5
4
3
2

1
9
8
7
6
5
4
3
2

1
9
8
7
6
5
4
3
2

1
9
8
7
6
5
4
3
2

1

9
8
7
6
5
4
3
2
1
9
8
7
6
5
4
3
2
1
9
8
7
6
5
4
3
2
1
9
8
7
6
5
4
3
2
1

9
8
7
6
5
4
3
2

1
9
8
7
6
5
4
3
2

1
9
8
7
6
5
4
3
2

1
9
8
7
6
5
4
3
2

1

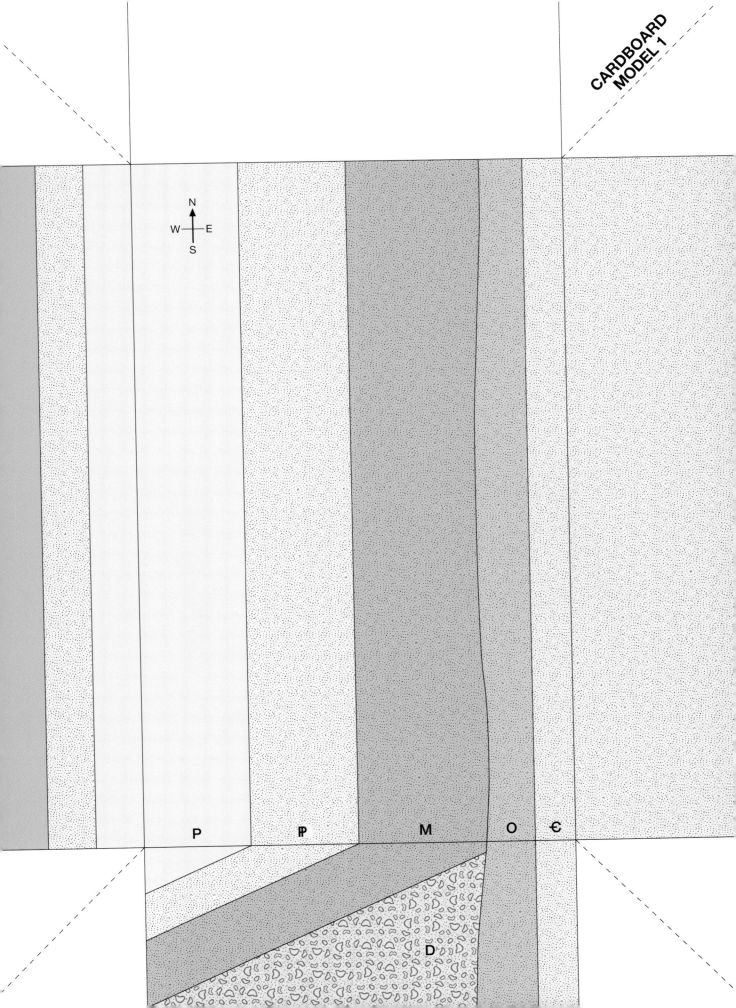

CARDBOARD
MODEL 1

N
W—E
S

P   ℙ   M   O   Є

D

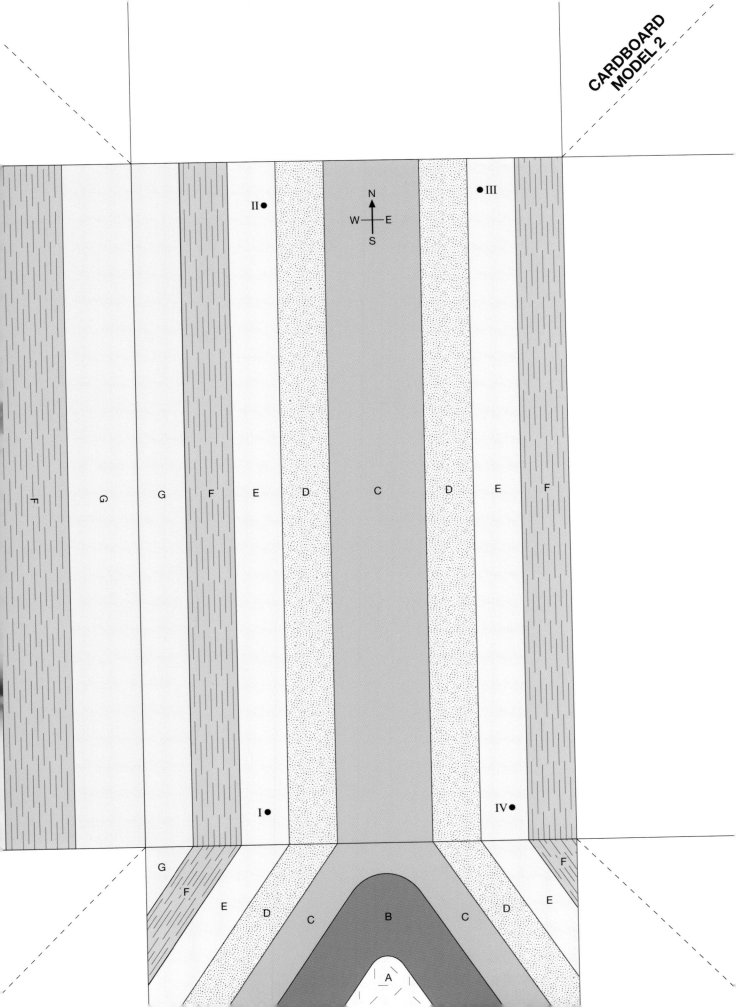

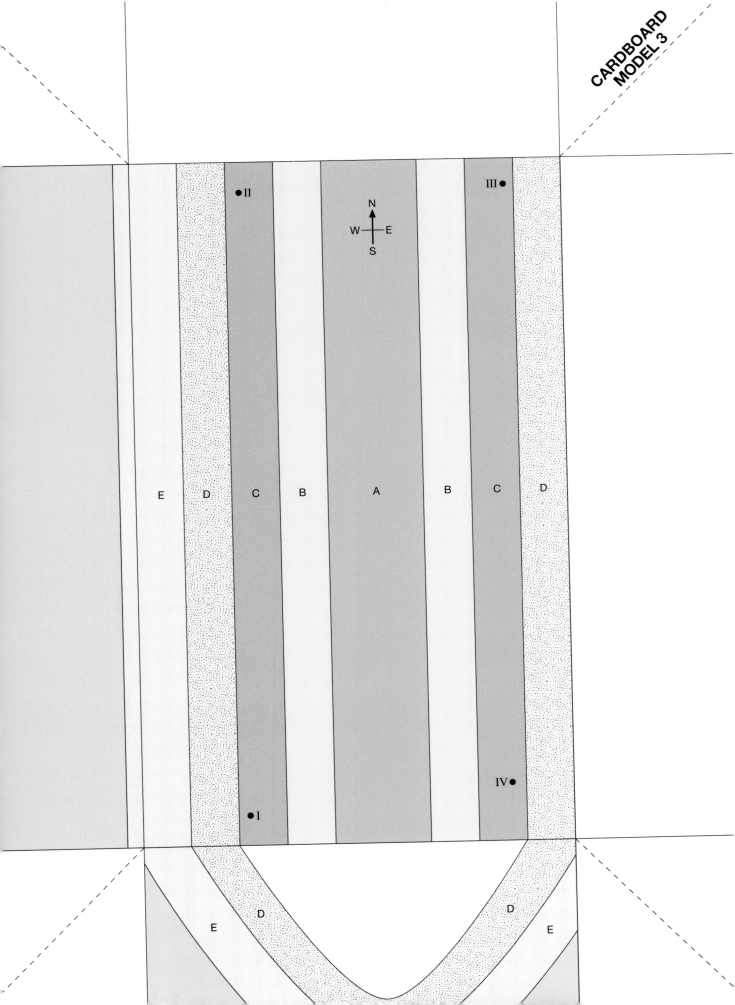

● II

III ●

N
W— —E
S

E   D   C   B   A   B   C   D

IV ●

● I

E   D   D   E

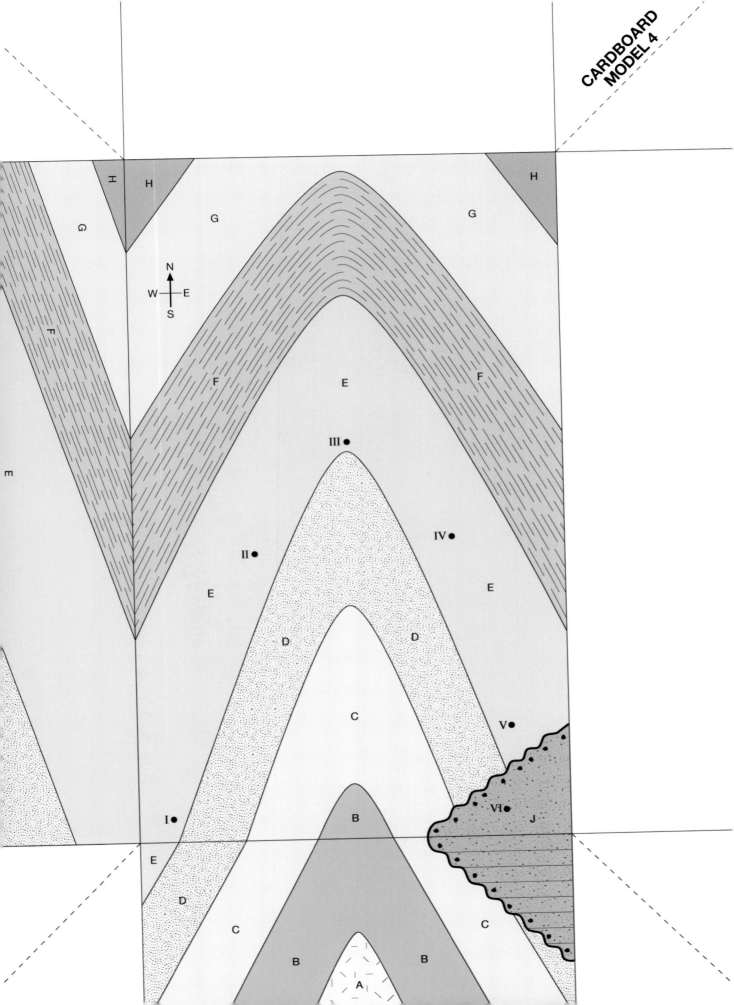

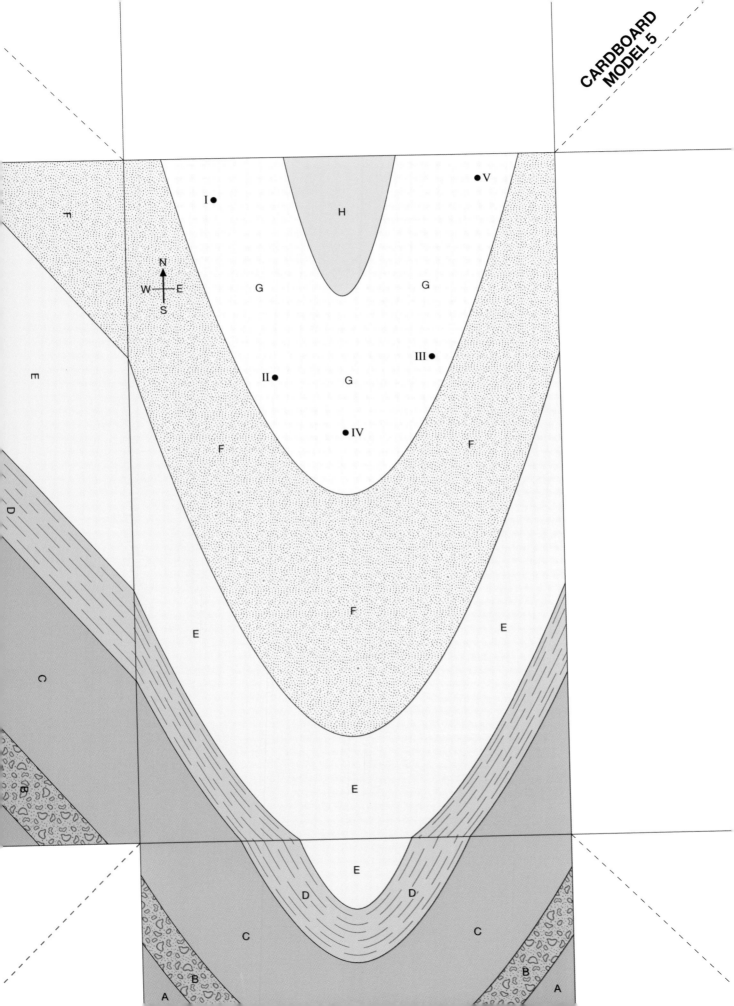

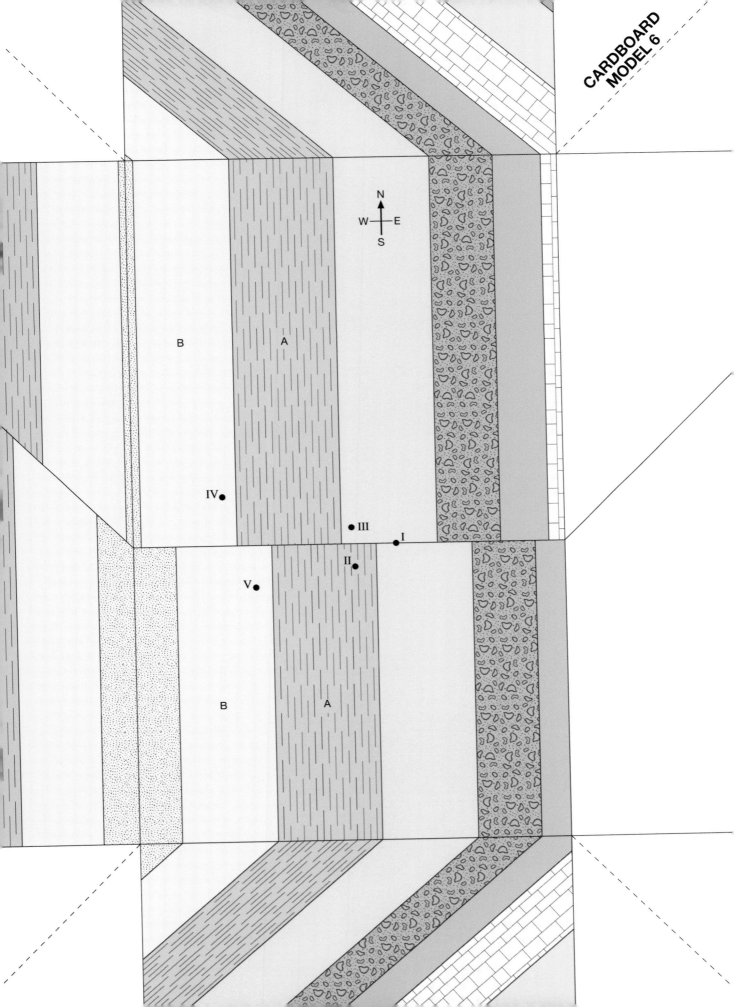

GEOTOOLS SHEET 1

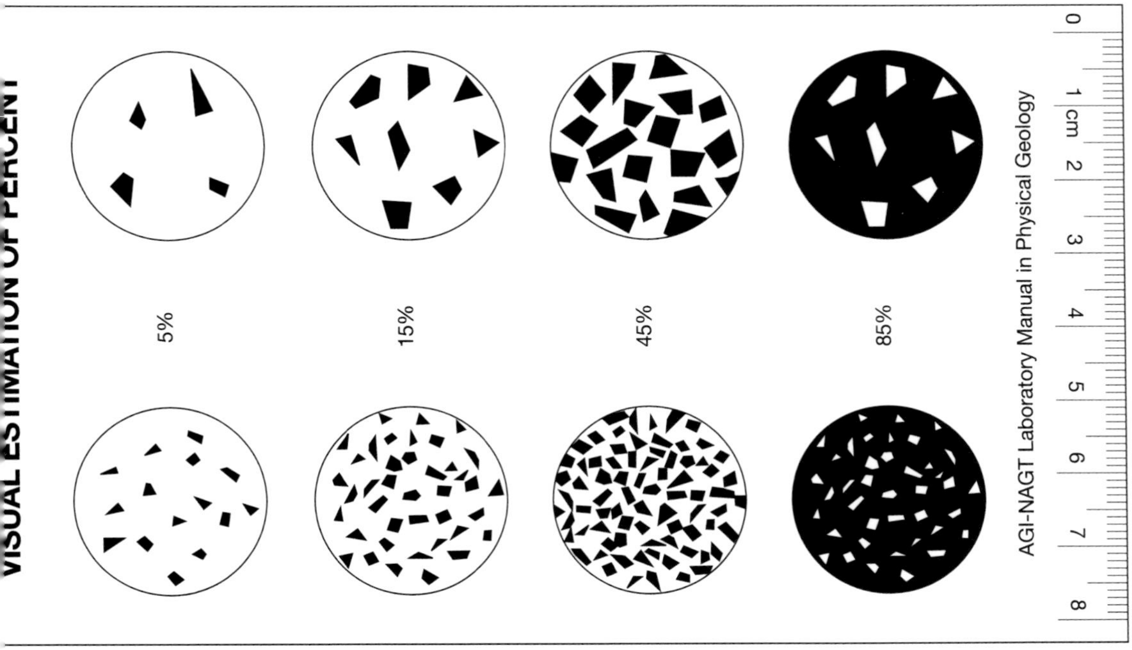
VISUAL ESTIMATION OF PERCENT
5%
15%
45%
85%
AGI-NAGT Laboratory Manual in Physical Geology
0 1 cm 2 3 4 5 6 7 8

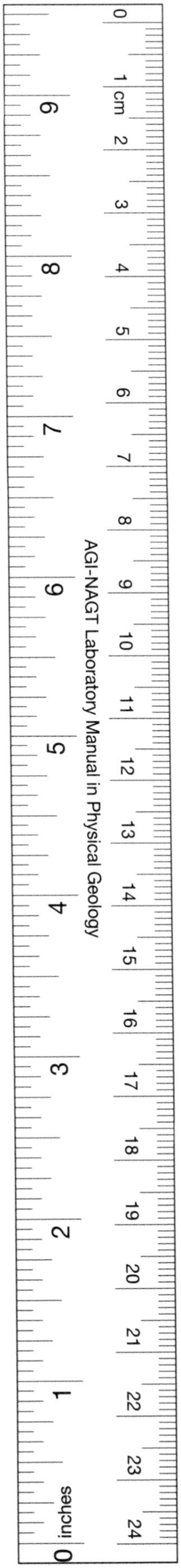
0 1 cm 2 3 4 5 6 7 8 9 10 11 12 13 14 15 16 17 18 19 20 21 22 23 24
AGI-NAGT Laboratory Manual in Physical Geology
0 inches 1 2 3 4 5 6 7 8 9

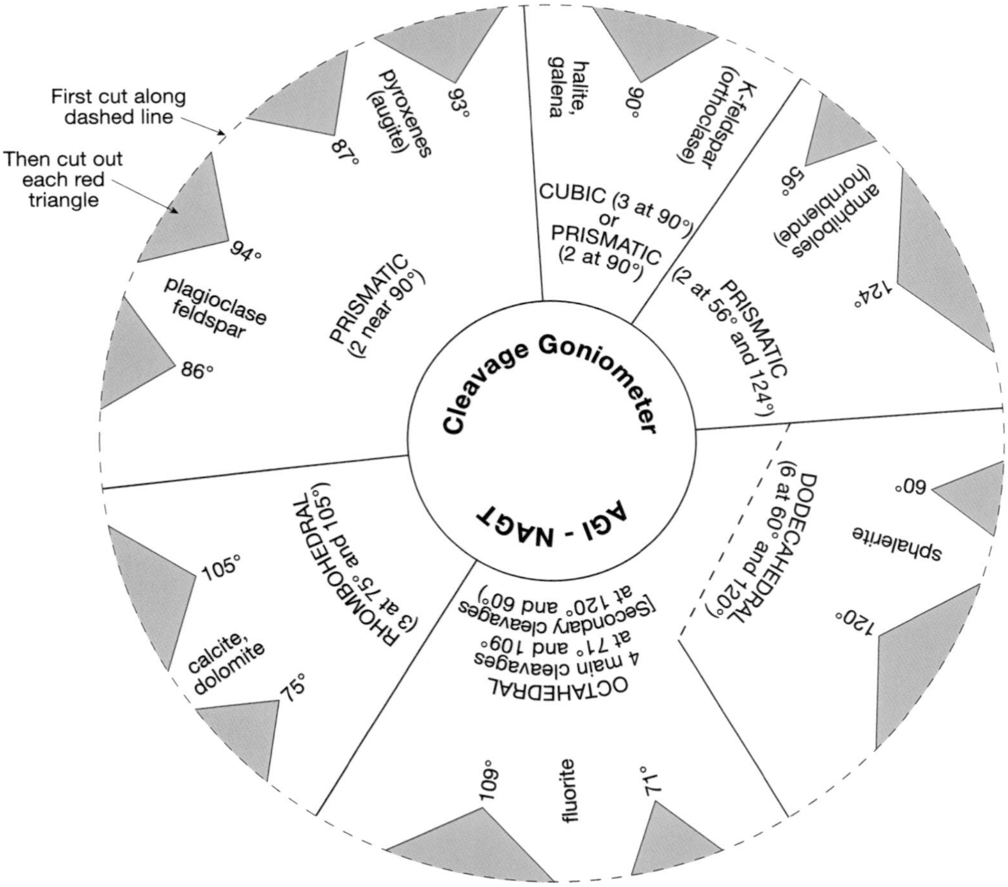
First cut along dashed line
Then cut out each red triangle
Cleavage Goniometer
AGI - NAGT
pyroxenes (augite)
93°
87°
halite, galena
90°
K-feldspar (orthoclase)
CUBIC (3 at 90°) or PRISMATIC (2 at 90°)
amphiboles (hornblende)
56°
124°
PRISMATIC (2 at 56° and 124°)
94°
plagioclase feldspar
PRISMATIC (2 near 90°)
86°
DODECAHEDRAL (6 at 60° and 120°)
60°
sphalerite
120°
105°
calcite, dolomite
75°
RHOMBOHEDRAL (3 at 75° and 105°)
OCTAHEDRAL 4 main cleavages at 71° and 109° [Secondary cleavages at 120° and 60°]
109°
fluorite
71°

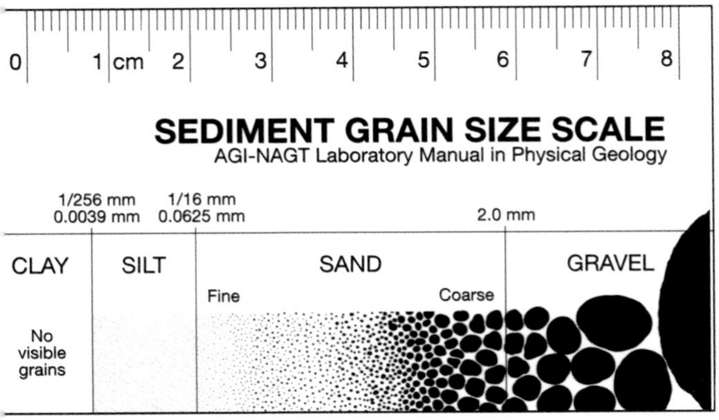
0 1 cm 2 3 4 5 6 7 8
SEDIMENT GRAIN SIZE SCALE
AGI-NAGT Laboratory Manual in Physical Geology
1/256 mm 0.0039 mm
1/16 mm 0.0625 mm
2.0 mm
CLAY
SILT
SAND
GRAVEL
Fine
Coarse
No visible grains

GeoTools

Cut out these tools or make copies of them for your personal use.